VOLUME ONE HUNDRED AND TWENTY THREE

ADVANCES IN PROTEIN CHEMISTRY AND STRUCTURAL BIOLOGY

Transport proteins

VOLUME ONE HUNDRED AND TWENTY THREE

ADVANCES IN PROTEIN CHEMISTRY AND STRUCTURAL BIOLOGY

Transport proteins

Edited by

Rossen Donev
Head of Research
MicroPharm Ltd
Newcastle Emlyn, Carmarthenshire
United Kingdom

ELSEVIER

ACADEMIC PRESS
An imprint of Elsevier

Academic Press is an imprint of Elsevier
125 London Wall, London EC2Y 5AS, United Kingdom
525 B Street, Suite 1650, San Diego, CA 92101, United States
50 Hampshire Street, 5th Floor, Cambridge, MA 02139, United States
The Boulevard, Langford Lane, Kidlington, Oxford OX5 1GB, United Kingdom

First edition 2021

ISBN: 978-0-12-822087-0
ISSN: 1876-1623

For information on all Academic Press publications visit our website at https://www.elsevier.com/books-and-journals

Publisher: Zoe Kruze
Acquisitions Editor: Ashlie Jackman
Editorial Project Manager: Leticia Lima
Production Project Manager: Abdulla Sait
Cover Designer: Greg Harris

Typeset by TNQ Technologies

Contents

Contributors *ix*

1. Biology of the iron efflux transporter, ferroportin **1**

Gautam Rishi and V. Nathan Subramaniam

1. Iron metabolism and the central role of ferroportin 1
2. Structure of ferroportin 3
3. Regulation of ferroportin 8
4. Disorders associated with ferroportin dysfunction 10
5. Ferroportin partners-molecules involved in efficient ferroportin functioning 11
6. Conclusion 12

References 13

2. The effect of sport and physical activity on transport proteins: implications for cancer prevention and control **17**

Maha Sellami and Nicola Luigi Bragazzi

1. Transport proteins 18
2. Muscles 18
3. Muscle and transport proteins 21
4. Sport and transport proteins: implications for cancer treatment and prevention 22
5. Conclusions 24

References 25

3. Cancer bioenergetics as emerging holistic cancer theory: the role of metabolic fluxes and transport proteins involved in metabolic pathways in the pathogenesis of malignancies. State-of-the-art and future prospects **27**

Nicola Luigi Bragazzi and Maha Sellami

1. Introduction 28
2. Cancer and bio-energetics: the emerging field of "onco-energetics" 34
3. Altered glycolysis and pyruvate oxidation metabolic fluxes in cancer 36
4. Altered glutamine metabolic fluxes in cancer 36
5. Altered serine metabolic fluxes in cancer 37
6. Altered TCA metabolism in cancer 38
7. Altered methionine metabolism in cancer 38

8. Altered tryptophan metabolism in cancer 39
9. Altered leucine metabolism in cancer 40
10. Altered ornithine metabolism in cancer 40
11. Altered proline metabolism in cancer 40
12. Altered acetate metabolism in cancer 41
13. Altered lipid metabolism in cancer 41
14. Mitochondrial structural and functional impairment 42
15. Tumor microenvironment 42
16. Nuclear DNA genes mutations 43
17. Cellular pathways mutations 43
18. Conclusions and future prospects 44
References 44

4. Investigating mutations at the hotspot position of the ERBB2 and screening for the novel lead compound to treat breast cancer – a computational approach 49

D. Thirumal Kumar, S. Udhaya Kumar, R. Magesh and C. George Priya Doss

1. Introduction 50
2. Materials and methods 54
3. Results 56
4. Discussion 64
5. Conclusion 66
Supplementary data 67
Acknowledgments 67
References 67

5. Autophagy proteins and its homeostasis in cellular environment 73

Vrushali Guhe, Bhavnita Soni, Prajakta Ingale and Shailza Singh

1. Introduction 74
2. Methodology 79
3. Results 83
4. Discussion 90
Acknowledgments 92
References 92

6. Expression of placental glucose transporter proteins in pregnancies complicated by fetal growth disorders 95

Paweł Jan Stanirowski, Michał Lipa, Dorota Bomba-Opoń and Mirosław Wielgoś

1. Introduction 96
2. Characteristics of GLUT proteins in human placenta 104

3. Placental glucose transport in pregnancies complicated by fetal macrosomia 112
4. Placental glucose transport in pregnancies complicated by fetal growth restriction 118
5. Placental GLUT-mediated interventions in pregnancies complicated by fetal growth disorders 122
6. Conclusions 123
References 123

7. The role of transferrins and iron-related proteins in brain iron transport: applications to neurological diseases 133

Camila Campos-Escamilla

1. Introduction 134
2. Brain iron transport 135
3. Iron and neurodegeneration 138
4. Iron transport related proteins 146
5. Transferrins 151
6. Perspectives on the use of transferrin as a drug delivery system 154
7. Concluding remarks 156
Acknowledgments 156
References 156

8. ABCG: a new fold of ABC exporters and a whole new bag of riddles! 163

Atanu Banerjee, Alexis Moreno, Jorgaq Pata, Pierre Falson and Rajendra Prasad

Abbreviations 164
1. ABC transporters and their relevance 164
2. Structure of type I and type II exporters 171
3. Yeast ABCG/PDR transporters: contribution toward understanding the enigma of non-catalytic nucleotide-binding sites 177
4. Concluding thoughts and future perspectives 182
Acknowledgments 183
References 183

9. Serum albumin: clinical significance of drug binding and development as drug delivery vehicle 193

Saad Tayyab and Shevin Rizal Feroz

1. Clinical implications of drug–serum albumin binding 194
2. HSA-based drug delivery systems 201

Acknowledgments 213
References 213

10. PTR2/POT/NPF transporters: what makes them tick? 219

Bala K. Prabhala, Moazur Rahman, Hussam H. Nour-eldin and Flemming Steen Jørgensen and Osman Mirza

1. Introduction 220
2. A plethora of natural substrates 221
3. 3D-structure and transport mechanism 230

References 234

11. Network analysis of transcriptomics data for the prediction and prioritization of membrane-associated biomarkers for idiopathic pulmonary fibrosis (IPF) by bioinformatics approach 241

Smriti Mishra, Mohammad Imran Shah, S. Udhaya Kumar, D. Thirumal Kumar, Chandrasekhar Gopalakrishnan, Abeer Mohammed Al-Subaie, R. Magesh, C. George Priya Doss and Balu Kamaraj

1. Introduction 242
2. Materials and methods 244
3. Results and discussion 248
4. Conclusion 267

Supplementary data 268
Acknowledgement 268
References 268

Contributors

Abeer Mohammed Al-Subaie
Department of Clinical Laboratory Sciences, College of Applied Medical Sciences, Imam Abdulrahman Bin Faisal University, Dammam, Saudi Arabia

Atanu Banerjee
Amity Institute of Biotechnology, Amity University Haryana, Gurgaon, Haryana, India

Dorota Bomba-Opoń
1st Department of Obstetrics and Gynecology, Medical University of Warsaw, Warsaw, Poland

Nicola Luigi Bragazzi
Postgraduate School of Public Health, Department of Health Sciences (DISSAL), University of Genoa, Genoa, Italy; Laboratory for Industrial and Applied Mathematics (LIAM), Department of Mathematics and Statistics, York University, Toronto, ON, Canada

Camila Campos-Escamilla
Programa de Doctorado en Ciencias Biomédicas, Universidad Nacional Autónoma de México. Unidad de Posgrado, Ciudad Universitaria, Ciudad de México, Mexico

Pierre Falson
Drug Resistance & Membrane Proteins Team, Molecular Microbiology and Structural Biochemistry Laboratory, CNRS-Lyon 1 University UMR5086, Institut de Biologie et Chimie des Protéines, Lyon, France

Shevin Rizal Feroz
Department of Biological Sciences and Biotechnology, Faculty of Science and Technology, Universiti Kebangsaan Malaysia, Bangi, Selangor, Malaysia

Chandrasekhar Gopalakrishnan
School of Biosciences and Technology, Vellore Institute of Technology, Vellore, Tamil Nadu, India

Vrushali Guhe
National Centre for Cell Science, NCCS Complex, Ganeshkhind, SP Pune University Campus, Pune, Maharashtra, India

Prajakta Ingale
National Centre for Cell Science, NCCS Complex, Ganeshkhind, SP Pune University Campus, Pune, Maharashtra, India

Flemming Steen Jørgensen
Department of Drug Design and Pharmacology, Faculty of Health and Medical Sciences, University of Copenhagen, Copenhagen, Denmark

Balu Kamaraj
Department of Neuroscience Technology, College of Applied Medical Sciences in Jubail, Imam Abdulrahman Bin Faisal University, Jubail, Saudi Arabia

Michał Lipa
1st Department of Obstetrics and Gynecology, Medical University of Warsaw, Warsaw, Poland; Club 35. Polish Society of Gynecologists and Obstetricians, Warsaw, Poland

R. Magesh
Faculty of Biomedical Sciences, Technology & Research, Department of Biotechnology, Sri Ramachandra University, Chennai, Tamil Nadu, India

Osman Mirza
Department of Drug Design and Pharmacology, Faculty of Health and Medical Sciences, University of Copenhagen, Copenhagen, Denmark

Smriti Mishra
Department of Biotechnology and Bioinformatics, Jaypee University of Information Technology, Waknaghat, Solan, Himachal Pradesh, India; Navipoint Health India Pvt Ltd, Moula-Ali, Hyderabad, Telangana, India

Alexis Moreno
Drug Resistance & Membrane Proteins Team, Molecular Microbiology and Structural Biochemistry Laboratory, CNRS-Lyon 1 University UMR5086, Institut de Biologie et Chimie des Protéines, Lyon, France

Hussam H. Nour-eldin
DynaMo Center, Copenhagen Plant Science Centre, Department of Plant and Environmental Sciences, University of Copenhagen, Frederiksberg, Denmark

Jorgaq Pata
Drug Resistance & Membrane Proteins Team, Molecular Microbiology and Structural Biochemistry Laboratory, CNRS-Lyon 1 University UMR5086, Institut de Biologie et Chimie des Protéines, Lyon, France

Bala K. Prabhala
Institute of Physics, Chemistry and Pharmacy, University of Southern Denmark, Odense, Denmark

Rajendra Prasad
Amity Institute of Biotechnology, Amity University Haryana, Gurgaon, Haryana, India; Amity Institute of Integrative Sciences and Health, Amity University Haryana, Gurgaon, Haryana, India

C. George Priya Doss
School of Biosciences and Technology, Vellore Institute of Technology, Vellore, Tamil Nadu, India

Moazur Rahman
School of Biological Sciences, University of the Punjab, Lahore, Punjab, Pakistan; Drug Discovery and Structural Biology Group, Health Biotechnology Division, National Institute for Biotechnology and Genetic Engineering (NIBGE), Faisalabad, Punjab, Pakistan

Gautam Rishi
Hepatogenomics Research Group, School of Biomedical Sciences, Faculty of Health, Queensland University of Technology (QUT), Brisbane, QLD, Australia

Maha Sellami
Sport Science Program (SSP), College of Arts and Sciences (CAS), Qatar University, Doha, Qatar

Mohammad Imran Shah
Department of Biotechnology and Bioinformatics, Jaypee University of Information Technology, Waknaghat, Solan, Himachal Pradesh, India; Navipoint Health India Pvt Ltd, Moula-Ali, Hyderabad, Telangana, India

Shailza Singh
National Centre for Cell Science, NCCS Complex, Ganeshkhind, SP Pune University Campus, Pune, Maharashtra, India

Bhavnita Soni
National Centre for Cell Science, NCCS Complex, Ganeshkhind, SP Pune University Campus, Pune, Maharashtra, India

Paweł Jan Stanirowski
1st Department of Obstetrics and Gynecology, Medical University of Warsaw, Warsaw, Poland; Club 35. Polish Society of Gynecologists and Obstetricians, Warsaw, Poland

V. Nathan Subramaniam
Hepatogenomics Research Group, School of Biomedical Sciences, Faculty of Health, Queensland University of Technology (QUT), Brisbane, QLD, Australia

Saad Tayyab
Biochemistry Programme, Institute of Biological Sciences, Faculty of Science, University of Malaya, Kuala Lumpur, Malaysia

D. Thirumal Kumar
School of Biosciences and Technology, Vellore Institute of Technology, Vellore, Tamil Nadu, India

S. Udhaya Kumar
School of Biosciences and Technology, Vellore Institute of Technology, Vellore, Tamil Nadu, India

Mirosław Wielgoś
1st Department of Obstetrics and Gynecology, Medical University of Warsaw, Warsaw, Poland

CHAPTER ONE

Biology of the iron efflux transporter, ferroportin

Gautam Rishi and V. Nathan Subramaniam*
Hepatogenomics Research Group, School of Biomedical Sciences, Faculty of Health, Queensland University of Technology (QUT), Brisbane, QLD, Australia
*Corresponding author: E-mail: nathan.subramaniam@qut.edu.au

Contents

1. Iron metabolism and the central role of ferroportin 1
   1.1 Identification of ferroportin 2
2. Structure of ferroportin 3
3. Regulation of ferroportin 8
4. Disorders associated with ferroportin dysfunction 10
5. Ferroportin partners-molecules involved in efficient ferroportin functioning 11
6. Conclusion 12
References 13

Abstract

Iron, the most common metal in the earth, is also an essential component for almost all living organisms. While these organisms require iron for many biological processes, too much or too little iron itself poses many issues; this is most easily recognized in human beings. The control of body iron levels is thus an important metabolic process which is regulated essentially by controlling the expression, activity and levels of the iron transporter ferroportin. Ferroportin is the only known iron exporter. The function and activity of ferroportin is influenced by its interaction with the iron-regulatory peptide hepcidin, which itself is regulated by many factors. Here we review the current state of understanding of the mechanisms that regulate ferroportin and its function.

1. Iron metabolism and the central role of ferroportin

Iron is essential for virtually all organisms. It plays an important role in many biological processes and acts as a co-factor of many proteins (Abbaspour, Hurrell, & Kelishadi, 2014; Rishi, Wallace, & Subramaniam, 2015). Its ability to exist in two redox states increases its utility as co-factor.

Advances in Protein Chemistry and Structural Biology, Volume 123
ISSN 1876-1623
https://doi.org/10.1016/bs.apcsb.2020.10.005

However, excess iron can also be injurious; at the center of this is the ease with which iron can change its oxidation state. Iron can undergo the Fenton reaction to generate free radical species including hydroxyl and superoxide iron which form a part of the collective reactive oxygen species (ROS). An accumulation of these reactive ions is related to several detrimental effects of iron accumulation resulting from damage to macromolecules including DNA, proteins, and lipids (Babitt & Lin, 2011). In the absence of a specific iron excretion mechanism the human body regulates physiological iron levels by modulating iron absorption, recycling and release from body stores (Rishi & Subramaniam, 2017). Two molecules have been identified as being crucial for maintaining iron homeostasis: the iron hormone, hepcidin and the iron exporter, ferroportin (FPN).

Human beings lack a defined mechanism for excreting excess iron, unlike other metals like copper; most of the iron lost by the body is lost by regular shedding of the epithelial cells lining the internal organs, and through blood loss during menstruation and childbirth in females (Ward & Kaplan, 2012). The absence of a defined mechanism to get rid of excess iron means that the levels of iron in the human body need to be tightly regulated; this regulation occurs primarily by modulating either iron absorption in the duodenum or its release from iron stores. The other challenge that multicellular organisms have to overcome is to develop mechanisms for efficient transfer of iron across tissues, which involves moving iron across the epithelia of the duodenum into the circulation (plasma) and then from the plasma to the iron-hungry tissues, and from the storage and recycling tissues (liver cells and splenic macrophages respectively) into the circulation. In mammals, FPN is at the center of these two regulatory processes (Ward & Kaplan, 2012). As an iron exporter, FPN is expressed on the surface of cells and is primarily involved in facilitating the release of iron from cells. This function of FPN is tightly regulated by a 25 amino acid peptide, hepcidin, also known as the master iron regulator (reviewed in Rishi et al., 2015). Hepcidin binds to FPN on the cell surface and results in its internalization and degradation (Nemeth et al., 2004). This interaction between hepcidin and FPN is central to the regulation of systemic iron homeostasis; dysregulation of either of these two molecules can lead to imbalances in body iron levels, leading to clinical complications as outlined previously.

1.1 Identification of ferroportin

Ferroportin was first identified in 2000 by three different research groups almost simultaneously using mice (Abboud & Haile, 2000; McKie et al.,

2000) and zebrafish models (Donovan et al., 2000). Each of the authors named the protein differently in each of these three studies: IREG1 (McKie et al., 2000), MTP1 (Abboud & Haile, 2000) or ferroportin1 (Donovan et al., 2000); however all three studies conclusively demonstrated that FPN was indeed an iron export protein. Mice with a deletion of *Fpn1* were embryonic lethal emphasizing the importance of iron and Fpn1 in early development (Donovan et al., 2005). However, in the same study a selective deletion of *Fpn1* in the visceral endoderm and placenta was not lethal but affected the growth and development of the knock-out mice, with knock-out mice developing severe neural tube defects (Donovan et al., 2005). Mutations in FPN lead to dysregulation of iron homeostasis (Montosi et al., 2001) providing another important piece of evidence supporting the essential role FPN plays in iron transport and homeostasis in general.

2. Structure of ferroportin

FPN is encoded by the Solute Carrier Family 40-member 1 (*SLC40A1*) gene. The most recent publicly available datasets suggest that the gene encodes for a 3330 bp mRNA encoded by 8 exons, which is then translated into a transmembrane protein made up of 571 amino acids. A comparison of the human, mouse and rat FPN protein sequences shows that the protein is more than 90% similar across these species, and that the molecular weight of the human and mouse protein is estimated to be approximately 62kD (Abboud & Haile, 2000).

The structure and topology of the protein itself have been controversial topics. Several studies have examined the number of transmembrane (TM) domains in the protein and suggested that it is between 9 and 12 TM domains (Devalia et al., 2002; Donovan et al., 2000; Goncalves et al., 2006; Liu, Yang, & Haile, 2005; McKie et al., 2000; Wallace, Harris, & Subramaniam, 2010). A detailed analysis of the mouse FPN protein using epitope tagging, mutagenesis and labeling experiments suggested the presence of either 10 or 12 TM domains (Liu, Yang, & Haile, 2005); Wallace et al. used functional analysis of FPN mutants identified in clinical studies and cell biology analysis as a basis for theoretical modeling to suggest that the protein has 12 TM domains (Wallace et al., 2010). The protein annotation on Pubmed (NP_055400.1) however lists the human FPN protein as having 10 TM domains. It is commonly accepted by experts in the field

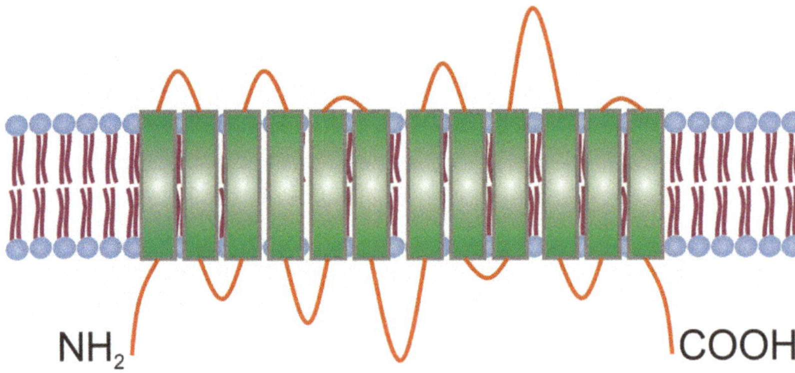

Fig. 1 Transmembrane structure of ferroportin. Schematic indicating the 12 transmembrane structure of ferroportin with amino and carboxy terminals of the transporter existing within the cytoplasm.

of iron metabolism that ferroportin consists of 12 transmembrane domains as indicated in Fig. 1.

The second controversy surrounding ferroportin structure involves the positioning of the carboxy terminal of the protein (Drakesmith, Nemeth, & Ganz, 2015; Pietrangelo, 2014). Several studies have suggested the carboxy terminal could either be extra- or intracellular using different epitope tags and detection techniques (indirect immunofluorescence and flow cytometry). Liu et al. (Liu, Yang, & Haile, 2005) and Rice et al. (Rice, Mendez, Hokanson, Rees, & Bjorkman, 2009) were unable to detect extracellular FPN expression in cells transfected with the carboxy terminal of the protein tagged with either Flag or green fluorescent protein (GFP), whereas in another study the hemagglutinin (HA) tag on the carboxy terminal of FPN could be detected in non-permeabilised cells, suggesting that the carboxy terminal of the protein is extracellular (Goncalves et al., 2006). These differences could be attributed to different tagged versions and different cell lines being used for these studies as all of these were overexpression systems (Wallace et al., 2010). In a recent publication using multiple amino and carboxy terminal epitope-tagged versions of FPN and specific antibodies Rishi et al. were able to demonstrate that both the amino and carboxy termini of the protein are indeed located intracellularly (Rishi, Secondes, Wallace, & Subramaniam, 2020).

Lastly, there are two schools of thought differing on the question whether FPN occurs as a monomer or multimer (Goncalves et al., 2006; Pietrangelo, 2014; Rice et al., 2009; Schimanski et al., 2008). This is particularly important in explaining the phenotype observed in patients with FPN mutations, as it could help explain whether the disease manifests because of haplo-insufficiency or from a dominant-negative effect. Most of our understanding of FPN structure and function does indeed come from analyzing the effect of identified mutations in hemochromatosis patients. Broadly, mutations affecting FPN function can be classified into either loss- or gain-of-function mutations (LOF/GOF). The LOF mutations impair the protein's ability to export iron; these mutations are mostly missense mutations and lead to the autosomal dominant form of the disease termed Ferroportin Disease (FD). In case of GOF mutations the FPN protein is essentially unresponsive to hepcidin and stays on the cell surface even in the presence of hepcidin. This results in an increased, unregulated absorption of iron in the duodenum leading to iron accumulation. These GOF mutations can either affect the hepcidin binding sites, or the sites essential for internalization and degradation of the protein. The existence of FPN as a multimeric protein would easily explain the dominant negative effect of the LOF mutations, however there are several studies which suggest otherwise. FPN appears to exist as a monomer in macrophages cultured from FD patients (Sabelli et al., 2017), where the results suggested that the LOF mutations may inhibit the effective localization of the protein by blocking the degradation or protein trafficking pathways, as more FPN protein could be detected in the lysosomes of macrophages from FD patients as compared to normal donors (Sabelli et al., 2017). The authors also showed that this phenomenon is related to body iron levels, the macrophages from FD patients cultured in lower iron-containing media were able to effectively express FPN on the surface and export iron; however this was impaired significantly when the iron levels were high. This contrasts with the previous studies where most of the LOF mutations have been shown to be causing a defect in protein trafficking to the membrane and hence affecting iron export (Wallace et al., 2010). However, all the previous studies had used transient overexpression models which may explain the aggregation of proteins in the endoplasmic reticulum. Some other studies have also suggested the presence of FPN monomers in both cultured cells and in vivo models, whereas studies from our laboratory and others suggest the existence FPN dimers or multimers (Rishi et al., 2020). The absence of a definitive crystal structure for FPN is one of the reasons for these shortfalls in our knowledge and a keen research question for several groups attempting to unravel the mechanisms of FPN function and its interactions with hepcidin.

The closet structure models based on comparisons to similar proteins in bacteria (Taniguchi et al., 2015) were initially identified in 2015. This study examined the crystal structure of a hypothetical bacterial iron transport protein (BbFPN, derived from *Bdellovibrio bacteriovorus*) with 24% identity and 40% similarity with human FPN and used it to model the structure of human FPN (Taniguchi et al., 2015). Functionally, this protein is capable of transporting several divalent cations including iron (Fe^{2+}) (Taniguchi et al., 2015). This structure revealed that BbFPN has 12 TM helices which are divided into the N lobe and C lobe, each comprising 6 TM helices. This model fits the previously proposed (Bonaccorsi di Patti et al., 2014) open-closed conformation model, where in the outward-facing structure the cavity between the amino and carboxy lobes opens outwards (extracellularly) and is not accessible from inside the cell, which then reverses in the inward-facing structure (Taniguchi et al., 2015). In the presence of sufficient intracellular iron the BbFPN undergoes a conformational change which allows iron to pass through the central cavity of the protein, these changes are prevented by hepcidin binding which arrests the molecule in the outward-facing state, thus restricting release of iron outside.

The latest study striving to understand the regulation of iron homeostasis and hepcidin-ferroportin interaction used a combination of cryo-electron microscopy, molecular simulations and biochemistry. These were performed on FPN and hepcidin using lipid nanodiscs (Billesbolle et al., 2020). The studies also analyzed FPN in the apo state and in complex with an iron mimetic, cobalt, instead of iron. This elegant analysis using human ferroportin identified metal binding sites in the amino and carboxy terminal domains of FPN. These studies confirm that hepcidin interaction with Fpn occurs in the outward conformation. Hepcidin inhibits iron transport by ferroportin by stoppering the efflux pathway (Billesbolle et al., 2020) (Fig. 2).

Additional analyses demonstrated that the hepcidin carboxy-terminus interacts directly with the divalent metal (in this case cobalt) in the FPN carboxy domain. Interestingly hepcidin binding to FPN is linked to iron binding with an 80-fold increased hepcidin affinity. Minimal binding was observed in the absence of metal (Billesbolle et al., 2020). These important findings suggest a new model of regulation of iron homeostasis where FPN-loaded with iron are the targets for hepcidin-mediated degradation. The FPN metal binding amino terminal domain appears to be the main site necessary for iron efflux while the metal binding carboxy-terminal domain plays an important role in hepcidin regulation of FPN activity.

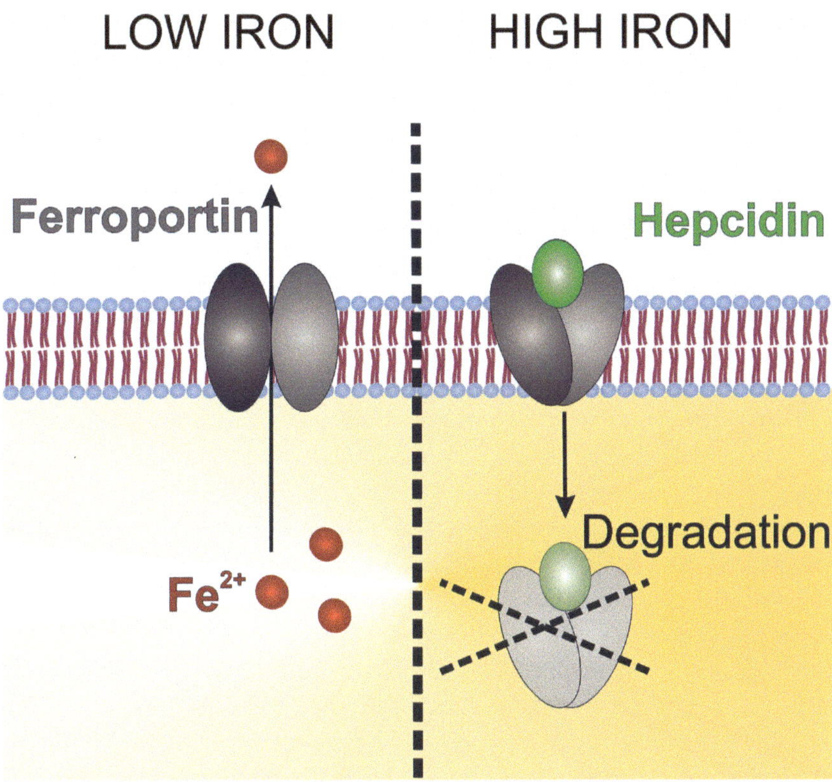

Fig. 2 Ferroportin structure and function. Schematic indicating the two lobes of ferroportin and the transport of iron through the ferroportin pore from the inside of the cell to exterior under conditions of low iron. In high iron conditions hepcidin acts as stopper to block iron transport through the pore. Hepcidin binding also results in the internalization and ubiquitination of ferroportin ending in its ultimate degradation.

In the most recent attempt to define the structure and function of ferroportin Li et al. generated and purified human recombinant ferroportin which was integrated into liposomes (Li, Yang, & Li, 2020). This study indicate that this recombinant ferroportin retains transport activity and transports ferrous iron preferentially and also transports protons. Using a series of known ferroportin mutants in these studies the authors indicate that iron is translocated through the pore region (Li et al., 2020) as opposed to studies using the bacterial ferroportin homolog in which iron binding is at the amino terminal lobe. Another key difference appears to be the inability of the bacterial protein to transport protons (Taniguchi et al., 2015).

In addition to the crystal structure comparisons and studies using recombinant proteins our other source of understanding the topology and function of FPN protein has been an examination of how the different mutations affect FPN function.

3. Regulation of ferroportin

FPN is expressed predominantly in cells which are involved in iron homeostasis; high levels of FPN have been reported in macrophages of the reticuloendothelial system, duodenal enterocytes, Kupffer cells in the liver, erythroblasts and placental syncytiotrophoblasts (Drakesmith et al., 2015). The enterocytes are responsible for transporting across the absorbed iron into the blood, whereas macrophages are involved in recycling the iron from senescent red blood cells; in both cases FPN plays the role of iron exporter. Given the different physiological roles of these cells, it is not surprising that FPN has cell-specific regulatory mechanisms. In addition, FPN can also be regulated at the transcriptional, post-transcriptional and post-translational levels.

Transcriptional control of FPN is prominent in cells exposed to heme, for example the macrophages recycling iron. FPN has an antioxidant response element (ARE) upstream of the transcription start site (Marro et al., 2010). The ARE's can be either bound by transcriptional repressors Btb and cnc homology 1 (BACH1) or transcriptional activators nuclear factor erythroid 2-like (NRF2). BACH1 is degraded in the presence of heme resulting in more NRF2 binding to the promoter leading to increased FPN transcription (Marro et al., 2010). This mechanism seems to play the role of an active control switch to prevent potentially damaging high iron levels in the macrophages. In addition, the FPN mRNA also contains an iron response element (IRE) in the 5′ untranslated region (Lymboussaki et al., 2003). This provides for another level of regulation where in the presence of iron-restricted conditions (iron deficiency), the IREs are bound by iron response proteins (IRPs) leading to translational repression and hence retention of iron in the macrophages (Lymboussaki et al., 2003). Together the heme-and iron-mediated regulation in the macrophages result in an efficient coordination of FPN expression which responds to decreased iron levels.

In the intestine, the regulatory mechanisms are poised toward controlling the absorption of iron rather than its release; the presence of an IRE in the 5′UTR of FPN mRNA is counter-intuitive in this case, as in iron deficiency it would lead to less FPN and hence less iron being absorbed. In iron-deficient conditions both FPN mRNA and protein levels have been shown to increase, at least in the intestine (Drakesmith et al., 2015). This is made possible by the overriding effect of hypoxia on the iron-mediated translational repression and partly due to the presence of an alternate transcript of FPN in the enterocytes which lacks the 5′ IRE in the UTR (Zhang, Hughes, Ollivierre-Wilson, Ghosh, & Rouault, 2009). In addition to the AREs, the FPN gene also contains hypoxia response elements (HREs), which are binding sites for hypoxia-inducible factors (HIFs). HIF1α and HIF2α both act as transcriptional regulators in response to hypoxia which responds to iron levels (Taylor et al., 2011). This regulation is mediated by HIF2α resulting in an increase in FPN mRNA levels in the intestine in conditions of iron deficiency (Taylor et al., 2011).

The most surprising finding about FPN was recently made using erythroblast-specific *Fpn* KO mice, which had serum iron levels similar to other Fpn cell-specific knockout mice, suggesting that erythroid FPN also plays a significant role in maintaining systemic body iron levels, at least in mice (Zhang, Ghosh, Ollivierre, Li, & Rouault, 2018). The erythroid cells also predominantly express the isoform lacking the 5′ IRE making it more responsive to iron deficiency conditions (Zhang et al., 2018). This discovery is an important step in understanding the delicately balanced process of iron homeostasis, as red blood cells (RBCs) are collectively the biggest reservoirs of body iron, and most of the iron in the body is utilized to synthesize heme for RBCs. In conditions of iron deficiency, the erythroblasts can keep pumping out emergency iron which would be essential for cells which may be more sensitive to low systemic iron levels, for example the cardiomyocytes and neurons (Zhang et al., 2018).

The inflammatory response acts as another post-transcriptional and an indirect post-translational regulator of FPN. Several studies have demonstrated a significant reduction in FPN mRNA levels in response to treatment with either the bacterial lipopolysaccharide (LPS) (Liu, Nguyen, Marquess, Yang, & Haile, 2005; Yang et al., 2002) or interferon gamma (Ludwiczek, Aigner, Theurl, & Weiss, 2003). The mechanisms of this transcriptional regulation are not clearly understood but seem to involve the toll-like receptors (TLRs) which act as receptors for a number of inflammation-induced cytokines and ligands including lipopolysaccharide (LPS) (Huang et al., 2012). This regulation appears to be another mechanism of restricting the availability of iron for the pathogens in response to infections in addition to the increase in hepcidin (inflammation is a positive regulator of hepcidin). The increased hepcidin limits the surface expression of FPN leading to reduced iron egress as well. How these two mechanisms fit, and the importance of the inflammation-mediated transcriptional regulation of FPN in hypoferremia of inflammation is unclear as this appears to have a relatively low effect on the serum iron levels (Arezes et al., 2015).

There have also been a few studies demonstrating the post-transcriptional regulation of FPN mRNA by micro RNAs (miRNA). miR-485-3p (Sangokoya, Doss, & Chi, 2013), miR17-5p (Kong et al., 2019), miR20a and miR20b (Jiang et al., 2019) have all been shown to bind to the 3′UTR of FPN mRNA leading to an increase in cellular iron content and resulting in decreased FPN expression.

The post-translational regulation of FPN is mediated by its biological ligand hepcidin. Hepcidin binding to FPN results in rapid ubiquitination of lysine residues; multiple ubiquitination sites are implicated in this process, however the lysines in the third TM loop of FPN seem to be the most important as mutating them resulted in a significant reduction in the internalization of FPN protein as compared to the wild type protein (Qiao et al., 2012). The discovery of this mechanism is relatively new and has opened up a new avenue for research. The ubiquitin-proteasome pathway (UPP) is one of the most important mechanisms to regulate protein turnover (Lecker, Goldberg, & Mitch, 2006). The system involves three steps catalyzed sequentially by three enzymes E1, E2 and E3. Although there is only one known functional E1 in mammalian organisms there are several known functional enzymes which can act as E2 and E3 (Lecker et al., 2006). This raises the question of cell-specific regulation of FPN degradation in response to hepcidin as well and would be an area of interest in the coming years.

4. Disorders associated with ferroportin dysfunction

Traditionally, mutations in FPN are classified under the heading of non-HFE hemochromatosis; however, as previously described these mutations can be classified into LOF or GOF and can lead to very different clinical presentations in patients (the known mutations in FPN have been listed elsewhere (Wallace & Subramaniam, 2016)). Broadly speaking, the GOF mutations lead to an iron overload phenotype similar to other hereditary hemochromatosis (HH) conditions, with increased parenchymal iron, increased transferrin saturation, increased serum ferritin levels and reduced hepcidin activity. The LOF mutations in FPN lead to a distinct form of iron overload with the iron predominantly detected in the Kupffer cells, hyperferritinemia is accompanied by low transferrin saturation; this form is usually referred to as the classical ferroportin disease (FD). Another distinction between the clinical effects of these mutations is the response of patients to phlebotomy, where the patients with GOF mutations generally respond favorably by normalizing their body iron levels as would be in HH cases, patients with FD have a tendency to develop anemia after phlebotomy.

These differences can be explained based on the effect of the mutations and considering the tissue specific consequences. The GOF mutations which are relatively rare, lead to a mutant FPN protein which is completely or partially capable of exporting iron but does not respond to hepcidin. These mutations may render the FPN molecule incapable of binding to hepcidin by interfering with the binding sites, or affect the internalization of the protein; both would result in an unregulated movement of iron from the gut into blood (increased and unregulated iron absorption), a characteristic common to other forms of HH, where there is usually insufficient amounts of hepcidin being produced in response to body iron levels.

Whereas in FD, FPN is incapable of exporting iron, these mutations could either affect protein trafficking or the iron transport ability of the protein. The result of these mutations would be an increase in iron levels in the cells and tissues involved in transporting iron across; an example is the macrophages in reticuloendothelial system leading to the Kupffer cell iron loading observed in these patients. Since enterocytes also need FPN in order to move iron across the duodenal enterocytes these mutations are also expected to affect iron absorption, and this becomes apparent in situations where FD patients undergo phlebotomy. Venesection removes blood and thus the major source of iron (recycling of senescent red blood cells), however, due to insufficient iron absorption through the gut due to defective iron export, this may result in anemia in FD patients.

5. Ferroportin partners-molecules involved in efficient ferroportin functioning

Ferroportin is the only known iron export protein. In the absence of a crystal structure of human FPN the mechanism of iron transport has been based on the mutation analysis and protein modeling, with the Asp39 and 181 position being determined as essential for substrate binding and transport (Bonaccorsi di Patti et al., 2014). Ferroportin can only export the ferric (Fe^{3+}) form of iron; however in the cell iron is stored in its ferrous form (Fe^{2+}). This oxidation to ferric iron is performed by two proteins ceruloplasmin (CP) and hephaestin. Both CP and hephaestin are multicopper enzymes capable of ferroxidase activity, thus essential for converting the ferrous form to transportable ferric form. Cp is primarily produced in the liver while hephaestin is highly expressed in the intestine, presumable adding another layer of cell specific mechanism of action and regulation of FPN activity (Fuqua et al., 2018).

Ferroxidase-active Cp stabilizes FPN at the plasma membrane supporting iron export; on the other hand, absence of Cp or presence of an inactive Cp lead to degradation of Fpn in specific cell types (Musci, Polticelli, & Bonaccorsi di Patti, 2014). However, no direct interaction between these two proteins has so far been detected. A similar role has been proposed for the amyloid precursor protein (APP) specially in the brain. Several studies (Dlouhy, Bailey, Steimle, Parker, & Kosman, 2019; McCarthy, Park, & Kosman, 2014; Wong et al., 2014) have shown that APP can stabilize the cell surface expression of labeled FPN and thus modify iron levels in the cells. Initially it was suggested that APP may also act as a ferroxidase in the brain thus assisting FPN in iron export in the brain (Duce et al., 2010), however later studies proved that this was not the case.

Hepcidin is the ligand for FPN. A functional analysis of FPN mutations suggests that hepcidin binds in the central cavity of the proposed FPN structure in between the amino and carboxy lobes. This binding induces a conformational change which then prevents the release of iron from FPN independent of the degradation of FPN. This suggest that hepcidin can regulate the egress of iron independent of FPN degradation in some cases (Aschemeyer et al., 2018). Inside the cavity hepcidin molecule is believed to interact with at least four helices, with Y501, D504 and A69 being the amino acids involved in hepcidin binding in addition to the previously known C326 residue (Preza et al., 2011).

6. Conclusion

So far FPN is the only known iron exporter molecule in humans and its interaction with hepcidin forms the central axis of iron metabolism making it one of the most suitable targets for interventions aimed at modulating body iron levels. In the last 20 years since its discovery, we have developed an understanding of the importance of FPN, mechanisms regulating FPN expression and what it does. However, there are a few gaps in this knowledge; until recently the structure of the FPN molecule has only been deduced using functional mutational analysis, protein modeling or extrapolated using BbFPN; in order to better understand how hepcidin interacts with FPN we will need the crystal structure of human FPN itself. There appear to be some tissue-specific isoforms of FPN, especially in the enterocytes and red blood cells; it is not known if they have a tissue/organ specific role and if there are tissue-specific E2 and E3 ubiquitination enzymes involved in FPN degradation. Several studies have suggested the possibility of FPN being a transporter for other metal ions (Troadec, Ward, Lo, Kaplan, & De Domenico, 2010; Yin et al., 2010), the biological importance of this function is unclear. What is the significance of inflammation-mediated FPN transcriptional repression and how does it occur? The answers to these questions would assist in designing better therapeutic interventions and enable us to understand the regulation of iron homeostasis in a tissue-specific context.

References

Abbaspour, N., Hurrell, R., & Kelishadi, R. (2014). Review on iron and its importance for human health. *Journal of Research in Medical Sciences: The Official Journal of Isfahan University of Medical Sciences, 19*(2), 164–174. https://pubmed.ncbi.nlm.nih.gov/24778671. https://www.ncbi.nlm.nih.gov/pmc/articles/PMC3999603/.

Abboud, S., & Haile, D. J. (2000). A novel mammalian iron-regulated protein involved in intracellular iron metabolism. *Journal of Biological Chemistry, 275*(26), 19906–19912. https://doi.org/10.1074/jbc.M000713200.

Arezes, J., Jung, G., Gabayan, V., Valore, E., Ruchala, P., Gulig, P. A., et al. (2015). Hepcidin-induced hypoferremia is a critical host defense mechanism against the siderophilic bacterium *Vibrio vulnificus*. *Cell Host & Microbe, 17*(1), 47–57. https://doi.org/10.1016/j.chom.2014.12.001.

Aschemeyer, S., Qiao, B., Stefanova, D., Valore, E. V., Sek, A. C., Ruwe, T. A., et al. (2018). Structure-function analysis of ferroportin defines the binding site and an alternative mechanism of action of hepcidin. *Blood, 131*(8), 899–910. https://doi.org/10.1182/blood-2017-05-786590.

Babitt, J. L., & Lin, H. Y. (2011). The molecular pathogenesis of hereditary hemochromatosis. *Seminars in Liver Disease, 31*(3), 280–292. https://doi.org/10.1055/s-0031-1286059.

Billesbolle, C. B., Azumaya, C. M., Kretsch, R. C., Powers, A. S., Gonen, S., Schneider, S., et al. (2020). Structure of hepcidin-bound ferroportin reveals iron homeostatic mechanisms. *Nature*. https://doi.org/10.1038/s41586-020-2668-z.

Bonaccorsi di Patti, M. C., Polticelli, F., Cece, G., Cutone, A., Felici, F., Persichini, T., et al. (2014). A structural model of human ferroportin and of its iron binding site. *FEBS Journal, 281*(12), 2851–2860. https://doi.org/10.1111/febs.12825.

Devalia, V., Carter, K., Walker, A. P., Perkins, S. J., Worwood, M., May, A., et al. (2002). Autosomal dominant reticuloendothelial iron overload associated with a 3-base pair deletion in the ferroportin 1 gene (SLC11A3). *Blood, 100*(2), 695–697. https://doi.org/10.1182/blood-2001-11-0132.

Dlouhy, A. C., Bailey, D. K., Steimle, B. L., Parker, H. V., & Kosman, D. J. (2019). Fluorescence resonance energy transfer links membrane ferroportin, hephaestin but not ferroportin, amyloid precursor protein complex with iron efflux. *Journal of Biological Chemistry, 294*(11), 4202–4214. https://doi.org/10.1074/jbc.RA118.005142.

Donovan, A., Brownlie, A., Zhou, Y., Shepard, J., Pratt, S. J., Moynihan, J., et al. (2000). Positional cloning of zebrafish ferroportin1 identifies a conserved vertebrate iron exporter. *Nature, 403*(6771), 776–781. https://doi.org/10.1038/35001596.

Donovan, A., Lima, C. A., Pinkus, J. L., Pinkus, G. S., Zon, L. I., Robine, S., et al. (2005). The iron exporter ferroportin/Slc40a1 is essential for iron homeostasis. *Cell Metabolism, 1*(3), 191–200. https://doi.org/10.1016/j.cmet.2005.01.003.

Drakesmith, H., Nemeth, E., & Ganz, T. (2015). Ironing out ferroportin. *Cell Metabolism, 22*(5), 777–787. https://doi.org/10.1016/j.cmet.2015.09.006.

Duce, J. A., Tsatsanis, A., Cater, M. A., James, S. A., Robb, E., Wikhe, K., et al. (2010). Iron-export ferroxidase activity of beta-amyloid precursor protein is inhibited by zinc in Alzheimer's disease. *Cell, 142*(6), 857–867. https://doi.org/10.1016/j.cell.2010.08.014.

Fuqua, B. K., Lu, Y., Frazer, D. M., Darshan, D., Wilkins, S. J., Dunn, L., et al. (2018). Severe iron metabolism defects in mice with double knockout of the multicopper ferroxidases hephaestin and ceruloplasmin. *Cellular and Molecular Gastroenterology and Hepatology, 6*(4), 405–427. https://doi.org/10.1016/j.jcmgh.2018.06.006.

Goncalves, A. S., Muzeau, F., Blaybel, R., Hetet, G., Driss, F., Delaby, C., et al. (2006). Wild-type and mutant ferroportins do not form oligomers in transfected cells. *Biochemical Journal, 396*(2), 265–275. https://doi.org/10.1042/BJ20051682.

Huang, Y. H., Yang, Y. L., Tiao, M. M., Kuo, H. C., Huang, L. T., & Chuang, J. H. (2012). Hepcidin protects against lipopolysaccharide-induced liver injury in a mouse model of obstructive jaundice. *Peptides, 35*(2), 212–217. https://doi.org/10.1016/j.peptides.2012.03.032.

Jiang, S., Fang, X., Liu, M., Ni, Y., Ma, W., & Zhao, R. (2019). MiR-20b down-regulates intestinal ferroportin expression in vitro and in vivo. *Cells, 8*(10). https://doi.org/10.3390/cells8101135.

Kong, Y., Hu, L., Lu, K., Wang, Y., Xie, Y., Gao, L., et al. (2019). Ferroportin downregulation promotes cell proliferation by modulating the Nrf2-miR-17-5p axis in multiple myeloma. *Cell Death & Disease, 10*(9), 624. https://doi.org/10.1038/s41419-019-1854-0.

Lecker, S. H., Goldberg, A. L., & Mitch, W. E. (2006). Protein degradation by the ubiquitin-proteasome pathway in normal and disease states. *Journal of the American Society of Nephrology, 17*(7), 1807–1819. https://doi.org/10.1681/ASN.2006010083.

Liu, X. B., Nguyen, N. B., Marquess, K. D., Yang, F., & Haile, D. J. (2005). Regulation of hepcidin and ferroportin expression by lipopolysaccharide in splenic macrophages. *Blood Cells Molecules and Diseases, 35*(1), 47–56. https://doi.org/10.1016/j.bcmd.2005.04.006.

Liu, X. B., Yang, F., & Haile, D. J. (2005). Functional consequences of ferroportin 1 mutations. *Blood Cells Molecules and Diseases, 35*(1), 33–46. https://doi.org/10.1016/j.bcmd.2005.04.005.

Li, S., Yang, Y., & Li, W. (2020). Human ferroportin mediates proton-coupled active transport of iron. *Blood Advances, 4*(19), 4758–4768. https://doi.org/10.1182/bloodadvances.2020001864.

Ludwiczek, S., Aigner, E., Theurl, I., & Weiss, G. (2003). Cytokine-mediated regulation of iron transport in human monocytic cells. *Blood, 101*(10), 4148–4154. https://doi.org/10.1182/blood-2002-08-2459.

Lymboussaki, A., Pignatti, E., Montosi, G., Garuti, C., Haile, D. J., & Pietrangelo, A. (2003). The role of the iron responsive element in the control of ferroportin1/IREG1/MTP1 gene expression. *Journal of Hepatology, 39*(5), 710–715. https://doi.org/10.1016/s0168-8278(03)00408-2.

Marro, S., Chiabrando, D., Messana, E., Stolte, J., Turco, E., Tolosano, E., et al. (2010). Heme controls ferroportin1 (FPN1) transcription involving Bach1, Nrf2 and a MARE/ARE sequence motif at position -7007 of the FPN1 promoter. *Haematologica, 95*(8), 1261–1268. https://doi.org/10.3324/haematol.2009.020123.

McCarthy, R. C., Park, Y. H., & Kosman, D. J. (2014). sAPP modulates iron efflux from brain microvascular endothelial cells by stabilizing the ferrous iron exporter ferroportin. *EMBO Reports, 15*(7), 809–815. https://doi.org/10.15252/embr.201338064.

McKie, A. T., Marciani, P., Rolfs, A., Brennan, K., Wehr, K., Barrow, D., et al. (2000). A novel duodenal iron-regulated transporter, IREG1, implicated in the basolateral transfer of iron to the circulation. *Molecular Cell, 5*(2), 299–309. http://www.ncbi.nlm.nih.gov/pubmed/10882071.

Montosi, G., Donovan, A., Totaro, A., Garuti, C., Pignatti, E., Cassanelli, S., et al. (2001). Autosomal-dominant hemochromatosis is associated with a mutation in the ferroportin (SLC11A3) gene. *Journal of Clinical Investigation, 108*(4), 619–623. https://doi.org/10.1172/JCI13468.

Musci, G., Polticelli, F., & Bonaccorsi di Patti, M. C. (2014). Ceruloplasmin-ferroportin system of iron traffic in vertebrates. *World Journal of Biological Chemistry, 5*(2), 204–215. https://doi.org/10.4331/wjbc.v5.i2.204.

Nemeth, E., Tuttle, M. S., Powelson, J., Vaughn, M. B., Donovan, A., Ward, D. M., et al. (2004). Hepcidin regulates cellular iron efflux by binding to ferroportin and inducing its internalization. *Science, 306*(5704), 2090–2093. https://doi.org/10.1126/science.1104742.

Pietrangelo, A. (2014). The ferroportin disease. *Clinics in Liver Disease, 3*(5), 98–100. https://doi.org/10.1002/cld.340.

Preza, G. C., Ruchala, P., Pinon, R., Ramos, E., Qiao, B., Peralta, M. A., et al. (2011). Mini-hepcidins are rationally designed small peptides that mimic hepcidin activity in mice and may be useful for the treatment of iron overload. *Journal of Clinical Investigation, 121*(12), 4880–4888. https://doi.org/10.1172/JCI57693.

Qiao, B., Sugianto, P., Fung, E., Del-Castillo-Rueda, A., Moran-Jimenez, M. J., Ganz, T., et al. (2012). Hepcidin-induced endocytosis of ferroportin is dependent on ferroportin ubiquitination. *Cell Metabolism, 15*(6), 918–924. https://doi.org/10.1016/j.cmet.2012.03.018.

Rice, A. E., Mendez, M. J., Hokanson, C. A., Rees, D. C., & Bjorkman, P. J. (2009). Investigation of the biophysical and cell biological properties of ferroportin, a multipass integral membrane protein iron exporter. *Journal of Molecular Biology, 386*(3), 717–732. https://doi.org/10.1016/j.jmb.2008.12.063.

Rishi, G., Secondes, E. S., Wallace, D. F., & Subramaniam, V. N. (2020). Evidence for dimerization of ferroportin in a human hepatic cell line using proximity ligation assays. *Bioscience Reports, 40*(5). https://doi.org/10.1042/BSR20191499.

Rishi, G., & Subramaniam, V. N. (2017). The liver in regulation of iron homeostasis. *American Journal of Physiology - Gastrointestinal and Liver Physiology, 313*(3), G157–G165. https://doi.org/10.1152/ajpgi.00004.2017.

Rishi, G., Wallace, D. F., & Subramaniam, V. N. (2015). Hepcidin: Regulation of the master iron regulator. *Bioscience Reports, 35*(3). https://doi.org/10.1042/BSR20150014.

Sabelli, M., Montosi, G., Garuti, C., Caleffi, A., Oliveto, S., Biffo, S., et al. (2017). Human macrophage ferroportin biology and the basis for the ferroportin disease. *Hepatology, 65*(5), 1512–1525. https://doi.org/10.1002/hep.29007.

Sangokoya, C., Doss, J. F., & Chi, J. T. (2013). Iron-responsive miR-485-3p regulates cellular iron homeostasis by targeting ferroportin. *PLoS Genetics, 9*(4), e1003408. https://doi.org/10.1371/journal.pgen.1003408.

Schimanski, L. M., Drakesmith, H., Talbott, C., Horne, K., James, J. R., Davis, S. J., et al. (2008). Ferroportin: Lack of evidence for multimers. *Blood Cells Molecules and Diseases, 40*(3), 360–369. https://doi.org/10.1016/j.bcmd.2007.09.007.

Taniguchi, R., Kato, H. E., Font, J., Deshpande, C. N., Wada, M., Ito, K., et al. (2015). Outward- and inward-facing structures of a putative bacterial transition-metal transporter with homology to ferroportin. *Nature Communications, 6*, 8545. https://doi.org/10.1038/ncomms9545.

Taylor, M., Qu, A., Anderson, E. R., Matsubara, T., Martin, A., Gonzalez, F. J., et al. (2011). Hypoxia-inducible factor-2alpha mediates the adaptive increase of intestinal ferroportin during iron deficiency in mice. *Gastroenterology, 140*(7), 2044–2055. https://doi.org/10.1053/j.gastro.2011.03.007.

Troadec, M. B., Ward, D. M., Lo, E., Kaplan, J., & De Domenico, I. (2010). Induction of FPN1 transcription by MTF-1 reveals a role for ferroportin in transition metal efflux. *Blood, 116*(22), 4657–4664. https://doi.org/10.1182/blood-2010-04-278614.

Wallace, D. F., Harris, J. M., & Subramaniam, V. N. (2010). Functional analysis and theoretical modeling of ferroportin reveals clustering of mutations according to phenotype. *American Journal of Physiology - Cell Physiology, 298*(1), C75–C84. https://doi.org/10.1152/ajpcell.00621.2008.

Wallace, D. F., & Subramaniam, V. N. (2016). The global prevalence of HFE and non-HFE hemochromatosis estimated from analysis of next-generation sequencing data. *Genetics in Medicine, 18*(6), 618–626. https://doi.org/10.1038/gim.2015.140.

Ward, D. M., & Kaplan, J. (2012). Ferroportin-mediated iron transport: Expression and regulation. *Biochimica et Biophysica Acta, 1823*(9), 1426–1433. https://doi.org/10.1016/j.bbamcr.2012.03.004.

Wong, B. X., Tsatsanis, A., Lim, L. Q., Adlard, P. A., Bush, A. I., & Duce, J. A. (2014). beta-Amyloid precursor protein does not possess ferroxidase activity but does stabilize the cell surface ferrous iron exporter ferroportin. *PLoS One, 9*(12), e114174. https://doi.org/10.1371/journal.pone.0114174.

Yang, F., Liu, X. B., Quinones, M., Melby, P. C., Ghio, A., & Haile, D. J. (2002). Regulation of reticuloendothelial iron transporter MTP1 (Slc11a3) by inflammation. *Journal of Biological Chemistry, 277*(42), 39786–39791. https://doi.org/10.1074/jbc.M201485200.

Yin, Z., Jiang, H., Lee, E. S., Ni, M., Erikson, K. M., Milatovic, D., et al. (2010). Ferroportin is a manganese-responsive protein that decreases manganese cytotoxicity and accumulation. *Journal of Neurochemistry, 112*(5), 1190–1198. https://doi.org/10.1111/j.1471-4159.2009.06534.x.

Zhang, D. L., Ghosh, M. C., Ollivierre, H., Li, Y., & Rouault, T. A. (2018). Ferroportin deficiency in erythroid cells causes serum iron deficiency and promotes hemolysis due to oxidative stress. *Blood, 132*(19), 2078–2087. https://doi.org/10.1182/blood-2018-04-842997.

Zhang, D. L., Hughes, R. M., Ollivierre-Wilson, H., Ghosh, M. C., & Rouault, T. A. (2009). A ferroportin transcript that lacks an iron-responsive element enables duodenal and erythroid precursor cells to evade translational repression. *Cell Metabolism, 9*(5), 461–473. https://doi.org/10.1016/j.cmet.2009.03.006.

CHAPTER TWO

The effect of sport and physical activity on transport proteins: implications for cancer prevention and control

Maha Sellami[a] and Nicola Luigi Bragazzi[b,c,*]

[a]Sport Science Program (SSP), College of Arts and Sciences (CAS), Qatar University, Doha, Qatar
[b]Postgraduate School of Public Health, Department of Health Sciences (DISSAL), University of Genoa, Genoa, Italy
[c]Laboratory for Industrial and Applied Mathematics (LIAM), Department of Mathematics and Statistics, York University, Toronto, ON, Canada
*Corresponding author: E-mail: bragazzi@yorku.ca

Contents

1. Transport proteins 18
2. Muscles 18
3. Muscle and transport proteins 21
4. Sport and transport proteins: implications for cancer treatment and prevention 22
5. Conclusions 24
References 25

Abstract

The present contribution briefly overviews the major biological functions of the plasma membrane and of the transport proteins (transporters), which enable the movement of different molecules and substrates (either charged or uncharged) by passive (facilitated diffusion) or active transport. In particular, transporters are overviewed at the level of the skeletal muscles, which represent a highly complex, heterogeneous, plastic and dynamic tissue and are one of the most abundant tissues in humans, accounting for up to 40% of their total weight and containing up to 50%–75% of all body proteins. Moreover, it is shown how sport and physical activity finely tune and modulate human proteome, especially in terms of structural and functional improvements concerning the density of the transport proteins. These changes are among the factors responsible for the positive outcomes of training, which involve mainly the cardiovascular and the endocrine/metabolic systems. Different kinds of training (strength and endurance) enable to achieve such improvements, even though there seems to exist a dose-relationship intensity-dependent effect, with responses after 6–8 weeks of exercise and disappearing in the chronic period (years of training). Finally, exercise-induced changes

Advances in Protein Chemistry and Structural Biology, Volume 123
ISSN 1876-1623
https://doi.org/10.1016/bs.apcsb.2020.07.001

at the level of transporters can play a role in terms of cancer prevention and management. Regular physical activity and exercise can, indeed, counteract the side-effects of chemotherapy drugs, including doxorubicin and other anthracycline derivatives, which may impair the functions of cardiac and skeletal muscles, probably modulating the expression of multidrug resistance proteins.

1. Transport proteins

The plasma membrane (variously referred to as cell or cytoplasmic membrane) exerts major biological functions, including separating and protecting the inner *milieu* from the outer environment and regulating the entering and exiting of molecules (Bernardino de la Serna, Schütz, Eggeling, & Cebecauer, 2016). Therefore, plasma membrane and its components play a vital and fundamental role, being instrumental in several cellular processes, like cell growth and functioning.

Transporters or transport proteins enable the movement of different molecules and substrates (either charged or uncharged) by passive (facilitated diffusion) or active transport, down or against their concentration gradient, exploiting, for instance, electrochemical gradient or ATP hydrolysis, respectively. According to the "Transporter Classification Database" (TCDB), approved by the "International Union of Biochemistry and Molecular Biology" (IUBMB), transport proteins can be basically divided into ion channels/pores and carriers (Table 1).

Whereas ion channels/pores include proteins differing in terms of structure (alpha-helical vs. beta-barrel porins, fusion pores, or membrane-bound channels, among others), there are, as well, several different kinds of carrier proteins. More in detail, carriers are integral membrane proteins involved in the movement of ions, small molecules, or greater compounds and substances, such as macromolecules, like proteins, moving through the cell membrane.

Carriers can be categorized into various groups: namely, (i) electrochemical potential-driven transporters (or porters, such as uniporters, and co-transporters, which include symporters and antiporters, facilitating the movement of specific small molecules and ions); (ii) primary active transporters; (iii) group translocators; and, (iv) transport electron carriers.

2. Muscles

Skeletal muscles represent a highly complex, heterogeneous, plastic and dynamic tissue and are one of the most abundant tissues in humans, accounting for up to 40% of their total weight and containing up to

Table 1 Classification of transport proteins.

Transporter family	Transporter sub-family	Examples
Ion channels/pores		Alpha-helical protein channels/alpha-type channels
		Beta-barrel protein porins
		Non-ribosomally synthesized channels
		Holins
		Vesicle fusion pores
		Viral fusion pores
		Paracellular channels
		Toxin channels/pore-forming toxins
		Peptide/membrane-bound channels
Carriers	Electrochemical potential-driven transporters	Porters (uniporters, symporters and antiporters)
	Primary active transporters	P—P-bond hydrolysis-driven transporters
		Decarboxylation-driven transporters
		Methyltransfer-driven transporters
		Oxidoreduction-driven transporters
		Light absorption-driven transporters
	Group translocators	Phosphotransfer-driven group translocators
		Nicotinamide ribonucleoside uptake transporters
		Acyl CoA ligase-coupled transporters
		Polysaccharide synthase/exporters
		Vacuolar polyphosphate polymerase-catalyzed group translocators
	Transport electron carriers	Transmembrane 1-electron transfer carriers
		Transmembrane 2-electron transfer carriers

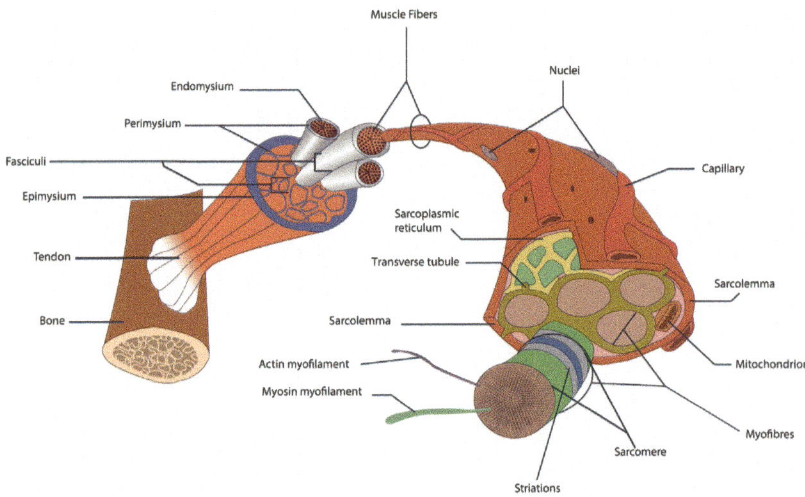

Fig. 1 Anatomic organization of the human skeletal muscle.

50%–75% of all body proteins (Frontera & Ochala, 2015; Weinert, 2009; Wolfe, 2006) (Fig. 1).

The heterogeneity of the human skeletal muscle reflects the variety of its tasks and functions, ranging from keeping posture to sub-maximal and maximal contractions (Bottinelli & Reggiani, 2000). Muscle is made up of multinucleated long and cylindrical cells called myofibres that are, in turn, composed of myofibrils (Fig. 1).

Myofibrils are made up of proteins termed as myofilaments, which include mainly actins and myosins. These contribute to forming the sarcomeres, the smallest contractile units (Fig. 2).

Myosin is the most commonly found muscle protein, playing a major role as fundamental component of the molecular motor of muscle contraction. Sarcomeric myosins belong to class II myosin and consist of two heavy

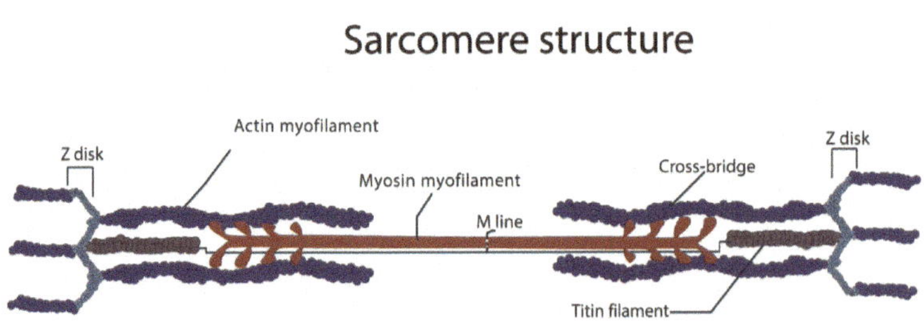

Fig. 2 Anatomic organization of the sarcomere.

chains (molecular weight of approximately 200 kD) and four light chains (molecular weight of around 20 kD) (Batters, Veigel, Homsher, & Sellers, 2014; Heissler & Sellers, 2016; Sellers, 2000).

The interaction between contractile proteins (termed as contraction–relaxation cycle) requires adequate amounts of energy that are mainly produced and released in an erobical fashion by mitochondria, which are particularly abundant in adult muscle fibers (Gonzalez-Freire et al., 2017). Besides biophysical and biochemical events, neuronal *stimuli* and coordination from spinal cord motor neurons *via* the neuromuscular junction are required. This results in acetylcholine release and subsequent activation of receptors, leading to the formation of an action potential, which spreads from the muscle cell surface to the muscle fibers through a specialized network of T tubules. The action potential stimulates the dihydropyridine receptor, enabling calcium (Ca^{2+}) influx into the cytoplasm, which results in the activation of ryanodine receptors. These trigger cascades that lead to further release of Ca^{2+} from the sarcoplasmic reticulum, the interaction between contractile proteins and the shortening of the muscle fiber.

Tendons are made up of collagen (mainly type I) and elastin embedded in a complex matrix consisting of proteoglycans and water with collagen and elastin accounting for up to 65%–80% and 1%–2% of the dry mass of the tendon, respectively. These elements are produced and released by specialized cells, termed as tenoblasts and tenocytes, which are the elongated fibroblasts and fibrocytes lying between collagen fibers.

Collagen fibers are the basic units of tendons, and consist of collagen fibrils, made up of soluble tropocollagen molecules cross-linked longitudinally, transversely and horizontally to form insoluble collagen and microfibrils. Collagen fibers are bound together by means of sheaths of connective tissues known as endotenon. A similar sheath of connective tissue termed as epitenon covers the tendon (Kannus, 2000).

3. Muscle and transport proteins

Muscle cells have different transport proteins (Bunprajun, Yuajit, Noitem, & Chatsudthipong, 2019; Holloway, Schwenk, Luiken, Glatz, & Bonen, 2010; Zorzano, Fandos, & Palacín, 2000), which are briefly overviewed in Table 2.

Table 2 An overview of transport proteins at the level of muscle cells.

Transport protein family	Example
Monocarboxylate transporters	MCT1, MCT2, MCT4
Sodium bicarbonate co-transporters	NBCe1, NBCe2
Sodium potassium pumps	α 1, α2, β1 Na,K-pumps
Sodium potassium chloride transporters	NKCC
Sodium-hydrogen exchangers/antiporters	NHE1
Glucose transporters	GLUT1, GLUT4
Fructose transporters	GLUT5, GLUT11
Potassium channels	K_{ATP}
Organic cation/carnitine transporter	OCTN2
Amino acid transporters	Taurine, glutamine transporters
Creatine transporters	CreaT/CRT1
Adenosine transporters	ATs
Ouabain binding transporters	
Water channels	AQP3, AQP4
Fatty acid transporters	FABPpm, FATP, and FAT/CD36

4. Sport and transport proteins: implications for cancer treatment and prevention

Sport and physical activity finely tune and modulate human proteome, especially in terms of structural and functional improvements concerning the density of the transport proteins. These changes are among the factors responsible for the positive outcomes of training, which involve the cardiovascular and the endocrine/metabolic systems (McGinley & Bishop, 2016). Different kinds of training (strength and endurance) enable to achieve such improvements, even though there seems to exist a dose-relationship intensity-dependent effect, with responses after 6–8 weeks of exercise and disappearing in the chronic period (years of training) (Juel, 2006).

From a cellular and molecular standpoint, different underlying mechanisms could explain these acute effects of remodeling of transport proteins: changes in substrate concentrations, Ca^{2+} levels, mitogen-activated protein kinases (MAPKs) and related pathways, adenosine monophosphate kinases (AMPKs) and related cascades, other kinases and networks, and interleukins such as interleukin-6 (IL-6) (Juel, 2006).

McGinley and Bishop (2016) investigated the impact of a four-week program of work-matched, high-intensity interval training exercise carried out above the lactate threshold or close to the peak aerobic power on human proteome. More in detail, the adaptive response was assessed in terms of changes in the levels of transport proteins belonging to the acid-base exchange and non bicarbonate muscle buffer capacity families. These included proteins such as the monocarboxylate transporters type 1 and 4 (MCT1 and MCT4), the chaperone protein basigin/CD147, the electrogenic sodium-bicarbonate co-transporter type e1 (NBCe1), the sodium/hydrogen exchanger type 1 (NHE1), and the carbonic anhydrases type II and XIV (CAII and CAXIV). Despite an overall improvement in each protein measured, based on training intensity, authors failed to detect changes for most proteins, except for CAIII and CAIV, which exhibited little differences, while MCT4 and basigin/CD147 varied according to the lactate threshold and peak aerobic power, respectively. These effects disappeared with detraining.

Sticka et al. (2018) explored the effects of conditioning programs on the expression of the glucose transporter 4 (GLUT4), the impairment of which is responsible for the development of type 2 diabetes. From the literature, it is well-known that exercise up-regulates GLUT4 levels, improving cellular glucose uptake and transport within skeletal muscle tissue. Authors recruited a sample of sixteen student athletes versus fifteen sedentary age- and sex-matched students. Authors found higher levels of GLUT4 after conditioning, especially among females.

Talanian et al. (2010) explored the effects of a six-week program of high-intensity interval training on the levels of fatty acid oxidation, fatty acid transport proteins (Bonen, Dyck, & Luiken, 1998; Jain et al., 2015; Luiken, Miskovic, Arumugam, Glatz, & Bonen, 2001) and their localization (namely, whole skeletal muscle, sarcolemmal, and mitochondrial membranes). A sample of ten untrained female subjects was recruited. After the training period, maximal oxygen uptake as well as the levels of FAT/CD36 at the levels of the whole muscle and mitochondrial membrane were found to be significantly increased, whereas remained unchanged at the sarcolemmal level. Concerning the fatty acid binding protein (FABPpm), its concentrations increased at the whole muscle and sarcolemmal levels, remaining unchanged at the mitochondrial level.

Millet et al. (2014) assessed the effects of a three-week intermittent hypoxic training (IHT) on anaerobic performance and skeletal muscle expression of transporters belonging to the family of monocarboxylate lactate transporters (namely, MCT1 and MCT4). For the purpose, the sample of recruited trained athletes was assigned to two interventional groups (normoxic and hypoxic groups). After training, authors found an increase in the average and peak power output, whereas VO_{2max} and MCT1 and MCT4 protein content remained unchanged. Normoxic and hypoxic training did not differ in terms of modulation and expression of muscle transport proteins.

Nielsen, Sabaratnam, Pedersen, Højlund, and Handberg (2019) have explored the impact of a bout of acute exercise (60 min of cycle ergometer) on long-chain fatty acid transporter mRNA content (CD36 and FATP4) in a sample of thirteen men with obesity and type 2 diabetes mellitus versus 14 obese but otherwise apparently healthy male controls. An overall exercise-induced change in transporter content was noted.

Whereas the positive outcomes of training affecting the cardiovascular and the endocrine/metabolic systems have been thoroughly investigated, less is known in terms of cancer prevention and management. From the existing scholarly literature, it is well-known that regular physical activity and exercise can counteract the cardiotoxicity of chemotherapy drugs, including doxorubicin and other anthracycline derivatives, which may impair the functions of cardiac and skeletal muscles. The precise mechanisms underlying these effects are yet to be elucidated. Some scholars hypothesize that exercise would lead to a decreased myocardial doxorubicin accumulation and reduced levels of oxidative stress biomarkers, like malondialdehyde (MDA) and 4-hydroxyalkenal (4-HDA). Quinn et al. (2016) and Gibson, Quinn, Pfannenstiel, Hydock, & Hayward (2014) speculated that exercise could finely tune and modulate the expression of multidrug resistance protein (MRP), such as MRP-1, MRP-2 and MRP-7. Utilizing an animal model, authors were able to detect lower left ventricle and *soleus* doxorubicin concentration rates, while cardiac MRP levels were increased. No significant changes in muscular MRP concentrations could be reported. Levels of oxidative stress biomarkers were found to be reduced both in heart and in skeletal muscles. Author concluded that the exercise-induced effects in terms of reduced side-effects of doxorubicin were site-specific, occurring at the cardiac but not muscular level.

5. Conclusions

Training induces modifications at the level of transport proteins. However, since data concerning the effects of exercise specific training programs (endurance, strength, etc.) on transporters in terms of cancer prevention and control are lacking, further research is needed in the field.

References

Batters, C., Veigel, C., Homsher, E., & Sellers, J. R. (March 11, 2014). To understand muscle you must take it apart. *Frontiers in Physiology, 5*, 90.

Bernardino de la Serna, J., Schütz, G. J., Eggeling, C., & Cebecauer, M. (September 29, 2016). There is No simple model of the plasma membrane organization. *Frontiers in Cell & Development Biology, 4*, 106.

Bonen, A., Dyck, D. J., & Luiken, J. J. (1998). Skeletal muscle fatty acid transport and transporters. *Advances in Experimental Medicine & Biology, 441*, 193–205.

Bottinelli, R., & Reggiani, C. (2000). Human skeletal muscle fibres: Molecular and functional diversity. *Progress in Biophysics & Molecular Biology, 73*(2–4), 195–262.

Bunprajun, T., Yuajit, C., Noitem, R., & Chatsudthipong, V. (March 2019). Exhaustive exercise decreases renal organic anion transporter 3 function. *The Journal of Physiological Sciences, 69*(2), 245–251.

Frontera, W. R., & Ochala, J. (March 2015). Skeletal muscle: A brief review of structure and function. *Calcified Tissue International, 96*(3), 183–195.

Gibson, N. M., Quinn, C. J., Pfannenstiel, K. B., Hydock, D. S., & Hayward, R. (2014). Effects of age on multidrug resistance protein expression and doxorubicin accumulation in cardiac and skeletal muscle. *Xenobiotica, 44*(5), 472–479.

Gonzalez-Freire, M., Semba, R. D., Ubaida-Mohien, C., Fabbri, E., Scalzo, P., Højlund, K., et al. (2017). The human skeletal muscle proteome project: A reappraisal of the current literature. *Journal of Cachexia Sarcopenia & Muscle, 8*(1), 5–18.

Heissler, S. M., & Sellers, J. R. (May 8, 2016). Various themes of myosin regulation. *Journal of Molecular Biology, 428*(9 Pt B), 1927–1946.

Holloway, G. P., Schwenk, R. W., Luiken, J. J., Glatz, J. F., & Bonen, A. (2010). Fatty acid transport in skeletal muscle: Role in energy provision and insulin resistance. *Clinical Lipidology, 5*(5), 731–745.

Jain, S. S., Luiken, J. J., Snook, L. A., Han, X. X., Holloway, G. P., Glatz, J. F., et al. (September 14, 2015). Fatty acid transport and transporters in muscle are critically regulated by Akt2. *FEBS Letters, 589*(19 Pt B), 2769–2775.

Juel, C. (2006). Training-induced changes in membrane transport proteins of human skeletal muscle. *European Journal of Applied Physiology, 96*(6), 627–635.

Kannus, P. (2000). Structure of the tendon connective tissue. *Scandinavian Journal of Medicine & Science in Sports, 10*(6), 312–320.

Luiken, J. J., Miskovic, D., Arumugam, Y., Glatz, J. F., & Bonen, A. (2001). Skeletal muscle fatty acid transport and transporters. *International Journal of Sport Nutrition and Exercise Metabolism, 11*(Suppl. l), S92–S96.

McGinley, C., & Bishop, D. J. (2016). Influence of training intensity on adaptations in acid/base transport proteins, muscle buffer capacity, and repeated-sprint ability in active men. *Journal of Applied Physiology, 121*, 1290–1305.

Millet, G., Bentley, D. J., Roels, B., Mc Naughton, L. R., Mercier, J., & Cameron-Smith, D. (2014). Effects of intermittent training on anaerobic performance and MCT transporters in athletes. *PLoS One, 9*(5), e95092.

Nielsen, M. H., Sabaratnam, R., Pedersen, A. J. T., Højlund, K., & Handberg, A. (2019). Acute exercise increases plasma levels of muscle-derived microvesicles carrying fatty acid transport proteins. *Journal of Clinical Endocrinology & Metabolism, 104*(10), 4804–4814.

Quinn, C. J., Gibson, N. M., Pfannenstiel, K. B., Bashore, A. C., Hayward, R., & Hydock, D. S. (2016). Effects of exercise on doxorubicin accumulation and multidrug resistance protein expression in striated muscle. *Global Journal of Medical Research, 6*, 1–13.

Sellers, J. R. (March 17, 2000). Myosins: A diverse superfamily. *Biochimica et Biophysica Acta, 1496*(1), 3–22.

Sticka, K. D., Schnurr, T. M., Jerome, S. P., Dajles, A., Reynolds, A. J., Duffy, L. K., et al. (2018). Exercise increases glucose transporter-4 levels on peripheral blood mononuclear cells. *Medicine & Science in Sports & Exercise, 50*(5), 938–944.

Talanian, J. L., Holloway, G. P., Snook, L. A., Heigenhauser, G. J., Bonen, A., & Spriet, L. L. (2010). Exercise training increases sarcolemmal and mitochondrial fatty acid transport proteins in human skeletal muscle. *American Journal of Physiology. Endocrinology and Metabolism, 299*(2), E180–E188.

Weinert, D. J. (August 2009). Nutrition and muscle protein synthesis: A descriptive review. *Journal of the Canadian Chiropractic Association, 53*(3), 186–193.

Wolfe, R. R. (September 2006). The underappreciated role of muscle in health and disease. *American Journal of Clinical Nutrition, 84*(3), 475–482.

Zorzano, A., Fandos, C., & Palacín, M. (August 1, 2000). Role of plasma membrane transporters in muscle metabolism. *Biochemical Journal, 349*(Pt 3), 667–688.

CHAPTER THREE

Cancer bioenergetics as emerging holistic cancer theory: the role of metabolic fluxes and transport proteins involved in metabolic pathways in the pathogenesis of malignancies. State-of-the-art and future prospects

Nicola Luigi Bragazzi[a,c,*,1] **and Maha Sellami**[b,1]
[a]Postgraduate School of Public Health, Department of Health Sciences (DISSAL), University of Genoa, Genoa, Italy
[b]Sport Science Program (SSP), College of Arts and Sciences (CAS), Qatar University, Doha, Qatar
[c]Laboratory for Industrial and Applied Mathematics (LIAM), Department of Mathematics and Statistics, York University, Toronto, ON, Canada
[*]Corresponding author: E-mail: bragazzi@yorku.ca

Contents

1. Introduction 28
2. Cancer and bio-energetics: the emerging field of "onco-energetics" 34
3. Altered glycolysis and pyruvate oxidation metabolic fluxes in cancer 36
4. Altered glutamine metabolic fluxes in cancer 36
5. Altered serine metabolic fluxes in cancer 37
6. Altered TCA metabolism in cancer 38
7. Altered methionine metabolism in cancer 38
8. Altered tryptophan metabolism in cancer 39
9. Altered leucine metabolism in cancer 40
10. Altered ornithine metabolism in cancer 40
11. Altered proline metabolism in cancer 40
12. Altered acetate metabolism in cancer 41
13. Altered lipid metabolism in cancer 41
14. Mitochondrial structural and functional impairment 42
15. Tumor microenvironment 42

[1] Both authors contributed equally.

Advances in Protein Chemistry and Structural Biology, Volume 123
ISSN 1876-1623
https://doi.org/10.1016/bs.apcsb.2020.09.001

16. Nuclear DNA genes mutations 43
17. Cellular pathways mutations 43
18. Conclusions and future prospects 44
References 44

Abstract

Cancer represents a global health concern, imposing a severe burden both from a societal and clinical perspective. Despite the latest advancements and achievements in the treatment and management of malignancy, cancer still imposes a dramatically high burden worldwide. Different theories (biophysical or biochemical, genetic or epigenetic) related to the origin of tumor cells have been put forth. These theories can also be subdivided into reductionist and emergentist/holistic theories. In the current overview, we will focus only on the cancer metabolic theory, one of the emergentist/holistic theories: it is holistic in that maintains that pathways, cascades and networks controlling energy metabolism, as well as those devoted to cell growth, cell cycle, replication, division and other cellular processes are highly interwoven and interconnected, and cannot be understood if not assuming a systems biology perspective. Cells should be seen as metabolic factories, in which metabolic fluxes and circuits (anabolic and catabolic) are plastically re-wired on the basis of the internal/external stimuli (cell make-up and genetic determinants, micro-environment, etc.). Complex regulatory and meta-regulatory systems exist that finely tune the functioning of cell, cell-cell communication and its interaction with the surrounding environment. At the tissue level, not all tissues share the same degree of metabolic plasticity (metabolic rigidity vs. metabolic flexibility), even though some metabolic coupling systems exist in order to guarantee an overall minimum extent of metabolic plasticity. The same broad picture of molecular events is necessary when describing the impairment and dysregulation of these processes, leading to multi-stage phenomena, including carcinogenesis.

1. Introduction

Cancer represents a global health concern, imposing a severe burden both from a societal and clinical perspective (Global Burden of Disease Cancer Collaboration et al., 2019).

Different types of treatment can be offered to mitigate against this burden in that cancer is a highly heterogeneous, complex and multi-factorial disease, with every patient suffering from a unique form of tumor: rather than being one-size-fits-it-all, each malignancy has its own molecular signature. Treatments can be prescribed alone or, more often, in combination. The choice of treatments should be adapted according to the specific needs of the patient (the so-called individualized or personalized oncology). The main types of cancer treatments are: surgery, chemotherapy, radiotherapy, hormone therapy, and targeted therapies, among others. All these treatments are aimed at eliminating and eradicating the disease. Depending also on the extension of the malignancy, they can act either locally, that is to say only at the level of the cancer cells of an affected organ (like surgery and radiotherapy), or systematically (such as chemotherapy or hormone therapy). Targeted therapies are selective and act against a specific molecular target at the level of the cancer cell itself.

Despite the latest advancements and achievements in the treatment and management of malignancy, cancer still imposes a dramatically high burden worldwide (Valle, Tramalloni, & Bragazzi, 2015). Different theories (biophysical or biochemical, genetic or epigenetic) related to the origin of tumor cells have been put forth, including the physio-mitotic theory (Hirata & Hirata, 2002a, 2002b, 2002c, 2003), the somatic mutation theory (SMT) (Kennedy, Loeb, & Herr, 2012; Sonnenschein & Soto, 2016), the tissue-organization field theory (TOFT) (Smythies, 2015), the two-hit theory (the so-called Knudson hypothesis) and its modification (the multi-hit theory or the Moolgavkar-Venzon-Knudson or MVK theory) (Bellacosa, 2003), the aneuploidy and paraneuploidy theories (Erenpreisa et al., 2014), the cancer stem cell theory (Allegra et al., 2014), the tumor microenvironment theory (Bizzarri & Cucina, 2014), and the multi-stage model of cancer development, among others (Paduch, 2015) (briefly overviewed in Table 1).

These theories can also be subdivided into reductionist and emergentist/holistic theories (Soto & Sonnenschein, 2005). SMT is considered the actual scientific dogma and is an example of the reductionist approach to cancer. However, despite its wide consensus, SMT has shown some inconsistencies and pitfalls, as it could have been obviously expected in explaining a phenomenon so complex, multi-faceted and multi-factorial, such as carcinogenesis (Baker & Kramer, 2007; Burgo & Migliore, 2015; Sonnenschein & Soto, 2000; Soto & Sonnenschein, 2004, pp. 15—18).

August Paul von Wassermann (February 21, 1866—March 16, 1925), a German bacteriologist and hygienist, in 1911, discovered that tumor cells consumed oxygen more avidly than normal cells and, utilizing murine models, hypothesized that selenium-eosin (*Eosin-Selen*), a disruptor of oxygen consumption, would have eventually reversed this condition (Brigelius-Flohe & Sies, 2015). Unfortunately, trials conducted among human subjects showed that selenium was toxic and, therefore, in 1929 the experimentations were discontinued. Otto Heinrich Warburg (October 8, 1883—August 1, 1970), a German physiologist, medical doctor and

Table 1 An overview of the major cancer theories.

Major cancer theory	Details/major premises
Physio-mitotic theory	Mitosis consists of two different phases: namely, maturation and duplication Cancerous tissue is a kind of organoid tissue
Somatic mutation theory (SMT)	Cancer, being a cellular proliferation problem, is derived from a single somatic cell that has accumulated multiple DNA mutations regarding genes controlling cell cycle Quiescence is the default status of all cells The theory relies on the Ame's axiom, which states that "substances act as carcinogens because they have mutagenic activity"
Tissue-organization field theory (TOFT)	Cancer represents a problem of tissue organization and not a mere cellular disorder Proliferation, and not quiescence, is the default state of all cells
Two-hit theory (the so-called Knudson hypothesis) and its modification (the multi-hit theory)	Cancer derives from two to eight "driver gene" mutations, leading to impaired pathways and networks and, finally, to the acquisition of a tumorigenic phenotype
Aneuploidy and paraneuploidy theories	Cancer arises aneuploid and aneuploidy/paraneuploidy leads to chromosomal instability and abnormality
Cancer stem cell theory	Epigenetic alterations and genetic events (mutations of genes responsible for signal transmission pathways and networks) may promote the formation of cancer stem cells
Tumor microenvironment theory	Microenvironment is not a secondary factor, but, instead, represents a major factor, being the physical-biochemical support of the "morphogenetic field", which guides cells toward differentiation and transformation
Multi-stage model of cancer development	Introduced by Armitage and Doll, after a thorough study of cancer registries, and modified by Pompei and Wilson, this theory states that cancer proceeds by a number of stages

Table 1 An overview of the major cancer theories.—cont'd

Major cancer theory	Details/major premises
Mitochondrial metabolic disease theory	Networks and pathways regulating the control of energy metabolism, cell growth, division and replication are highly interwoven and connected
Delayed hyper-sensitivity reaction theory	Chronic, long-standing inflammation caused by immune disruptors is characterized by changes that lead to increased immune suppression and multistep disease processes including cancer

Nobel laureate in Physiology or Medicine in 1931, has made important discoveries in the field of the metabolism of tumors, the fermentation and the respiration of cancer cells (Koppenol, Bounds, & Dang, 2011; Warburg, 1956). Since his investigations and the subsequent, pioneering works of Albert Lester Lenhinger (February 17, 1917—March 4, 1986), sir Hans Adolf Krebs (August 25, 1900—November 22, 1981), and other biochemists (including Chance, Petersen, Weinhouse, Vaupel, Williams, and Chi Van Dang, among others) (Obre & Rossignol, 2015), in the last decades a huge body of scientific evidence has accumulated to support the theory that cancer may be a metabolic disorder (Seyfried & Shelton, 2010; Seyfried, Flores, Poff, & D'Agostino, 2014; Verschoor et al., 2013), even though the initial observations of Warburg were at least partially wrong in that (i) mitochondrial impairment could be still reversible, and not irreversible as postulated by Warburg, (ii) the Warburg effect is not directly linked to respiration damage, since tumor cells can retain mitochondrial respiration, and (iii) not only glucose metabolism is altered in cancer as disruption of other metabolic pathways contribute to the pathogenesis of malignancy (Koppenol, Bounds, & Dang, 2011).

In the current overview, we will focus only on the cancer metabolic theory, one of the emergentist/holistic theories together with TOFT, tumor microenvironment theory and delayed hyper-sensitivity theory (Khatami, 2016). It is holistic in that, as we will see in the following paragraphs, maintains that pathways, cascades and networks controlling energy metabolism, as well as those devoted to cell growth, cell cycle, replication, division and other cellular processes are highly interwoven and interconnected, and cannot be understood if not assuming a systems biology perspective.

Cells should be seen as metabolic factories, in which metabolic fluxes and circuits (anabolic and catabolic) are plastically re-wired on the basis of the internal/external stimuli (cell make-up and genetic determinants, micro-environment, etc.). Complex regulatory and meta-regulatory systems exist that finely tune the functioning of cell, cell-cell communication and its interaction with the surrounding environment. At the tissue level, not all tissues share the same degree of metabolic plasticity (metabolic rigidity vs. metabolic flexibility), even though some metabolic coupling systems exist in order to guarantee an overall minimum extent of metabolic plasticity (Dart, 2016).

The same broad picture of molecular events is necessary when describing the impairment and dysregulation of these processes, leading to multi-stage phenomena, including carcinogenesis (Table 2 for an overview of the main new concepts introduced by the metabolic theory of cancer).

Table 2 A brief "metabolic dictionary", providing an overview of the new concepts introduced by the metabolic theory of cancer adapted from (DeBerardinis & Chandel, 2016; Martinez-Outschoorn et al., 2010, 2011; Zhou, Huang, & Wei, 2010).

New concept	Definition
Autophagic tumor stroma model of cancer metabolism, also known as battery-operated tumor growth model	Cancer cell is a "parasite", which utilizes oxidative stress as a "weapon" in order to extract recycled nutrients from cancer associated fibroblasts, *via* the induction of autophagy (a process termed "stromal autophagy"), that in its turn, provides energy-rich recycled nutrients (such as lactate, ketones, and glutamine) to fuel oxidative mitochondrial metabolism in cancer cells. Hypoxia, acidosis and other feature of the microenviroment can help driving autophagy. This scenario allows for the vectorial and unilateral transfer of energy from the tumor stroma (catabolism) to cancer cells (anabolism)
Lactate-producing tumor cells	Stromal cells and tumor-associated fibroblasts exhibit a reverse Warburg phenotype
Lactate-using tumor cells	Tumor cells exhibiting the classical Warburg phenotype
Metabolic coupling	In absence of tumor, it is a system that ensures a minimum, universal degree of metabolic uniformity among different tissues and metabolic compartments (mediating between metabolic rigidity and metabolic flexibility)

Table 2 A brief "metabolic dictionary", providing an overview of the new concepts introduced by the metabolic theory of cancer adapted from (DeBerardinis & Chandel, 2016; Martinez-Outschoorn et al., 2010, 2011; Zhou, Huang, & Wei, 2010).—cont'd

New concept	Definition
Metabolic oncogene	The unified concept emerging from the convergence of the metabolic and the genetic theories on cancer origin
Metabolic reprogramming/remodeling/rewiring	"Conventional metabolic pathways whose activities are enhanced or suppressed in tumor cells relative to benign tissues as a consequence of tumorigenic mutations and/or other factors" (DeBerardinis & Chandel, 2016)
Metabolic rigidity	Tissue-specific metabolic dependence on given substrates for energy synthesis and on given storage systems for energy stocking; metabolic rigidity is lost during cancer transformation
Metabolic switch	"Aberrant activation of aerobic glycolysis, de novo lipid biosynthesis and glutamine-dependent anaplerosis to fuel robust cell growth and proliferation" (Zhou et al., 2010)
Non-oncogene addiction	Survival of tumor cells depends not exclusively on the activation of oncogenes and inhibition of oncosuppressors, but also on non-tumorigenic genes (such as metabolic genes)
Oncometabolite	"Metabolites whose abundance increases markedly in tumors" (DeBerardinis & Chandel, 2016)
Oncometabolomics/oncometabonomics	Metabolomics/metabonomics applied in the field of cancer research
Reverse Warburg effect	The opposite of the conventional Warburg effect, theorized by the autophagic tumor stroma model of cancer metabolism (typical for example of lactate-producing stromal cells)
Warburg effect/Warburg phenotype	Enhanced glycolytic activity and impaired mitochondrial respiration/oxidative phosphorylation

2. Cancer and bio-energetics: the emerging field of "onco-energetics"

Cancer is a disease characterized by abnormal cell proliferation within normal body tissue. These cells are all derived from the same clone, the cancer-initiating cell that has acquired certain characteristics allowing it to divide indefinitely and to be able to form metastases. Cancer is a general term for any disease in which certain cells of the human body divide in an uncontrolled manner. The resulting new cells can contribute to the formation, development of a malignancy and its spread through the body *via* metastases.

The hallmarks of a cancer cell are, indeed:

1. the skill to promote and maintain proliferation *via ad hoc* signaling pathways and cascades;
2. insensitivity to anti-proliferative signals;
3. resistance to apoptosis;
4. unlimited proliferation (loss of senescence);
5. ability to induce angiogenesis; and,
6. ability to invade tissues and to spread.

In addition to these six characteristics, four new ones have been recently added:

7. genomic instability and mutation;
8. inflammatory microenvironment;
9. escape to the immune system; and,
10. change in energy metabolism.

From a bio-energetic standpoint, tumor is characterized by "metabolic reprogramming" (Dart, 2016; López-Ríos et al., 2007), a situation in which tumor cells switch from mitochondrial oxidative phosphorylation (OXPHOS), using glucose, fatty acids, and other metabolic intermediates as the energy sources, to glycolysis, a very low-energy providing and less efficient pathway to produce adenosine triphosphate (ATP), even when sufficient oxygen is present, that is to say even in aerobic conditions (the so-called "Warburg phenotype" or aerobic enhanced glycolysis) (Baffy, 2017; Ferreira, Hebrant, & Dumont, 2012; Formentini, Martínez-Reyes, & Cuezva, 2010). Metabolic reprogramming, as previously mentioned, is, indeed, the seventh hallmark of cancer (Corbert & Feron, 2017; Hanahan & Weinberg, 2000).

However, some recent metabolomic/metabonomic studies have shown that cancer cells are able to fully oxidize glucose (Hanahan & Weinberg,

2000). As briefly mentioned, the initial Warburg's theory has been then reframed and modified to take into account the recent discoveries. The current view is that tumor cell metabolism depends on an array of factors including (i) the bioavailability of metabolic fuels, (ii) their metabolic plasticity and flexibility, which is generally reduced due to mitochondrial impairments, and (iii) the micro-environmental conditions (hypoxia, acidosis, etc.). As such, the Warburg effect varies according to the kind of tumor, the type of tissue/organ involved and other individual features (Corbert & Feron, 2017; Hanahan & Weinberg, 2000).

As already mentioned, metabolic reprogramming/re-wiring of metabolic fluxes and networks is, indeed, highly complex and dynamic, and different factors can contribute to the Warburg effect (Table 3), as outlined in the following paragraphs.

This nuanced picture is emerging from the convergence of different molecular, cellular approaches and fields, resulting into a new super-specialized strain, named as onco-energetics.

Table 3 An overview of the main mechanisms/factors underlying the Warburg effect, adapted from (Carew & Huang, 2002).

Mechanisms/Factors of the Warburg phenotype
Altered glucose metabolism
Altered glutamine metabolism
Altered serine metabolism
Altered tricarboxylic acid (TCA) cycle
Altered methionine metabolism
Altered tryptophan metabolism
Altered lipid metabolism
Mitochondrial DNA (mtDNA) alterations, mutations, amplifications and deletions (at the level of complex I and complex IV of the respiratory chain; non-coding D-loop; 12S rRNA and 16S rRNA; ND1, ND2, ND4, ND4L, ND5, ND6; COX I, COX II and COX III; ATPase 6 and ATPase 8; cytochrome B and cytochrome C)
Mitochondrial alterations and impairments (at the level of membrane composition; network dynamics, motility, fusion and fission; electron transport system; biogenesis and biodegradation)
Nuclear DNA mutations, alterations and deletions (genes such as c-myc, Ras, Src, Akt, PTEN, cyclin D1, Notch, and p53)
Tumor micro-environment (hypoxia, acidosis)

3. Altered glycolysis and pyruvate oxidation metabolic fluxes in cancer

Probably because of the rapid cellular turnover, there is an imbalance between maximum rates of glycolysis and pyruvate oxidation. The latter is tightly regulated by the pyruvate dehydrogenase (PDH) complex, which transforms pyruvate into acetyl-CoA. Due to cell proliferation, the PDH flux cannot match the glycolytic flux, forcing cells to use other high-capacity metabolic systems, like lactate dehydrogenase isoform A (LDHA), which oxidizes nicotinamide adenine dinucleotide reduced form (NADH) into NAD^+ and transforms private into lactate, in order to avoid pyruvate accumulation (Corbet & Feron, 2017; DeBerardinis, 2008; Hanahan & Weinberg, 2000).

In addition, Warburg described an increase in lactate production and thus hypothesized that cancer can occur when the mitochondrial oxidative metabolism is altered. Oncogenes stimulating glycolysis have been discovered, including fetal isoforms re-expressed in cancer, such as PKM2 which is slower than PKM1 thus allowing the accumulation of glycolysis intermediates that will be used by the anabolic pathways to allow the proliferation (Israelsen & Vander Heiden, 2015). The Warburg effect also raises the hypothesis of dysfunctional mitochondria to explain why pyruvate is not degraded in the Krebs cycle. In the 1990s, it was shown that this mitochondrial dysfunction may be due to inhibition of PDH by over-expression of PDK1 (controlled by HIF1α), reduction of mitochondrial biogenesis (due to inactivation of p53) or inhibition of the activity of the respiratory chain (triggered by p53 mutated *via* SCO2 and HIF1α *via* a COX4A-1/2 isoform switch). Activation of the PI3K/Akt channel also promotes glycolysis and the entry of glucose into the cell during the Warburg effect. Indeed, Akt allows the increase of the number of glucose transporters to the membrane which allows a greater glucose entry into the cell. Akt promotes glycolysis (and thus the Warburg effect) by stimulating certain enzymes such as hexokinase and phosphofructokinase 1. Akt also intervenes in the synthesis of lipids from glucose (Robey & Hay, 2009).

4. Altered glutamine metabolic fluxes in cancer

Gutamine, a highly abundant amino-acid in human plasma (approximately 500 μM), plays a major role in different metabolic pathways and networks, in that it is a fundamental source of nitrogen for the de novo biosynthesis of nonessential amino-acids, nucleotides (purines and pyrimidines), and hexosamines. Moreover, it is a precious source of glutamate used for the synthesis of glutathione (DeBerardinis, 2008; Newman & Maddocks, 2017a,b), a tripeptide (Glu-Cys-Gly) involved in the neutralization of peroxide free radicals. Glutamine is a source also for the fatty acid synthesis, since the carbon derived from the uptaken glucose is excreted as lactate, rather than entering the tricarboxylic acid (TCA) cycle. Glutamine undergoes reductive carboxylation: glutamine-derived α-ketoglutarate is being reduced through the consumption of NADPH by the enzyme isocitrate dehydrogenase (IDH) in a non-canonical reverse reaction, in order to form citrate (Altman, Stine, & Dang, 2016; Zielinski et al., 2017).

While the enzyme glutaminase (GLS) was identified in the thirties in its two main isoforms ("kidney-type" or GLS1 and "liver-type" or GLS2), the interest in glutamine within the framework of cancer metabolism is quite recent (DeBerardinis, 2008; Newman & Maddocks, 2017a,b).

Summarizing glutamine metabolism is highly interconnected with different cellular functions and networks, including protein biosynthesis and trafficking, nucleotide biosynthesis, stress pathway inhibition and suppression, lipogenesis, and autophagy (DeBerardinis, 2008; Newman & Maddocks, 2017a,b).

5. Altered serine metabolic fluxes in cancer

Another metabolic pathway altered in tumor cells is the metabolism of serine (Alberghina et al., 2012; DeBerardinis, 2008; Newman & Maddocks, 2017a,b). Serine metabolism has been shown to be dysregulated in cancers; however, the result that this confers to tumors remains controversial and conflicting (Mattaini, Sullivan, & Vander Heiden, 2016). Serine plays a central role in anabolism, directly related to glucose conversion. Indeed, serine can be synthesized from an intermediate of glycolysis: 3-phosphoglycerate (3-PG). About 10% of this 3-PG is deflected for serine synthesis. Cancer cells are particularly avid of serine as a major source of one-carbon units, which are required for the production and assembly of different cellular components, such as nucleotides, and enable the biosynthesis and recycling of functional metabolites, such as ATP, S-adenosyl-methionine (SAM), NADH, and NADPH, among others (Alberghina et al., 2012; Smolková & Ježek, 2012; Yang & Vousden, 2016). Furthermore, serine metabolism is tightly interconnected with the liver kinase B1 isoform (LKB1)/5' AMP-activated protein kinase (AMPK)/mTOR signaling axis (Newman & Maddocks, 2017a,b), which regulates tumor growth and replication (Gao & Locasale, 2016).

Serine restriction interferes with tumor growth, whereas glycine restriction does not impede cancer cells proliferation (Kottakis et al., 2016), underlining the differential role of some metabolites. Up-regulation of the serine, glycine, and one-carbon (SGOC) metabolic pathway is required for the development of several malignancies, including neuroendocrine prostate cancer.

6. Altered TCA metabolism in cancer

As mentioned in the introduction, interest about TCA alterations in cancer metabolism is recent, in that according to the dogma of Warburg's initial hypothesis tumor cells were thought to bypass TCA cycle and, as such, the role of this metabolic pathway was considered to be marginal, not to say null (Anderson et al., 2017). Succinate dehydrogenase (SDH), fumarase (FH) and IDH are the enzymes of the citric acid (CAC) or TCA or Krebs cycle most frequently altered in tumor cells. FH converts fumarate to malate and acts as tumor suppressor; it is usually inactivated in cancer cells. The same holds for SDH, which is responsible of succinate oxidation producing fumarate (Shackelford & Shaw, 2009).

Genetic alteration in TCA cycle enzymes have been well demonstrated through decades. To date, the literature has found only a few recessive mutations whose clinical consequences were similar to changes in the electron transport chain (ETC) and oxidative phosphorylation. These changes were associated with multisystemic alterations and neurological lesions, but no predisposition to cancer. This is induced because of a very strong alteration of ATP synthesis. Over the last ten years, dominant defects associated with oncogenesis have been demonstrated in the cytoplasmic and mitochondrial isoforms of three nuclear-encoding enzymes, SDH, FH and IDH (Cardaci & Ciriolo, 2012), however, further research is need to better understand the extrametabolic roles of TCA cycle metabolites and their signaling toward the formation of tumors (King, Selak, & Gottlieb, 2006).

7. Altered methionine metabolism in cancer

Methionine, essential amino-acid first isolated in 1921 by John Howard Mueller, plays an important role in tumor cells. Methionine is an essential amino acid that can serve as a precursor to the synthesis of S-adenosylmethionine (SAM). It can also serve as a precursor to cysteine synthesis by condensing into serine to give cysteine and α-ketobutyrate. Propionyl-CoA arises from the oxidation of α-ketobutyrate, which can give succinyl-CoA. Cancer cells become, indeed, methionine-dependent and eventually follow apoptosis due to methionine restriction (Labuschagne, van den Broek, Mackay, Vousden, & Maddocks, 2014).

Methionine also plays an indispensable role in gene fine tuning, modulation, activation and inactivation due to hypermethylation and/or hypomethylation of a number of genes (Anderson, Mucka, Kern, & Feng, 2017). Methionine (MET) restriction (MR) has been demonstrated to arrest cancer developement and make tumors more sensitive to chemotherapy, together with other metabolites. Several cancer studies have suggested that methionine may be one of these signaling metabolites, as many cancers appear to depend on it for growth (Sharma, Singh, & Kanwar, 2014). However, the precise process through which methionine controls cell growth remains poorly understood from a molecular/cellular perspective (Ables & Johnson, 2017).

8. Altered tryptophan metabolism in cancer

L-Tryptophan (L-Trp) is an essential amino acid having different important biological functions. In addition to participate in the biosynthesis of proteins, tryptophan is metabolized in mammals according to various pathways, namely those of serotonin and kynurenine. In this last pathway, tryptophan is catabolized into end products such as acid Nicotine, niacin (vitamin B3) and nicotinamide adenine dinucleotide (NAD +). Although tryptophan enzymes 2,3-dioxygenase (TDO) and indoleamine 2,3-dioxygenase (IDO) do not have the same cellular location and are not induced by the same partners, these two enzymes both catalyze, similarly, the first step of the degradation of tryptophan; therefore, tryptophan, is catabolized in the tumor tissue by the rate-limiting enzyme IDO expressed in tumor cells or antigen-presenting cells (APCs), and plays a major role in cancer metabolism, contributing to create an immunosuppressive *milieu* by inducing T-cell anergy and apoptosis. Together with IDO, TDO represents another interesting druggable target (Ananieva, 2015; Platten, Wick, & Van den Eynde, 2012).

9. Altered leucine metabolism in cancer

Leucine, an essential amino-acid, is involved in several cellular processes, including cell activation and regulation, as well as in the process of carcinogenesis: more specifically, cancer cells compete with T cells for some metabolites, including leucine, and manage to escape immune surveillance. The loss of some enzymes deputed to leucine metabolism, as cytosolic or mitochondrial branched chain aminotransferases (BCATc and BCATm) leads to increased levels of intracellular leucine and fosters the activation of the mTORC1 pathway (Ananieva, 2015).

The role of leucine in cancer was noticed for the first time in a work by Waterhouse and Mason (Waterhouse & Mason, 1981). Some molecules like sirtuin 4 isoform (Sirt4) can control leucine metabolism.

10. Altered ornithine metabolism in cancer

Ornithine decarboxylase (ODC) is the key rate reducing enzyme in the pathway of polyamine synthesis and it is overexpressed in a numerous type of cancers, including endometrial cancer (Kim et al., 2017). Ornithine is a non-proteinogenic amino acid that plays a major role in the urea cycle. One of the first reports concerning ornithine is the investigation by Webber and colleagues (Webber et al., 1980).

In some endometrial cancers cells are sensitive to difluoromethylornithine (DFMO), which is the specific inhibitor of ODC. Successful results in tumor growth have been obtained in both humans and animals after administration of diets enriched with micro-nutrients and amino acids stimulating the immune system (including glutamine, arginine, ornithine and a-ketoglutarate, among others). Ornithine decarboxylase has been therefore considered as marker in biochemistry of tumor transformation of target cells. However, its mechanisms of induction of carcinogenesis and especially its modulation by xenobiotic substances still require further studies.

11. Altered proline metabolism in cancer

Proline is an amino acid that is not essential to the body. It can be synthesized by the latter from glutamic acid. It has different functions within the body: the synthesis of collagen (fundamental for the skin, joints, tendons) is the major one. Contributing to the healing of wounds, and being the precursor of hydroxyproline, Proline has an important role in the maintenance of connective tissues. Proline plays a major role also in apoptosis, autophagy and in response to nutrient and oxygen deprivation. Noteworthy, proline-derived reactive oxygen species can serve as an important driving signal for reprogramming. Since Proline is a major component of collagen (1/4 of the molecule), the latter could represent a proline reservoir for tumor cells. PYCR1, pyrroline-5-carboxylate reductase 1, is a crucial catalyzer enzyme in the synthesis pathway of proline synthesis which, when reversed, inhibits tumor growth. The proline pathway synthesis interacts with the glucose metabolism network. All these findings make Proline an important druggable molecule for the new targeted, personalized therapy of cancer (Phang, Liu, Hancock, & Fischer, 2015).

12. Altered acetate metabolism in cancer

Acetate, within the context of malignancies, acts as both an epigenetic regulator of posttranslational protein modification, and as a major carbon source for cancer cell biomass accumulation (Jaworski, Namboodiri, & Moffett, 2016).

Decomposition of the lipid is a main source of acetyl-CoA, and during hypoxic condition, acetate becomes a major source of acetyl-CoA. It is transported through the monocarboxylate transporter family. Acetyl-CoA synthetases (ACSS) acts as a catalyzer in the reaction of acetyl-CoA synthesis. This enzyme has been found in mitochondria and cytosol and for some researchers it has an important role in tumor growth.

Acetyl-CoA formation through acetate pathway is sometimes observed in hypoxic cells. It is poorly understood if this condition is the result of the increased/decreased rate of ACSS activity in situations characterized by deprivation of oxygen.

13. Altered lipid metabolism in cancer

Cancer cells get their energy from carbohydrates, proteins and lipids. In addition, they also need oxygen that they collect in the blood to optimize the extraction of energy from these nutrient sources. Cancer cells have very high energy needs, because of their very active metabolism: it is therefore always at the expense of the cells of the tissue that hosts them, that they capture the "food" they need. Cancer cells have different options for getting enough nutrients and oxygen. The cancer cells, however, reveal a metabolic fragility: the more a tumor grows, the more it becomes acidic, the more it becomes dependent on lipids as an energy source. Therefore, lipid metabolism plays a major role in conferring cancer a malignant and particularly aggressive feature. Lipid metabolism clinically correlates with tumor invasion and metastases, as well as with survival (Luo et al., 2017).

Acidity induces a large consumption of fatty acids by the tumors, but not only: they also use a lot of glutamine, as previously overviewed.

14. Mitochondrial structural and functional impairment

The OXPHOS system is divided into two segments: the oxidative system (OS), formed of substrate transporters, Krebs cycle enzymes and electron transport system (ETS), and the phosphorylating system (PS), consisting of the ATP/adenosine diphosphate (ADP) translocase (ANT) and Pi (PiT) transporters and ATP synthase.

Tumor mitochondria are characterized by altered features in morphology and ultra-structure, including protein and lipid composition and increased mass, and are, as such, responding differently to growth media. Moreover, the pathways and networks regulating calcium homeostasis are impaired. Some metabolic enzymes are altered, like the surface-bound, fetal-like hexokinase, insensitive to glucose-6-phosphate (G6P) mediated inhibition. Further, a down-regulation of the catalytic subunit of the mitochondrial H^+-ATP synthase (beta-F1-ATPase) usually occurs in most tumors. This down-regulation is associated with evasion of apoptosis and cell-death and reduced clinical survival. For its relevance, this molecular signature has been termed as a major "bio-energetic signature of cancer" (BEC).

Other features include increased amount in the natural inhibitor protein (IF1) of the mitochondrial ATP synthase.

15. Tumor microenvironment

The rapid proliferation of cancer cells often generates a hypoxic microenvironment triggering the stabilization of the hypoxia inducible factor 1 alpha (HIF-1a). HIF-1a, aided by c-myc, c-ras, c-src, induces the expression of many glycolytic enzymes, including LDHA, GLUT1 and GLUT3, pyruvate dehydrogenase kinase 1 (PDK1), PFKFB3, PFKFB4,

PKM2, and hexokinase II (HXK2). Moreover, HIF-1a contributes to the repression of the biogenesis of mitochondria by inhibiting c-myc activity and also affects mitochondrial respiration by switching the molecular composition of cytochrome *c* oxidase.

16. Nuclear DNA genes mutations

Genes, such as c-myc, regulate and modulate the expression of other genes related to ribosomal biosynthesis and biogenesis as well as mitochondria, and can trans-activate glycolytic genes, like LDHA binding the promoter cis-acting elements, which resemble the core carbohydrate response element (ChoRE), CACGTG, as well as the glucose transporter 1 (GLUT1), enhancing both glucose uptake and lactate production.

Other genes include Ras, Src, Akt, Bcr-abl and p53. Ras enhances glucose uptake and metabolism as well as mitochondrial bio-energetics, besides increasing the entry of glucose carbon into the TCA cycle and different cataplerotic pathways, including a singular process termed as "autophagic cell-death", in which macropinocitosis and cell-death are activated in order to provide tumor cells with glutamine derived from autophagic degradation of intracellular proteins, among others.

p53, the master guardian of the genome, can directly activate genes such as TP53-inducible glycolysis and apoptosis regulator (TIGAR), a PFKFB family member that inhibits glycolysis, shunting glucose into the pentose phosphate pathway (PPP). p53 also activates genes such as SCO2 that enhance more efficient mitochondrial respiration, as well as the gene coding for phosphate-activated glutaminase (GLS2), a p53-inducible regulator of glutamine metabolism and reactive oxygen species (ROS). Finally, p53 has also a direct apoptogenic role at the level of mitochondria (Mihara et al., 2003).

17. Cellular pathways mutations

Altered interactions between some cascades or pathways, such as the canonical WNT/beta-catenin pathway and peroxisome proliferator-activated receptor gamma (PPAR gamma) (Lecarpentier, Claes, Vallée, & Hébert, 2017) have an impact on the metabolic and bio-energetic capacity of tumors. The role of other cellular pathways mutations and their impact on onco-metabolism has yet to be fully understood.

18. Conclusions and future prospects

Metabolic targets appear promising in the efforts of treating cancer. However, further larger high-quality experimentations are urgently needed. With this regard, it is anticipated that the application of omics sciences, such as oncometabolomics/oncometabonomics, utilizing sophisticated approaches, like pulsed stable isotope-resolved metabolomics (pSIRM) and metabolic flux analysis (MFA), will contribute to further dissecting the molecular basis of tumor metabolism and individuating new promising pharmacological targets.

References

Ables, G. P., & Johnson, J. E. (August 2017). Pleiotropic responses to methionine restriction. *Experimental Gerontology, 94*, 83–88.

Alberghina, L., Gaglio, D., Gelfi, C., Moresco, R. M., Mauri, G., Bertolazzi, P., et al. (September 12, 2012). Cancer cell growth and survival as a system-level property sustained by enhanced glycolysis and mitochondrial metabolic remodeling. *Frontiers in Physiology, 3*, 362.

Allegra, A., Alonci, A., Penna, G., Innao, V., Gerace, D., Rotondo, F., et al. (November 2014). The cancer stem cell hypothesis: A guide to potential molecular targets. *Cancer Investigation, 32*(9), 470–495.

Altman, B. J., Stine, Z. E., & Dang, C. V. (October 2016). From Krebs to clinic: Glutamine metabolism to cancer therapy. *Nature Reviews Cancer, 16*(10), 619–634.

Ananieva, E. (November 26, 2015). Targeting amino acid metabolism in cancer growth and anti-tumor immune response. *World Journal of Biological Chemistry, 6*(4), 281–289.

Anderson, N. M., Mucka, P., Kern, J. G., & Feng, H. (July 26, 2017). The emerging role and targetability of the TCA cycle in cancer metabolism. *Protein Cell*. https://doi.org/10.1007/s13238-017-0451-1 [Epub ahead of print].

Baffy, G. (August 2017). Mitochondrial uncoupling in cancer cells: Liabilities and opportunities. *Biochimica et Biophysica Acta, 1858*(8), 655–664.

Baker, S. G., & Kramer, B. S. (August 6, 2007). Paradoxes in carcinogenesis: New opportunities for research directions. *BMC Cancer*, 7, 151.

Bellacosa, A. (December 2003). Genetic hits and mutation rate in colorectal tumorigenesis: Versatility of Knudson's theory and implications for cancer prevention. *Genes Chromosomes and Cancer, 38*(4), 382–388.

Bizzarri, M., & Cucina, A. (2014). Tumor and the microenvironment: A chance to reframe the paradigm of carcinogenesis? *BioMed Research International, 2014*, 934038.

Brigelius-Flohe, R., & Sies, H. (2015). *Diversity of selenium functions in health and disease*. CRC Press.

Burgio, E., & Migliore, L. (April 2015). Towards a systemic paradigm in carcinogenesis: Linking epigenetics and genetics. *Molecular Biology Reports, 42*(4), 777–790.

Cardaci, S., & Ciriolo, M. R. (2012). TCA cycle defects and cancer: when metabolism tunes redox state. *International Journal of Cell Biology, 2012*, 161837.

Carew, J. S., & Huang, P. (2002). Mitochondrial defects in cancer. *Molecular Cancer, 1*, 9.

Corbet, C., & Feron, O. (August 2017). Cancer cell metabolism and mitochondria: Nutrient plasticity for TCA cycle fueling. *Biochimica et Biophysica Acta, 1868*(1), 7–15.

Dart, A. (September 23, 2016). Tumour metabolism: Metabolic flexibility. *Nature Reviews Cancer, 16*(10), 617.
DeBerardinis, R. J. (November 2008). Is cancer a disease of abnormal cellular metabolism? New angles on an old idea. *Genetics in Medicine, 10*(11), 767–777.
DeBerardinis, R. J., & Chandel, N. S. (2016). Fundamentals of cancer metabolism. *Science Advances, 2*(5), e1600200.
Erenpreisa, J., Salmina, K., Huna, A., Jackson, T. R., Vazquez-Martin, A., & Cragg, M. S. (December 17, 2014). The "virgin birth", polyploidy, and the origin of cancer. *Oncoscience, 2*(1), 3–14.
Ferreira, L. M., Hebrant, A., & Dumont, J. E. (September 6, 2012). Metabolic reprogramming of the tumor. *Oncogene, 31*(36), 3999–4011.
Formentini, L., Martínez-Reyes, I., & Cuezva, J. M. (July 2010). The mitochondrial bioenergetic capacity of carcinomas. *IUBMB Life, 62*(7), 554–560.
Gao, X., & Locasale, J. W. (December 13, 2016). Serine metabolism links tumor suppression to the epigenetic landscape. *Cell Metabolism, 24*(6), 777–779.
Global Burden of Disease Cancer Collaboration, Fitzmaurice, C., Abate, D., Abbasi, N., Abbastabar, H., et al. (September 27, 2019). Global, regional, and national cancer incidence, mortality, years of life lost, years lived with disability, and disability-adjusted life-years for 29 cancer groups, 1990 to 2017: A systematic analysis for the global burden of disease study. *JAMA Oncology, 5*(12), 1749–1768.
Hanahan, D., & Weinberg, R. A. (January 7, 2000). The hallmarks of cancer. *Cell, 100*(1), 57–70.
Hirata, Y., & Hirata, S. (May 2002a). Physio-mitotic theory and a new concept of cancer development. *Medical Hypotheses, 58*(5), 361–364.
Hirata, Y., & Hirata, S. (May 2002b). Physio-mitotic theory and a new concept of embryological differentiation. *Medical Hypotheses, 58*(5), 365–368.
Hirata, Y., & Hirata, S. (April 2002c). Physio-mitotic theory and a new concept of leukemia development. *Medical Hypotheses, 58*(4), 251–253.
Hirata, Y., & Hirata, S. (October 2003). Fundamental physio-mitotic theory. *Medical Hypotheses, 61*(4), 449–457.
Israelsen, W. J., & Vander Heiden, M. G. (July 2015). Pyruvate kinase: Function, regulation and role in cancer. *Seminars in Cell and Developmental Biology, 43*, 43–51.
Jaworski, D. M., Namboodiri, A. M., & Moffett, J. R. (2016). Acetate as a metabolic and epigenetic modifier of cancer therapy. *The Journal of Cellular Biochemistry, 117*(3), 574–588.
Kennedy, S. R., Loeb, L. A., & Herr, A. J. (April 2012). Somatic mutations in aging, cancer and neurodegeneration. *Mechanism of Ageing and Development, 133*(4), 118–126.
Khatami, M. (December 2016). Is cancer a severe delayed hypersensitivity reaction and histamine a blueprint? *Clinical and Translational Medicine, 5*(1), 35.
Kim, H. I., Schultz, C. R., Buras, A. L., Friedman, E., Fedorko, A., & Seamon, L. (2017). Ornithine decarboxylase as a therapeutic target for endometrial cancer. *PLoS One, 12*(12), e0189044.
King, A., Selak, M. A., & Gottlieb, E. (2006). Succinate dehydrogenase and fumarate hydratase: linking mitochondrial dysfunction and cancer. *Oncogene, 25*(34), 4675–4682.
Koppenol, W. H., Bounds, P. L., & Dang, C. V. (May 2011). Otto Warburg's contributions to current concepts of cancer metabolism. *Nature Reviews Cancer, 11*(5), 325–337.
Kottakis, F., Nicolay, B. N., Roumane, A., Karnik, R., Gu, H., Nagle, J. M., et al. (November 17, 2016). LKB1 loss links serine metabolism to DNA methylation and tumorigenesis. *Nature, 539*(7629), 390–395.
Labuschagne, C. F., van den Broek, N. J., Mackay, G. M., Vousden, K. H., & Maddocks, O. D. (May 22, 2014). Serine, but not glycine, supports one-carbon metabolism and proliferation of cancer cells. *Cell Reports,* 7(4), 1248–1258.

Lecarpentier, Y., Claes, V., Vallée, A., & Hébert, J. L. (2017). Interactions between PPAR Gamma and the canonical Wnt/Beta-catenin pathway in type 2 diabetes and colon cancer. *PPAR Research, 2017*, 5879090.

López-Ríos, F., Sánchez-Aragó, M., García-García, E., Ortega, A. D., Berrendero, J. R., Pozo-Rodríguez, F., et al. (October 1, 2007). Loss of the mitochondrial bioenergetic capacity underlies the glucose avidity of carcinomas. *Cancer Research, 67*(19), 9013–9017.

Luo, X., Cheng, C., Tan, Z., Li, N., Tang, M., Yang, L., et al. (2017). Emerging roles of lipid metabolism in cancer metastasis. *Molecular Cancer, 16*(1), 76.

Mattaini, K. R., Sullivan, M. R., & Vander Heiden, M. G. (2016). The importance of serine metabolism in cancer. *The Journal of Cell Biology, 214*(3), 249–257.

Martinez-Outschoorn, U. E., Whitaker-Menezes, D., Pavlides, S., Chiavarina, B., Bonuccelli, G., Casey, T., et al. (2010). The autophagic tumor stroma model of cancer or "battery-operated tumor growth": a simple solution to the autophagy paradox. *Cell Cycle, 9*(21), 4297–4306.

Martinez-Outschoorn, U. E., Pavlides, S., Howell, A., Pestell, R. G., Tanowitz, H. B., Sotgia, F., et al. (2011). Stromal-epithelial metabolic coupling in cancer: integrating autophagy and metabolism in the tumor microenvironment. *The International Journal of Biochemistry and Cell Biology, 43*(7), 1045–1051.

Mihara, M., Erster, S., Zaika, A., Petrenko, O., Chittenden, T., & Pancoska, P. (2003). p53 has a direct apoptogenic role at the mitochondria. *Molecular Cell, 11*(3), 577–590.

Newman, A. C., & Maddocks, O. D. K. (June 6, 2017a). One-carbon metabolism in cancer. *British Journal of Cancer, 116*(12), 1499–1504.

Newman, A. C., & Maddocks, O. D. K. (June 7, 2017b). Serine and functional metabolites in cancer. *Trends in Cell Biology*. https://doi.org/10.1016/j.tcb.2017.05.001. pii: S0962-8924(17)30075-2. [Epub ahead of print].

Obre, E., & Rossignol, R. (February 2015). Emerging concepts in bioenergetics and cancer research: Metabolic flexibility, coupling, symbiosis, switch, oxidative tumors, metabolic remodeling, signaling and bioenergetic therapy. *The International Journal of Biochemistry and Cell Biology, 59*, 167–181.

Paduch, R. (January 2015). Theories of cancer origin. *European Journal of Cancer Prevention, 24*(1), 57–67.

Phang, J. M., Liu, W., Hancock, C. N., & Fischer, J. W. (2015). Proline metabolism and cancer: emerging links to glutamine and collagen. *Current Opinion in Clinical Nutrition & Metabolic Care, 18*(1), 71–77.

Platten, M., Wick, W., & Van den Eynde, B. J. (November 1, 2012). Tryptophan catabolism in cancer: Beyond Ido and tryptophan depletion. *Cancer Research, 72*(21), 5435–5440.

Robey, R. B., & Hay, N. (2009). Is Akt the "Warburg kinase"?-Akt-energy metabolism interactions and oncogenesis. *Seminars in Cancer Biology, 19*(1), 25–31.

Seyfried, T. N., Flores, R. E., Poff, A. M., & D'Agostino, D. P. (March 2014). Cancer as a metabolic disease: Implications for novel therapeutics. *Carcinogenesis, 35*(3), 515–527.

Seyfried, T. N., & Shelton, L. M. (January 27, 2010). Cancer as a metabolic disease. *Nutrition and Metabolism, 7*, 7.

Shackelford, D. B., & Shaw, R. J. (August 2009). The LKB1-AMPK pathway: Metabolism and growth control in tumour suppression. *Nature Reviews Cancer, 9*(8), 563–575.

Sharma, B., Singh, S., & Kanwar, S. S. (2014). L-methionase: A therapeutic enzyme to treat malignancies. *BioMed Research International, 2014*, 506287.

Smolková, K., & Ježek, P. (2012). The role of mitochondrial NADPH-dependent isocitrate dehydrogenase in cancer cells. *International Journal of Cell Biology, 2012*, 273947.

Smythies, J. (May 23, 2015). Intercellular signaling in cancer-the SMT and TOFT hypotheses, exosomes, telocytes and metastases: Is the messenger in the message? *Journal of Cancer, 6*(7), 604–609.

Sonnenschein, C., & Soto, A. M. (October 2016). Carcinogenesis explained within the context of a theory of organisms. *Progress in Biophysics and Molecular Biology, 122*(1), 70–76.
Sonnenschein, C., & Soto, A. M. (December 2000). Somatic mutation theory of carcinogenesis: Why it should be dropped and replaced. *Molecular Carcinogenesis, 29*(4), 205–211.
Soto, A. M., & Sonnenschein, C. (February 2005). Emergentism as a default: Cancer as a problem of tissue organization. *Journal of Bioscience, 30*(1), 103–118.
Soto, A. M., & Sonnenschein, C. (October 2004). The somatic mutation theory of cancer: Growing problems with the paradigm? *BioEssays, 26*(10), 1097–1107.
Valle, I., Tramalloni, D., & Bragazzi, N. L. (June 10, 2015). Cancer prevention: State of the art and future prospects. *Journal of Preventive Medicine and Hygiene, 56*(1), E21–E27.
Verschoor, M. L., Ungard, R., Harbottle, A., Jakupciak, J. P., Parr, R. L., & Singh, G. (2013). Mitochondria and cancer: Past, present, and future. *BioMed Research International, 2013*, 612369.
Warburg, O. (February 24, 1956). On the origin of cancer cells. *Science, 123*(3191), 309–314.
Webber, M. M., Buffkin, D. C., Juillard, G. J., Schwabe, A. D., Verma, R. C., & Bennet, L. R. (1980). Ornithine metabolism in normal subjects and patients with cancer. *The Journal of Nuclear Medicine, 21*(12), 1194–1196.
Waterhouse, C., & Mason, J. (1981). Leucine metabolism in patients with malignant disease. *Cancer, 48*(4), 939–944.
Yang, M., & Vousden, K. H. (October 2016). Serine and one-carbon metabolism in cancer. *Nature Reviews Cancer, 16*(10), 650–662.
Zhou, S., Huang, C., & Wei, Y. (2010). The metabolic switch and its regulation in cancer cells. *Science China. Life Sciences, 53*(8), 942–958.
Zielinski, D. C., Jamshidi, N., Corbett, A. J., Bordbar, A., Thomas, A., & Palsson, B. O. (January 25, 2017). Systems biology analysis of drivers underlying hallmarks of cancer cell metabolism. *Scientific Reports, 7*, 41241.

CHAPTER FOUR

Investigating mutations at the hotspot position of the ERBB2 and screening for the novel lead compound to treat breast cancer — a computational approach

D. Thirumal Kumar[a], S. Udhaya Kumar[a], R. Magesh[b,*] and C. George Priya Doss[a,*]

[a]School of Biosciences and Technology, Vellore Institute of Technology, Vellore, Tamil Nadu, India
[b]Faculty of Biomedical Sciences, Technology & Research, Department of Biotechnology, Sri Ramachandra University, Chennai, Tamil Nadu, India
*Corresponding authors: E-mail: georgepriyadoss@vit.ac.in, magesh.r@sriramachandra.edu.in

Contents

1. Introduction 50
2. Materials and methods 54
   2.1 Database for prevalence mutations in breast cancer 54
   2.2 Pathogenicity and stability prediction tools 54
   2.3 Amino acid conservation analysis 54
   2.4 Lipinski's rule 55
   2.5 Molecular docking 55
   2.6 Protein dynamics simulation 56
3. Results 56
   3.1 Mutation prevalence 56
   3.2 Pathogenicity and stability analysis 56
   3.3 Conservation analysis 57
   3.4 Protein dynamics simulations 57
   3.5 Virtual screening 58
   3.6 Molecular interaction studies 60
   3.7 Protein-ligand dynamics simulations 62
4. Discussion 64
5. Conclusion 66
Supplementary data 67
Acknowledgments 67
References 67

Advances in Protein Chemistry and Structural Biology, Volume 123
ISSN 1876-1623
https://doi.org/10.1016/bs.apcsb.2020.10.001

Abstract

Membrane proteins are the most common types of cancer that are active in the prognosis. Membrane proteins are a distinguishing characteristic of a cancer cell. In tumor cell therapy, the overexpressed membrane proteins are becoming ever more relevant. The 3-kinase (PI3K)/AKT phosphatidylinositol pathway is downstream triggered by different extracellular signals, and this signaling pathway activation impacts a variety of proliferation of the cellular processes like cell growth and surviving. Frequent PI3K/AKT dysregulation in human cancer has rendered proteins of this pathway desirable for diagnostic markers. Members of the ERBB family-like ERBB2 and ERBB3 activate intracellular signaling pathways such as PI3K/AKT. The mutations in these proteins dysfunctions the proteins in the downstream. Considering this importance, we have developed a computational pipeline to identify the mutation position with a highest number of mutations and to screen them for pathogenicity, stability, conservation, and structural changes using PredictSNP, iStable, ConSurf, and GROMACS simulation software respectively. Further, a virtual screening approach was initiated to find the most similar non-toxic lead compound, which could be an alternative to the currently used lapatinib. To conclude, protein-ligand dynamics were undertaken to study the actions of native and mutants with the lapatinib and the lead compound. From the overall analysis, we identified position 755 with leucine in the native condition is prone to frequent mutations. The leucine at 755th position is more prone to mutate as serine and tryptophan. Further from the computational analysis, we identified that the mutation L755S is more significant than the L755W mutation. We have witnessed CID140590176 be a potential lead compound with no toxicity. The behavior of the lead compound has shown more compactness with an increased number of intermolecular hydrogen bonds in the ERBB2 with L755S. This lead compound can be further taken for experimental validations, and we believe that this lead compound could be a potent ERBB2 inhibitor.

1. Introduction

A transmembrane protein is a form of an integral membrane protein that covers the whole of the cell membrane (Fowler et al., 2016). Several transmembrane proteins act as doorways to allow the movement of particular constituents through the membrane. It also undergoes substantial conformation changes to transfer materials across the membrane (Chipot et al., 2018). Membrane proteins are the most common forms of cancer that are involved in the prognosis. Membrane proteins are a distinguishing characteristic of a cancer cell. In cancer cell therapy, the overexpressed membrane receptors are becoming ever more critical (Kampen, 2011). Cancer is a term given to a group of diseases that are related to each other.

Cancer is formed when the abnormal cells in the body start to divide rapidly without any control leading to the formation of malignant tumors (Lengauer, Kinzler, & Vogelstein, 1998). However, leukemia is an exceptional case where tumors or lumps are not formed. NCI provides a detailed list of cancer-based on its location in the body (https://www.cancer.gov/types). According to the American Cancer Society, among the three people, it is said that one among them succumbs to cancer, making it one of the deadliest diseases (Siegel, Miller, & Jemal, 2019).

The common symptoms of cancer include a lump, unexplained weight loss, abnormal bleeding, etc. (Lengauer et al., 1998). Until date, several types of cancers have been identified to affect humans (Stern, 2000). Based on the estimated new cases, breast cancer, lung cancer, and prostate cancer rank the top positions (Bray et al., 2018). There are various types of cancers, among which breast cancer affects most of the women around us. It has even surpassed overtaken cervix cancer, which was the most prevalent cancer a decade ago. Even though breast cancer occurs most commonly in women, however, it can also develop in a few men (Giordano, Buzdar, & Hortobagyi, 2002). It happens when the tissues inside the breast start to divide rapidly without any uncontrolled growth leading to the formation of tumors. The number of cases reported each year in the world is 1.1 million, and more than four lakh women succumb to this disease (Ferlay et al., 2015). A study by Schonfeld et al. in 2011 showed that nulliparous women had a 2.2-fold higher risk of developing breast cancer than parous women (Schonfeld et al., 2011). Women who had first-degree relatives with breast cancer have 2 to 3 times the risk of getting breast cancer (Brewer, Jones, Schoemaker, Ashworth, & Swerdlow, 2017). The risk further increases if the relative was affected by their younger age. Also, the person will be at higher risk if she/he has more than one close relative affected with breast cancer (Brewer et al., 2017).

The significant pathways that are complicated in causing breast cancer are the PI3K-AKT signaling pathway, MAPK signaling pathway, the Estrogen signaling pathway, the Notch signaling pathway, and Wnt Signaling Pathway. A cell has to pass through many hurdles to become a cancerous cell. In other words, we can say that there is a large number of machinery such as apoptosis, DNA checkpoints, repair mechanisms that are present to save the cell from becoming a malignant cancerous cell (Hahn & Weinberg, 2002). There are regular mechanisms that happen now and then for the survival of healthy cells, sometimes mislead to becoming a cancerous cell. With the vast amount of research in the field of abnormal cell growth have led to the establishment that cancers may also have a tremendous genetic factor and that the somatic mutations of these oncogenes and also the tumor suppressor genes may be the initiators of the carcinogenic process (Barrett, 1993). For this form, PI3K performances as a trigger in the development for additional messengers at the cell membrane, including PIP3 (lipid second messenger). It helps to activate cellular processes that support the cell to perform normal functions, including AKT. This signaling mechanism of activation of AKT via PI3K is called the signaling pathway PI3K-AKT, which regulates cell development, multiplication, motility, and survival (Cantley, 2002; Luo, Manning, & Cantley, 2003). There is also information uttering that the activation of this AKT-PI3K initiation is detected by several additional mechanisms; likewise, that might also develop tumorigenesis (Yuan & Cantley, 2008). The significant proteins and their primary functions are represented in Fig. 1.

This Epidermal Growth factor receptor is a member of the ErBb family of receptors, a subfamily of four structurally related receptors EGFR (ErbB1, HER1), ErbB2 (HER 2, neu in rodents), ErbB3 (HER3), ErbB4 (HER4) (Riese, Komurasaki, Plowman, & Stern, 1998). Many mechanisms can trigger an oncogene to over-express. The ERBB2 protein, which acts as the initiator of the PI3K-AKT-mTOR pathway, is transmembrane. In normal cells, activation of the ERBB2 signaling pathway controls normal cell growth, differentiation, motility, and adhesion in several cell lineages (Gassmann et al., 1995; Lee et al., 1995; Meyer & Birchmeier, 1994). In

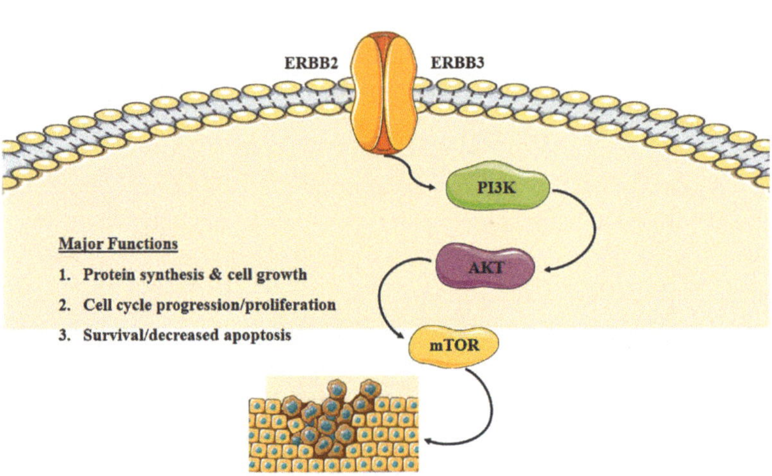

Fig. 1 The pictorial representation of the PI3K-AKT-mTOR activation signaling pathway.

normal cells, there are two copies of the gene per cell. During ERBB2 amplification, there may be as many as 100 c-erbB2 genes per tumor cell. Breast cancers might well take up to 25—50 clones of the ERBB2 and close to 40—100-fold rise in the expression of ERBB2 proteins, ranging in up to 2 million receptors on the tumor cell surface (Kallioniemi et al., 1992; Lohrisch & Piccart, 2001). The ERBB2 receptors function as monomers; the ERBB2 proteins experience dimerization and transphosphorylation in their intracellular regions after ligand binding to the extracellular domains. Heterodimerization fallouts in autophosphorylation of tyrosine residues within the cytoplasmic field of the receptors. These inductees three different signal transduction pathways, such as PI3K, MAPK, and phospholipase C-γ (PLCγ) pathways (Katz, Amit, & Yarden, 2007). The ERBB2-ERBB3 heterodimer is the most effective enhancer of the PI3K/AKT pathway (Carmona et al., 2016; Toomey et al., 2017). All ERBB involved dimerization (ERBB1/ERBB2, ERBB2/ERBB3, and ERBB2/ERBB4) can stimulate the MAPK pathway (Iqbal & Iqbal, 2014). PI3K/AKT pathways and MAPK pathways are important signaling pathways to stimulate cell proliferation and avoid apoptosis. A receptor tyrosine kinase ERBB2 on chromosome 17 is the most frequently mutated and overexpressed oncogene in breast cancer. Amplification of ERBB2 happens in 15%—30% of breast cancers due to the mutations (SNPs) and results in overexpression of the protein drug ERBB2 (Burstein, 2005). The SNPs are broadly classified into two distinct classes. (1) Coding synonymous (csSNPs), (2) Nonsynonymous SNPs (nsSNPs). These non-synonymous SNPs are further classified into missense mutations and nonsense mutations. The coding synonymous SNPs have a low impact on protein structure, whereas the non-synonymous SNPs have a great influence over the protein structure and have a higher risk of cancer. Non-synonymous SNPs are capable of altering the physicochemical property of the protein and alter protein stability and structure (George Priya Doss, Chakraborty, Narayan, & Kumar, 2014; Mosaeilhy et al., 2017). nsSNP can be broadly divided into four categories based on their mode of action: (1) Flexibility, stability, folding and aggregation of protein (2) Reaction kinetics, functional sites, and dependence on the environment parameters like pH, temperature, (3) Subcellular localization and protein expressions (4) Protein-Protein, Protein-small molecule, (5) protein-membrane and protein-DNA interaction (David & Sternberg, 2015; Herrgard et al., 2003).

In this study, we have identified the most prevalent mutation position and the corresponding mutations in the position: the pathogenicity, destabilizing properties, strength of conservation. The protein simulations were executed to identify the most deleterious mutation among the mutations occurring in the same hotspot position. Further, a virtual screening method was used to find a potential alternative to the currently used drug lapatinib.

2. Materials and methods

2.1 Database for prevalence mutations in breast cancer

Several biological databases report breast cancer-causing proteins and mutations. Among them, public databases such as UniProt and COSMIC are highly curated (Apweiler, 2004; Tate et al., 2019). The 3D structure of proteins was retrieved from Protein Data Bank (Joosten et al., 2011). The ERBB2 sequence in the FASTA format was taken from the UniProt with ID P04626. The frequency of each mutation is obtained from the COSMIC database. The mutation position with the highest number of mutations was considered as the hotspot mutations. The 3D structure of the ERBB2 was obtained from the PDB database with PDB ID 3PP0 (Aertgeerts et al., 2011).

2.2 Pathogenicity and stability prediction tools

Pathogenicity and stability were examined by means of in silico approaches: PredictSNP and iStable servers. PredictSNP is a harmonized classifier that integrates several other prediction tools (Bendl et al., 2014). These algorithms accord evidence concerning the category of mutation based on various parameters and thus help in classifying the mutations as neutral or deleterious. The FASTA sequences attained from the UniProt database, and the hotspot mutations were provided as the data for the prediction. iStable uses an SVM for prediction of the stability of proteins, which changes once an SNP follows. Outcomes from more than a few predictors improve the influence as meta predictions, which has improved performance than any solo implement (Chen, Lin, & Chu, 2013). The FASTA sequences got from the UniProt database, and the hotspot mutations were provided as the data for the prediction.

2.3 Amino acid conservation analysis

The website ConSurf was used to measure the protein's hotspot mutation location conservation pattern (Glaser et al., 2003). Through multiple sequence alignment, it defines conserved locations and then estimates the evolutionary survival rate using analytical Bayesian inference. Additionally, it includes the evolutionary survival profiles dependent on the protein structure or sequence. The ConSurf score varies from 1 to 9, with 1 indicating constantly developing sites, 5 representing the average sites, and 9 describing gradually evolving sites. Both the protein's exposed and buried regions are given together with the conservation pattern. This test also estimates the amino acid's structural/functional effect through the protein. The FASTA protein sequences that were extracted from the UniProt repository were supplied as the prediction information.

2.4 Lipinski's rule

The molecules screened from the PubChem database were subjected to ADME (absorption, distribution, metabolism, and excretion) analysis using the SwissADME server (Daina, Michielin, & Zoete, 2017). The server predicts different properties of the molecules based on LogP, total polar surface area, number of atoms, molecular weight, etc. (Lipinski, Lombardo, Dominy, & Feeney, 1997). The SMILES were given as the input, and the structural properties were calculated for all small molecules. The toxicity of the identified lead compound was predicted using the online mcule toxicity prediction server (Robert Kiss, 2012).

2.5 Molecular docking

The mutations in the native protein structure were introduced employing the SwissPDB viewer (Guex & Peitsch, 1997). Molecular docking of the selected drug and small molecules were executed with the native ERBB2 and the proteins with hotspot mutations employing AutoDock 4.2.3 (Morris et al., 2009). Subsequently, the calculation of hydrogen, the required charges, and AutoDock 4 type force field parameters, the molecules were improved to pdbqt file format. Subsequently, after making the required modifications to ligands, the ligands were also improved to pdbqt format. The ligand was set flexible, and the protein was set rigid. The grid was placed at 60 along the cubical axis by means of the auto grid constraints. Lamarckian Genetic Algorithm was employed in order to understand and analyze the ligand-receptor docking (Morris et al., 1998). GA helps to handle a large degree of freedom. The best protein-ligand confirmation was taken for protein dynamics simulations. Similarly, molecular docking was conducted using the ligands obtained from virtual screening with the native protein and the proteins with mutants, and the best confirmation was subjected to GROMACS dynamics simulations (Abraham et al., 2015).

2.6 Protein dynamics simulation

The simulations were carried out in two methods. (1) Molecular Dynamics of protein without ligand and (2) Molecular Dynamics with the selected ligand. The Simulations was accomplished with the assistance of the GROMACS using the GROMOS53a6 force field. The ligand constraints were taken from the Automated Topology Builder server (Malde et al., 2011; Oostenbrink, Villa, Mark, & van Gunsteren, 2004). To ascertain the solubility of the system, counter ions were added using the genion. To confirm that the system is without unrestricted simulations, the equilibration procedure was accomplished in two stages. Stage 1, the NVT, was kept constant. Here the bonds were controlled through 50,000 steps, besides the temperature was retained at 300 K utilizing the Berendsen thermostat (Lemak & Balabaev, 1994). Phase II, i.e., the NPT, was kept constant (Parrinello & Rahman, 1980). The simulations were performed with the well-equilibrated system. The detailed simulations methodology were followed from our previous works (Thirumal Kumar et al., 2017, 2018; Kumar, Jain, et al., 2020; Kumar, Kumar, et al., 2020; Udhayakumar, Kumar, Siva, Doss, & Zayed, 2020).

3. Results

3.1 Mutation prevalence

The prevalence data of the mutations in the ERBB2 protein, causing breast cancer were retrieved from the COSMIC database (https://cancer.sanger.ac.uk/cosmic). The leucine at 755th position of the ERRB2 protein was found to be a highly prevalent mutation position causing the breast cancer. The leucine at this position was found to mutate to tryptophan and serine. L755S mutation was found to be in 18 samples, and L755W mutation was found to be in 1 sample (Fig. 2). These two mutations (L755S and L755W) were taken for further studies.

3.2 Pathogenicity and stability analysis

The pathogenicity of the L755S and L755W mutations of the ERBB2 was analyzed using the PredictSNP. Both the mutations were predicted to be deleterious by PredictSNP in-built tools. Further stability study using the

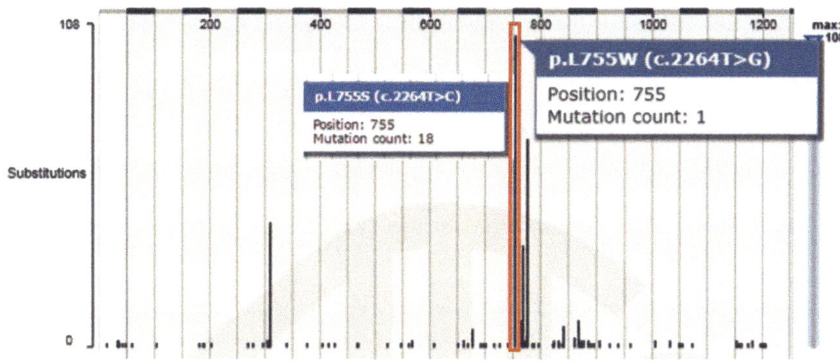

Fig. 2 ERBB2 mutations prevalence data obtained from the COSMIC database.

Table 1 Pathogenicity and stability predictions of hotspot mutations (L755S and L755W) in ERBB2 protein using PredictSNP and iStable.

Tool predictions	L755S	L755W
PhD-SNP	Deleterious	Deleterious
PolyPhen-1	Deleterious	Deleterious
PolyPhen-2	Deleterious	Deleterious
SIFT	Deleterious	Deleterious
SNAP	Deleterious	Deleterious
nsSNPAnalyzer	Deleterious	Deleterious
PredictSNP	Deleterious	Deleterious
i-MUTANT	Decrease	Decrease
MUpro	Decrease	Decrease
i-stable	Decrease	Decrease

iStable server, the mutations L755S and L755W were found to be decreasing the protein stability by all the three predictors (Table 1).

3.3 Conservation analysis

ConSurf online algorithm was employed to compute the conservation of respective amino acids in the ERBB2 protein. Through the results, we observed that the leucine at hotspot 755th position mutated to serine and tryptophan was found to be highly conserved with a scale of 9 (Fig. 3).

3.4 Protein dynamics simulations

The native structure and mutations were employed for the analysis of the dynamic behavior of the protein in an environment encompassing water. The conformational changes were analyzed through molecular dynamic

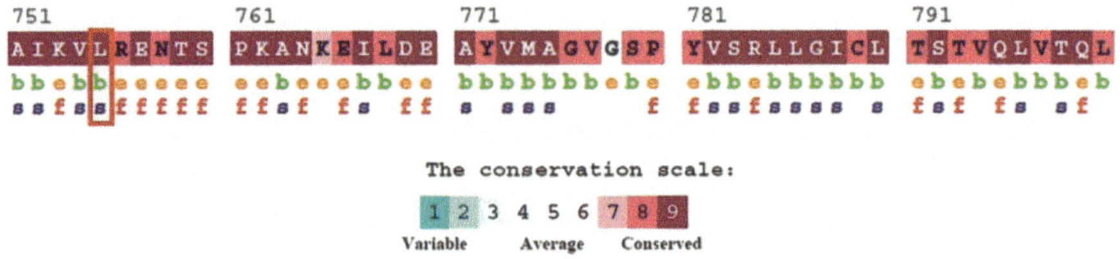

Fig. 3 Evolutionary conservation analysis of ERBB2 using ConSurf server.

simulation studies to check their RMSD (Root Mean Square Deviation), gyrate (Rg), and H-bonds. The stability of the proteins was analyzed for 30 ns simulation using the GROMACS software package. The RMSD was observed and is depicted in Fig. 4A, which shows that there are no much deviations of more significant than ~0.5 nm for native and mutated structure and initially, the mutated structure remains unsteady until ~20 ns and later convergence toward the end of the simulation period. Lower RMSD values of the protein−compound complexes specify the stability of the system (Fig. 4A). Fig. 4B represents the Rg of the native protein and shows that the protein is compact and does not lose its interactions. The gyration of the native remains stable when compared to the mutations. The L775W stabilizes its Rg to less than ~2.0 nm after the ~20th ns, whereas the L755S mutation shows high Rg deviations of ~2.05 nm to ~2.1 nm throughout the simulation (Fig. 4B). The hydrogen bonds have also been shown to decrease highly in the L755S mutation in comparison to the native protein and L755W (Fig. 4C).

3.5 Virtual screening

A virtual screening pipeline was initiated through the PubChem database search for the compounds that are 90% similar to the control drug lapatinib. Our results yielded a set of 1099 molecules (Table S1). Out of these 1099, 10 compounds were found to satisfy the Lipinski rule of 5 (Table 2). These ten compounds were finally taken for molecular docking screening using Auto-Dock vina, and the compound with PubChem ID 140590176 (CID140590176) (CID) have shown to express the significant binding affinity of −11 kcal/mol. This compound was taken for further investigations (Table 3). The toxicity of the identified lead compound was checked using the online mcule toxicity prediction server.

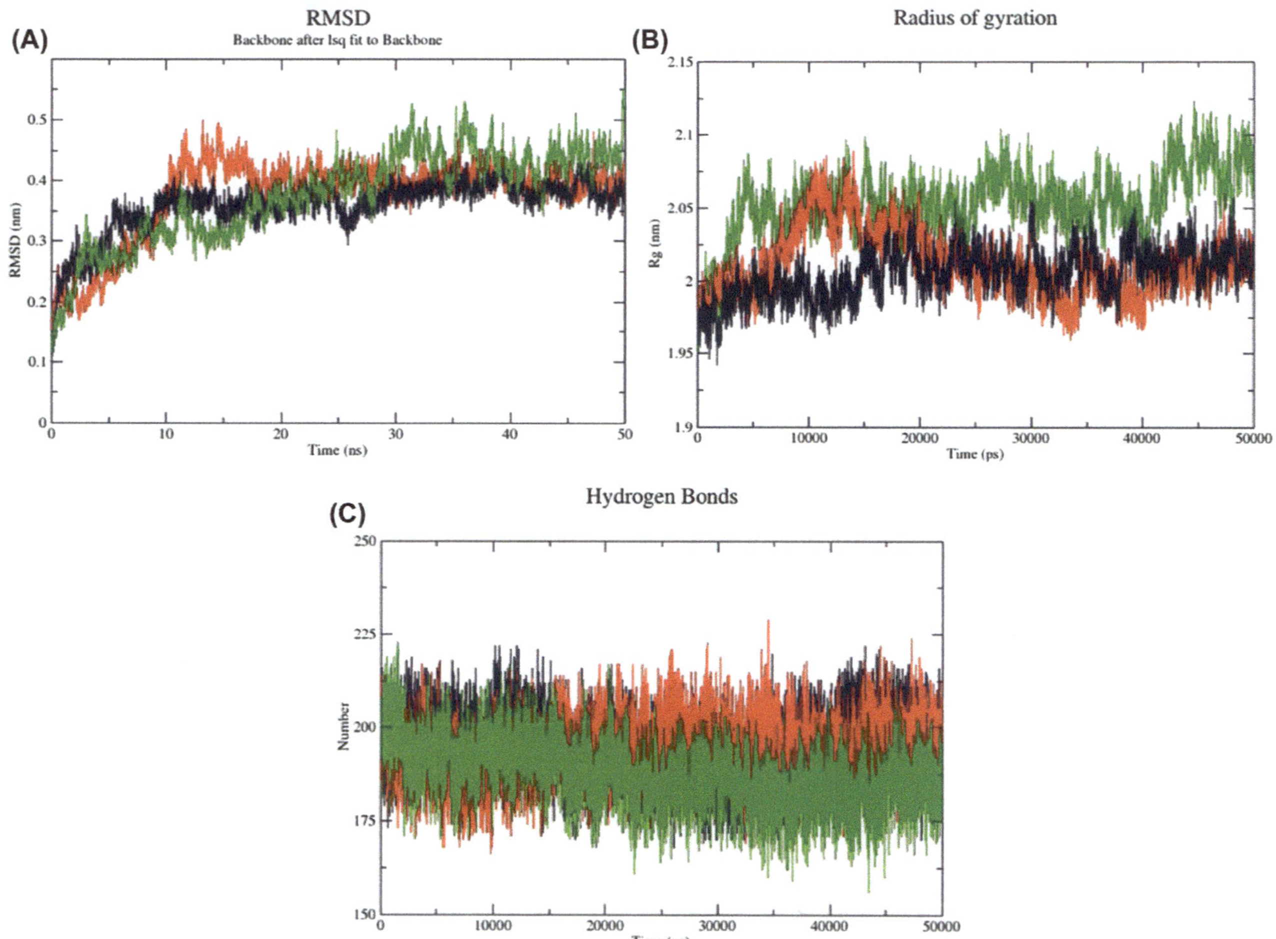

Fig. 4 GROMACS analysis of native and mutant ERBB2 proteins for 50 ns. (A) RMSD (B) Rg (C) Intramolecular hydrogen bonds. Color scheme: native — *black*, L755S — *green*, L755W — *red*.

Table 2 List of compounds satisfying Lipinski rule of 5 and their molecular properties.

S.No	PubChem ID	MW	nHBD	nHBA	LogP
1	145633426	455	2	7	3.6
2	142763491	471.9	2	7	4.6
3	142763490	486	2	7	5
4	140590176	486	1	7	5
5	137631553	474.9	2	8	3.7
6	118753421	488.9	4	9	2.6
7	118500438	471	2	7	3.9
8	71582332	470.9	2	8	3.7
9	71315271	472.9	3	8	3.2
10	58172657	408.9	3	6	4.7

MW – Molecular weight in gram; nHBA – Number of hydrogen bond acceptor; nHBD – Number of hydrogen bond donor; LogP – logarithm of partition coefficient.

Table 3 List of binding affinities obtained from the AutoDock vina screening between the ERBB2 with L755S and PubChem compounds.

S.No	Protein	PubChem ID	Binding affinity (kcal/mol)
1	ERBB2 with L755S	140590176	−11
2	ERBB2 with L755S	142763490	−10.9
3	ERBB2 with L755S	71582332	−10.7
4	ERBB2 with L755S	142763491	−10.4
5	ERBB2 with L755S	137631553	−10.3
6	ERBB2 with L755S	118753421	−9.9
7	ERBB2 with L755S	118500438	−9.9
8	ERBB2 with L755S	71315271	−9.8
9	ERBB2 with L755S	145633426	−9.5
10	ERBB2 with L755S	58172657	−9.3

3.6 Molecular interaction studies

Molecular dockings were performed to compare the binding efficacy between the control drug lapatinib and the lead compound CID140590176 C against the native and mutant (L755S) ERBB2 proteins. The binding energy between the native ERBB2 against the lapatinib and CID140590176 was found to be −9.62 kcal/mol and −9.83 kcal/mol, respectively. The binding energy between the mutant ERBB2 against the lapatinib and CID140590176 was established to be −7.59 and −7.67 kcal/mol correspondingly. The binding energy with the lead compound is comparatively higher than the control drug. The lapatinib was found to interact with the native ERBB2 through 23 residues, namely, LEU755, LYS753, ALA751, GLU770, ARG868, GLY865, PHE864, ASP863, ILE767, GLU766, ALA763, ALA730, GLY729, SER728, GLY727, ASP808, CYS805, VAL734, ARG849, ASN850, THR798, LEU796, and THR862 (Fig. 5A). The CID140590176 was found to interact with the native ERBB2 through 25 residues, namely, GLU770, ALA771, MET774, LEU796, LEU785, LYS753, SER783, THR798, ALA751, GLN799, MET801, LEU852, GLY804, CYS805, ARG849, VAL734, LEU807, ASP808, SER728, GLY727, LEU726, THR862, ASP863, PHE864, and ILE767 (Fig. 5B).

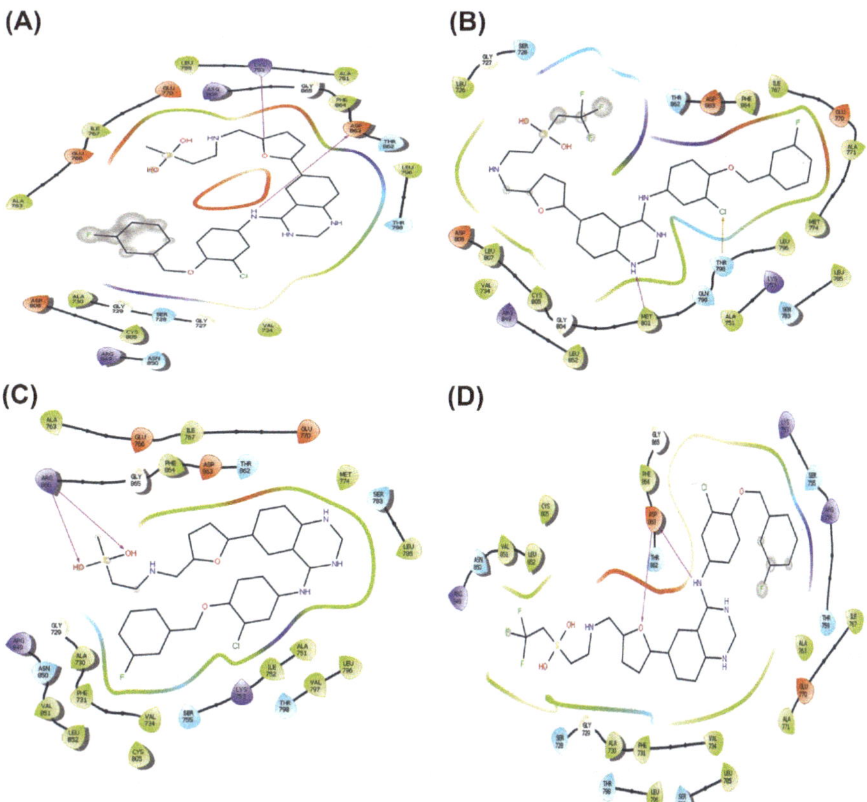

Fig. 5 2D visualization of interacting residues with the control drug and test lead compound. (A) native ERBB2 complexed with lapatinib (B) native ERBB2 complexed with CID140590176 (C) mutant ERBB2 complexed with lapatinib (D) mutant ERBB2 complexed with CID140590176.

The lapatinib was found to interact with the mutant ERBB2 through 28 residues, namely, ALA763, GLU766, ILE767, GLU770, ARG868, GLY865, PHE864, ASP863, THR862, MET774, SER783, LEU785, GLY729, ALA730, PHE731, VAL734, SER755, LYS753, ILE752, ALA751, ARG849, ASN850, VAL851, LEU852, CYS805, THR798, VAL797, and LEU796 (Fig. 5C). The CID140590176 was found to interact with the mutant ERBB2 through 28 residues, namely, LYS753, SER755, ARG756, THR759, ILE767, ALA763, GLU770, ALA771, LEU785, VAL734, SER783, LEU796, PHE731, ALA730, GLY729, SER728, CYS805, LEU852, VAL851, ASN850, ARG849, GLY865, PHE864, ASP863, and THR862 (Fig. 5D).

3.7 Protein-ligand dynamics simulations

The compound lapatinib docked with the native showed stability after 15 ns around 0.4 nm and had slight deviations (Fig. 6A). The L775W complexed with lapatinib showed that the stability is acquired after 10 ns simulation, whereas at 25 ns again occurs the deviation and thereafter, stability is maintained (Fig. 6A). On comparison of the native ERBB2 and L755S and L755W complexes, higher deviations were observed in the mutant complex. Similarly, the stability of both proteins with a compound with CID140590176 is unsteady throughout the simulation, and the complex of the mutant converges to 0.35 nm at around 18 ns (Fig. 6B). Gyration analysis shows that the compound binding with both proteins possesses stability, and targeting the protein paves the way for a potential drug discovery approach also, the protein is compact and does not lose its interactions with the compound (Fig. 6C). The plot results represent the radius of gyration of the native and mutant protein and complex with CID140590176. This shows that the compound binding to the native protein shows that it is compact and does not lose its interactions (Fig. 6D). The hydrogen bond analysis plot for MD simulation shows the number of H-bonds for each complex. In both complexes, H-bonds remain until the end of the simulation. This shows that the compound binding to the protein shows that the protein is compact and does not lose its interactions. The result shows that the compound binding to the native protein has two hydrogen bonds, whereas, with the mutant, the interaction is strong with four hydrogen bonds. The hydrogen bond analysis shows that there is no loss of hydrogen bond in the complex during the simulation period, and Fig. 6 represents the same. A constant range of 4–6 H-bond is seen throughout the simulation period (Fig. 6E and F).

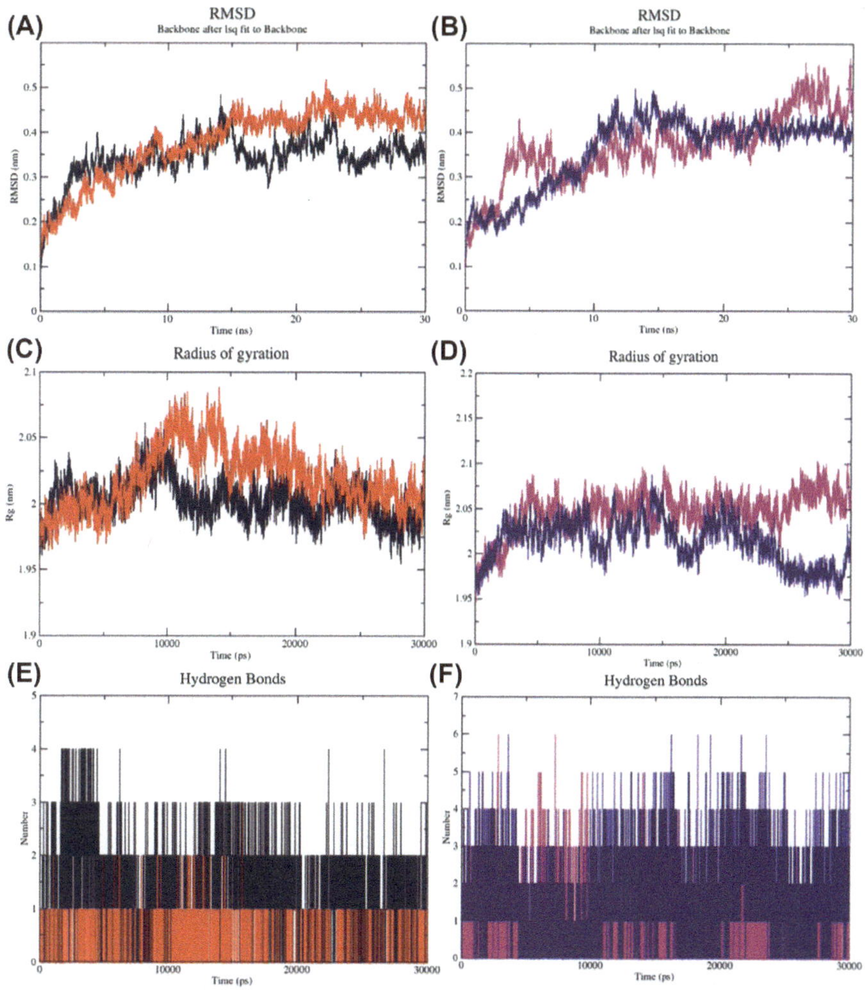

Fig. 6 GROMACS analysis of native and mutant ERBB2 proteins complexed with lapatinib and CID140590176 for 30 ns. (A) RMSD of native and mutant ERBB2 complexed with lapatinib (B) RMSD of native and mutant ERBB2 complexed with CID140590176 (C) Radius of gyration of native and mutant ERBB2 complexed with lapatinib (D) Radius of gyration of native and mutant ERBB2 complexed with CID140590176 (E) Number of intermolecular hydrogen bonds between native and mutant ERBB2 complexed with lapatinib (F) Number of intermolecular hydrogen bonds between native and mutant ERBB2 complexed with CID140590176. Color Scheme: native with lapatinib – *red*, mutant with lapatinib – *black*, native with CID140590176 – *pink*, mutant with CID140590176 – *blue*.

4. Discussion

A transmembrane protein is a protein that extends across biological membranes. Some of them pass across the plasma membrane's lipid bilayer, while some are found on the organelle membrane. The transmembrane protein family collects proteins with roles that are largely unclear (Schmit & Michiels, 2018). Many experiments have shown that the expression of a transmembrane protein in tumor tissues can be down- or up-regulated when opposed to neighboring healthy tissues. Most transmembrane proteins are up in cancer control. Any of these are involved in tumor growth, invasion, and metastasis development, while some are linked to poor prognosis, which can be used as a biomarker for prognosis (Lin et al., 2019). Breast cancer is the second most prevalent cause of cancer-related deaths in women and accounts for 30% of all current general cancer diagnosis (Sudhakar et al., 2016). The protein mutations in the PI3K/AKT pathway plays a crucial role in the breast cancer causative (Thirumal Kumar & George Priya Doss, 2016, 2017). In protein simulation, RMSD of the protein calculates the stability of the protein and the convergence (George Priya Doss, Nagasundaram, Chakraborty, Chen, & Zhu, 2013; Thirumal Kumar & George Priya Doss, 2016; Saralamba et al., 2011; Thirumal Kumar et al., 2019; Kumar et al., 2020; UdhayaKumar, Doss, & Zayed, 2020). The radius of the gyration parameter helps in understanding the compactness of the protein structure. An increased Rg elucidates the compactness is lost or lower and vice-versa (Lobanov et al.). The number of intra and intermolecular hydrogen bonds determines the strength of the molecules (Pace et al., 2011). Virtual screening methods have become one of the critical drug discovery methods because of the immediate need to find medicines for the several recently identified target mutants in many diseases (Mahajan et al., 2017; Nagasundaram et al., 2016; Thirumal Kumar, Jain, Udhaya Kumar, George Priya Doss, & Zayed, 2020).

With these considerations, we have developed a computational flowchart to find the noteworthy pathogenic mutation occurring in the hotspot position of the ERBB2 protein involved in the PI3K/AKT/mTOR pathway. We identified that leucine at 755th position to be hotspot mutation positions of ERBB2, which can frequently mutate to serine and tryptophan (L755S and L755W) (Fig. 2). This mutation position L755 was found to be highly conserved from the conservation analysis (Fig. 3). In our present study, we employed the PredictSNP tool, which comprises the algorithms of other popular deleteriousness prediction algorithms to examine the pathogenicity of L755S and L755W occurring in the hotspot position of the proteins. From the ERBB2 protein analysis, we observed that both L755S and L755W were identified to be deleterious by all the tools (Table 1). On the stability change analysis due to mutation, we interesting observed that both the mutations are capable of decreasing the stability of the protein (Table 1). The native structure of ERBB2 was retrieved from the PDB with ID 3PP0 was taken for structural studies. The mutant models were built employing SwissPDB viewer to get the 3D model of the mutants. In ERBB2, the leucine at the 755th position was mutated to tryptophan and serine. Protein dynamics simulation was further initiated to study the alterations in the structural pattern of the proteins owing to the destabilizing mutations.

From the initial protein simulations in the ERBB2 protein to compare the L755S and L755W mutations, we observed that the L755S showed higher deviation, lower compactness, and decreased the number of intramolecular hydrogen bonds in ERBB2 protein with L755S mutation on comparison to the native ERBB2 protein and ERBB2 protein with L755W mutation (Fig. 4). Thus L755S mutation was further taken for the virtual screening. A virtual screening method was implemented to find novel lead compounds to treat L755S mutation of ERBB2 protein. In the case of ERBB2, the drugs lapatinib, which is currently in use, possess drug resistance. So we have attempted to identify a similar lead compound with the same efficacy and lesser toxicity. In the attempt of virtual screening, from 90% similarity search against the PubChem for the lapatinib drug yielded 1099 molecules (Table S1). For those 1099, 10 compounds have been found to follow Lipinski Rule 5 (Table 2). These ten compounds were finally taken using AutoDock vina for molecular docking screening, and the compound with PubChem ID 140590176 exhibited the topmost binding affinity with −11 kcal/mol (Table 3). This compound has been retained for further investigation. The molecular dockings were initiated to compare this identified novel lead compound against the protein molecules in the native and mutated condition in comparison to the control drugs. From the complete molecular interaction results, we observed that the binding affinity, number of hydrogen bonds, and number of residual interactions were best suitable in the identified novel lead compound when compared to the control drug (Fig. 5). Finally, from the toxicity analysis, we observed that the proposed lead compound CID140590176 for ERBB2 had shown nil toxicity from the mcule toxicity prediction server. Protein-ligand dynamics were initiated to comprehend the actions of novel lead compounds with the protein molecules. The identified lead compounds showed a better compactness and hydrogen bond interaction when compared to the control drugs (Fig. 6). This result sheds a light that the novel lead compounds CID140590176 can be promisingly taken for the further experimental testing to treat the breast cancer caused by the hotspot mutation positions in the ERBB2 transporter protein.

5. Conclusion

Membrane proteins are the most common cancer types and are involved in the prognosis. Membrane proteins are a distinctive feature in cancer cells. The overexpressed membrane receptors are becoming increasingly crucial in cancer-cell therapy. The cascade of 3-kinase (PI3K)/AKT phosphatidylinositol is activated downstream by various extracellular signals, and this signaling pathway activation affects a range of cellular activity proliferation such as cell growth and survival. Frequent dysregulation of PI3K/AKT in human cancer has made proteins from this pathway attractive for diagnostic markers. ERBB family members such as ERBB2 and ERBB3 stimulate intracellular signaling pathways such as PI3K/AKT. The mutations in specific protein dysfunctions in the downstream proteins. Considering this value, we built a computational pipeline to classify the location of the mutation with a higher number of mutations and to test them for pathogenicity, stability, survival, and structural changes using the simulation tools PredictSNP, iStable, ConSurf, and GROMACS, respectively. In addition, a simulated screening technique was introduced to identify the most comparable non-toxic lead compound, which could be an alternative to the lapatinib currently used. Finally, protein-ligand simulations were performed with the lapatinib and the lead compound to consider the behavior of native and mutant proteins. We defined position 755 with leucine in the native condition, which is vulnerable to repeated mutations from the general study. In the 755th position, the leucine is more likely to mutate as serine and tryptophan. In addition to the statistical study, we have found that the L755S mutation is more important than the L755W mutation. We have observed CID140590176 being a highly toxic lead compound. The lead compound behavior has demonstrated more compactness in the ERBB2 with L755S mutation, with an expanded number of hydrogen bonds. This lead compound can also be used for laboratory validations, and we suspect this lead compound may be a potent inhibitor of ERBB2.

Supplementary data

Supplementary data to this chapter can be found online at https://doi.org/10.1016/bs.apcsb.2020.10.001.

Acknowledgments

The authors acknowledge the management of the Vellore Institute of Technology (VIT), Vellore, India, Sri Ramachandra Institute of Higher Education and Research (SRIHER), Chennai, India, and BRAF @ CDAC for providing the facilities required to perform this work.

Conflict of Interest

We have no conflicts of interest to disclose

References

Abraham, M. J., Murtola, T., Schulz, R., Páll, S., Smith, J. C., Hess, B., et al. (2015). GROMACS: High performance molecular simulations through multi-level parallelism from laptops to supercomputers. *SoftwareX, 1–2,* 19–25. https://doi.org/10.1016/j.softx.2015.06.001.

Aertgeerts, K., Skene, R., Yano, J., Sang, B.-C., Zou, H., Snell, G., et al. (2011). Structural analysis of the mechanism of inhibition and allosteric activation of the kinase domain of HER2 protein. *Journal of Biological Chemistry, 286*(21), 18756–18765. https://doi.org/10.1074/jbc.M110.206193.

Apweiler, R. (2004). UniProt: The universal protein knowledgebase. *Nucleic Acids Research, 32*(90001), 115D–119. https://doi.org/10.1093/nar/gkh131.

Barrett, J. C. (1993). Mechanisms of multistep carcinogenesis and carcinogen risk assessment. *Environmental Health Perspectives, 100,* 9–20. https://doi.org/10.1289/ehp.931009.

Bendl, J., Stourac, J., Salanda, O., Pavelka, A., Wieben, E. D., Zendulka, J., et al. (2014). PredictSNP: Robust and accurate consensus classifier for prediction of disease-related mutations. *PLoS Computational Biology, 10*(1), e1003440. https://doi.org/10.1371/journal.pcbi.1003440.

Bray, F., Ferlay, J., Soerjomataram, I., Siegel, R. L., Torre, L. A., & Jemal, A. (2018). Global cancer statistics 2018: GLOBOCAN estimates of incidence and mortality worldwide for 36 cancers in 185 countries. *CA: A Cancer Journal for Clinicians, 68*(6), 394–424. https://doi.org/10.3322/caac.21492.

Brewer, H. R., Jones, M. E., Schoemaker, M. J., Ashworth, A., & Swerdlow, A. J. (2017). Family history and risk of breast cancer: An analysis accounting for family structure. *Breast Cancer Research and Treatment, 165*(1), 193–200. https://doi.org/10.1007/s10549-017-4325-2.

Burstein, H. J. (2005). The distinctive nature of HER2-positive breast cancers. *New England Journal of Medicine, 353*(16), 1652–1654. https://doi.org/10.1056/NEJMp058197.

Cantley, L. C. (2002). The phosphoinositide 3-kinase pathway. *Science, 296*(5573), 1655–1657. https://doi.org/10.1126/science.296.5573.1655.

Carmona, F. J., Montemurro, F., Kannan, S., Rossi, V., Verma, C., Baselga, J., et al. (2016). AKT signaling in ERBB2-amplified breast cancer. *Pharmacology & Therapeutics*. https://doi.org/10.1016/j.pharmthera.2015.11.013.

Chen, C.-W., Lin, J., & Chu, Y.-W. (2013). iStable: off-the-shelf predictor integration for predicting protein stability changes. *BMC Bioinformatics, 14*(Suppl. 2), S5. https://doi.org/10.1186/1471-2105-14-S2-S5.

Chipot, C., Dehez, F., Schnell, J. R., Zitzmann, N., Pebay-Peyroula, E., Catoire, L. J., et al. (2018). Perturbations of native membrane protein structure in alkyl phosphocholine detergents: A critical assessment of NMR and biophysical studies. *Chemical Reviews, 118*(7), 3559–3607. https://doi.org/10.1021/acs.chemrev.7b00570.

Daina, A., Michielin, O., & Zoete, V. (2017). SwissADME: A free web tool to evaluate pharmacokinetics, drug-likeness and medicinal chemistry friendliness of small molecules. *Scientific Reports, 7*, 42717. https://doi.org/10.1038/srep42717.

David, A., & Sternberg, M. J. E. (2015). The contribution of missense mutations in core and rim residues of protein-protein interfaces to human disease. *Journal of Molecular Biology, 427*(17), 2886–2898. https://doi.org/10.1016/j.jmb.2015.07.004.

Ferlay, J., Soerjomataram, I., Dikshit, R., Eser, S., Mathers, C., Rebelo, M., et al. (2015). Cancer incidence and mortality worldwide: Sources, methods and major patterns in GLOBOCAN 2012. *International Journal of Cancer, 136*(5), E359–E386. https://doi.org/10.1002/ijc.29210.

Fowler, P. W., Hélie, J., Duncan, A., Chavent, M., Koldsø, H., & Sansom, M. S. P. (2016). Membrane stiffness is modified by integral membrane proteins. *Soft Matter, 12*(37), 7792–7803. https://doi.org/10.1039/c6sm01186a.

Gassmann, M., Casagranda, F., Orioli, D., Simon, H., Lai, C., Klein, R., et al. (1995). Aberrant neural and cardiac development in mice lacking the ErbB4 neuregulin receptor. *Nature, 378*(6555), 390–394. https://doi.org/10.1038/378390a0.

George Priya Doss, C., Chakraborty, C., Narayan, V., & Kumar, D. T. (2014). Computational approaches and resources in single amino acid substitutions analysis toward clinical research. *Advances in Protein Chemistry and Structural Biology, 94*, 365–423. https://doi.org/10.1016/B978-0-12-800168-4.00010-X.

George Priya Doss, C., Nagasundaram, N., Chakraborty, C., Chen, L., & Zhu, H. (2013). Extrapolating the effect of deleterious nsSNPs in the binding adaptability of flavopiridol with CDK7 protein: A molecular dynamics approach. *Human Genomics, 7*(1), 10. https://doi.org/10.1186/1479-7364-7-10.

Giordano, S. H., Buzdar, A. U., & Hortobagyi, G. N. (2002). Breast cancer in men. *Annals of Internal Medicine, 137*(8), 678. https://doi.org/10.7326/0003-4819-137-8-200210150-00013.

Glaser, F., Pupko, T., Paz, I., Bell, R. E., Bechor-Shental, D., Martz, E., et al. (2003). ConSurf: Identification of functional regions in proteins by surface-mapping of phylogenetic information. *Bioinformatics, 19*(1), 163–164. https://doi.org/10.1093/bioinformatics/19.1.163.

Guex, N., & Peitsch, M. C. (1997). SWISS-MODEL and the Swiss-PdbViewer: An environment for comparative protein modeling. *Electrophoresis, 18*(15), 2714–2723. https://doi.org/10.1002/elps.1150181505.

Hahn, W. C., & Weinberg, R. A. (2002). Rules for making human tumor cells. *New England Journal of Medicine, 347*(20), 1593–1603. https://doi.org/10.1056/NEJMra021902.

Herrgard, S., Cammer, S. A., Hoffman, B. T., Knutson, S., Gallina, M., Speir, J. A., et al. (2003). Prediction of deleterious functional effects of amino acid mutations using a library of structure-based function descriptors. *Proteins: Structure, Function, and Genetics, 53*(4), 806–816. https://doi.org/10.1002/prot.10458.

Iqbal, N., & Iqbal, N. (2014). Human epidermal growth factor receptor 2 (HER2) in cancers: Overexpression and therapeutic implications. *Molecular Biology International, 2014*, 852748. https://doi.org/10.1155/2014/852748.

Joosten, R. P., te Beek, T. A. H., Krieger, E., Hekkelman, M. L., Hooft, R. W. W., Schneider, R., et al. (2011). A series of PDB related databases for everyday needs. *Nucleic Acids Research, 39*(Database issue), D411–D419. https://doi.org/10.1093/nar/gkq1105.

Kallioniemi, A., Kallioniemi, O. P., Sudar, D., Rutovitz, D., Gray, J. W., Waldman, F., et al. (1992). Comparative genomic hybridization for molecular cytogenetic analysis of solid tumors. *Science, 258*(5083), 818–821. http://www.ncbi.nlm.nih.gov/pubmed/1359641.

Kampen, K. R. (2011). Membrane proteins: The key players of a cancer cell. *Journal of Membrane Biology, 242*(2), 69–74. https://doi.org/10.1007/s00232-011-9381-7.

Katz, M., Amit, I., & Yarden, Y. (2007). Regulation of MAPKs by growth factors and receptor tyrosine kinases. *Biochimica et Biophysica Acta, 1773*(8), 1161–1176. https://doi.org/10.1016/j.bbamcr.2007.01.002.

Kumar, D. T., Jain, N., Kumar, S. U., Jena, P. P., Ramamoorthy, S., Doss, C. G., & Zayed, H. (2020). Molecular dynamics simulations to decipher the structural and functional consequences of pathogenic missense mutations in the galactosylceramidase (GALC) protein causing Krabbe's disease. *Journal of Biomolecular Structure & Dynamics*, 1–36. https://doi.org/10.1080/07391102.2020.1742790.

Kumar, D. T., Kumar, S. U., Laeeque, A. S., Abhay, S. A., Bithia, R., Magesh, R., … Doss, C. G. (2020). Computational model to analyze and characterize the functional mutations of NOD2 protein causing inflammatory disorder - Blau syndrome. *Advances in Protein Chemistry and Structural Biology, 120*, 379–408. https://doi.org/10.1016/bs.apcsb.2019.11.005.

Kumar, S. U., Kumar, D. T., Mandal, P. D., Sankar, S., Haldar, R., Kamaraj, B., … Zayed, H. (2020). Comprehensive in silico screening and molecular dynamics studies of missense mutations in Sjogren-Larsson syndrome associated with the ALDH3A2 gene. *Advances in Protein Chemistry and Structural Biology, 120*, 349–377. https://doi.org/10.1016/bs.apcsb.2019.11.004.

Lee, K.-F., Simon, H., Chen, H., Bates, B., Hung, M.-C., & Hauser, C. (1995). Requirement for neuregulin receptor erbB2 in neural and cardiac development. *Nature, 378*(6555), 394–398. https://doi.org/10.1038/378394a0.

Lemak, A. S., & Balabaev, N. K. (1994). On the berendsen thermostat. *Molecular Simulation, 13*(3), 177–187. https://doi.org/10.1080/08927029408021981.

Lengauer, C., Kinzler, K. W., & Vogelstein, B. (1998). Genetic instabilities in human cancers. *Nature, 396*(6712), 643–649. https://doi.org/10.1038/25292.

Lin, C. Y., Lee, C. H., Chuang, Y. H., Lee, J. Y., Chiu, Y. Y., Wu Lee, Y. H., et al. (2019). Membrane protein-regulated networks across human cancers. *Nature Communications, 10*(1). https://doi.org/10.1038/s41467-019-10920-8.

Lipinski, C. A., Lombardo, F., Dominy, B. W., & Feeney, P. J. (1997). Experimental and computational approaches to estimate solubility and permeability in drug discovery and development settings. *Advanced Drug Delivery Reviews, 23*(1), 3–25. https://doi.org/10.1016/S0169-409X(96)00423-1.

Lobanov, M. I., Bogatyreva, N. S., & Galzitskaia, O. V. (2008). Radius of gyration is indicator of compactness of protein structure. *Molecular Biology (Molekuliarnaia Biologiia), 42*(4), 701–706. http://www.ncbi.nlm.nih.gov/pubmed/18856071.

Lohrisch, C., & Piccart, M. (2001). HER2/neu as a predictive factor in breast cancer. *Clinical Breast Cancer, 2*(2), 129–135. https://doi.org/10.3816/CBC.2001.n.017.

Luo, J., Manning, B. D., & Cantley, L. C. (2003). Targeting the PI3K-akt pathway in human cancer: Rationale and promise. *Cancer Cell, 4*(4), 257–262. http://www.ncbi.nlm.nih.gov/pubmed/14585353.

Mahajan, P., Suri, N., Mehra, R., Gupta, M., Kumar, A., Singh, S. K., et al. (2017). Discovery of novel small molecule EGFR inhibitory leads by structure and ligand-based virtual screening. *Medicinal Chemistry Research, 26*(1), 74–92. https://doi.org/10.1007/s00044-016-1728-2.

Malde, A. K., Zuo, L., Breeze, M., Stroet, M., Poger, D., Nair, P. C., et al. (2011). An automated force field Topology builder (ATB) and repository: Version 1.0. *Journal of Chemical Theory and Computation, 7*(12), 4026–4037. https://doi.org/10.1021/ct200196m.

Meyer, D., & Birchmeier, C. (1994). Distinct isoforms of neuregulin are expressed in mesenchymal and neuronal cells during mouse development. *Proceedings of the National Academy of Sciences, 91*(3), 1064–1068. https://doi.org/10.1073/pnas.91.3.1064.

Morris, G. M., Goodsell, D. S., Halliday, R. S., Huey, R., Hart, W. E., Belew, R. K., et al. (1998). Automated docking using a Lamarckian genetic algorithm and an empirical binding free energy function. *Journal of Computational Chemistry, 19*(14), 1639–1662. https://doi.org/10.1002/(SICI)1096-987X(19981115)19:14<1639::AID-JCC10>3.0.CO;2-B.

Morris, G. M., Huey, R., Lindstrom, W., Sanner, M. F., Belew, R. K., Goodsell, D. S., et al. (2009). AutoDock4 and AutoDockTools4: Automated docking with selective receptor flexibility. *Journal of Computational Chemistry, 30*(16), 2785–2791. https://doi.org/10.1002/jcc.21256.

Mosaeilhy, A., Mohamed, M. M., George Priya Doss, C., El Abd, H. S. A., Gamal, R., Zaki, O. K., et al. (2017). Genotype-phenotype correlation in 18 Egyptian patients with glutaric acidemia type I. *Metabolic Brain Disease, 32*(5), 1417–1426. https://doi.org/10.1007/s11011-017-0006-4. Epub 2017 Apr 7.

Nagasundaram, N., Doss, G. P. C., Chakraborty, C., Karthick, V., Kumar, D. T., Balaji, V., et al. (2016). Mechanism of artemisinin resistance for malaria PfATP6 L263 mutations and discovering potential antimalarials: An integrated computational approach. *Scientific Reports, 6*(January), 1–12. https://doi.org/10.1038/srep30106.

Oostenbrink, C., Villa, A., Mark, A. E., & van Gunsteren, W. F. (2004). A biomolecular force field based on the free enthalpy of hydration and solvation: The GROMOS force-field parameter sets 53A5 and 53A6. *Journal of Computational Chemistry, 25*(13), 1656–1676. https://doi.org/10.1002/jcc.20090.

Pace, C. N., Fu, H., Fryar, K. L., Landua, J., Trevino, S. R., Shirley, B. A., et al. (2011). Contribution of hydrophobic interactions to protein stability. *Journal of Molecular Biology, 408*(3), 514–528. https://doi.org/10.1016/j.jmb.2011.02.053.

Parrinello, M., & Rahman, A. (1980). Crystal structure and pair potentials: A molecular-dynamics study. *Physical Review Letters, 45*(14), 1196–1199. https://doi.org/10.1103/PhysRevLett.45.1196.

Riese, D. J., Komurasaki, T., Plowman, G. D., & Stern, D. F. (1998). Activation of ErbB4 by the bifunctional epidermal growth factor family hormone epiregulin is regulated by ErbB2. *Journal of Biological Chemistry, 273*(18), 11288–11294. https://doi.org/10.1074/jbc.273.18.11288.

Robert Kiss, M. S. F. A. S. (2012). http://Mcule.com: a public web service for drug discovery. *Journal of Cheminformatics, 4*(Suppl. 1), P17. BioMed Central. https://doi.org/10.1186/1758-2946-4-S1-P17.

Saralamba, S., Pan-Ngum, W., Maude, R. J., Lee, S. J., Tarning, J., Lindegårdh, N., et al. (2011). Intrahost modeling of artemisinin resistance in Plasmodium falciparum. *Proceedings of the National Academy of Sciences of the United States of America, 108*(1), 397–402. https://doi.org/10.1073/pnas.1006113108.

Schmit, K., & Michiels, C. (2018). TMEM proteins in cancer: A review. *Frontiers in Pharmacology, 9*(1345). https://doi.org/10.3389/fphar.2018.01345.

Schonfeld, S. J., Pfeiffer, R. M., Lacey, J. V., Berrington de González, A., Doody, M. M., Greenlee, R. T., et al. (2011). Hormone-related risk factors and postmenopausal breast cancer among nulliparous versus parous women: An aggregated study. *American Journal of Epidemiology, 173*(5), 509–517. https://doi.org/10.1093/aje/kwq404.

Siegel, R. L., Miller, K. D., & Jemal, A. (2019). Cancer statistics, 2019. *CA: A Cancer Journal for Clinicians, 69*(1), 7–34. https://doi.org/10.3322/caac.21551.

Stern, D. F. (2000). Tyrosine kinase signalling in breast cancer. ErbB family receptor tyrosine kinases. *Breast Cancer Research, 2*(3), 176–183. https://doi.org/10.1186/bcr51.

Sudhakar, N., George Priya Doss, C., Kumar, D. T., Chakraborty, C., Anand, K., & Suresh, M. (2016). Deciphering the impact of somatic mutations in exon 20 and exon

9 of PIK3CA gene in breast tumors among Indian women through molecular dynamics approach. *Journal of Biomolecular Structure and Dynamics, 34*(1), 29–41. https://doi.org/10.1080/07391102.2015.1007483.

Tate, J. G., Bamford, S., Jubb, H. C., Sondka, Z., Beare, D. M., Bindal, N., et al. (2019). COSMIC: The catalogue of somatic mutations in cancer. *Nucleic Acids Research, 47*(D1), D941–D947. https://doi.org/10.1093/nar/gky1015.

Thirumal Kumar, D., & George Priya Doss, C. (2016). Investigating the inhibitory effect of wortmannin in the hotspot mutation at codon 1047 of PIK3CA kinase domain: A molecular docking and molecular dynamics approach. *Advances in Protein Chemistry and Structural Biology, 102*, 267–297. https://doi.org/10.1016/bs.apcsb.2015.09.008.

Thirumal Kumar, D., & George Priya Doss, C. (2017). Role of E542 and E545 missense mutations of PIK3CA in breast cancer: A comparative computational approach. *Journal of Biomolecular Structure and Dynamics, 35*(12), 2745–2757. https://doi.org/10.1080/07391102.2016.1231082.

Thirumal Kumar, D., George Priya Doss, C., Sneha, P., Tayubi, I. A., Siva, R., Chakraborty, C., & Magesh, R. (2017). Influence of V54M mutation in giant muscle protein titin: a computational screening and molecular dynamics approach. *Journal of Biomolecular Structure and Dynamics, 35*(5), 917–928. https://doi.org/10.1080/07391102.2016.1166456.

Thirumal Kumar, D., Jain, N., Udhaya Kumar, S., George Priya Doss, C., & Zayed, H. (2020). Identification of potential inhibitors against pathogenic missense mutations of PMM2 using a structure-based virtual screening approach. *Journal of Biomolecular Structure and Dynamics*. https://doi.org/10.1080/07391102.2019.1708797.

Thirumal Kumar, D., Jerushah Emerald, L., Priya Doss, GeorgeC., Sneha, P., Siva, R., Charles Emmanuel Jebaraj, W., & Zayed, H. (2018). Computational approach to unravel the impact of missense mutations of proteins (D2HGDH and IDH2) causing D-2-hydroxyglutaric aciduria 2. *Metabolic Brain Disease, 33*(5), 1699–1710. https://doi.org/10.1007/s11011-018-0278-3.

Thirumal Kumar, D., Umer Niazullah, M., Tasneem, S., Judith, E., Susmita, B., George Priya Doss, C., et al. (2019). A computational method to characterize the missense mutations in the catalytic domain of GAA protein causing Pompe disease. *Journal of Cellular Biochemistry, 120*(3), 3491–3505. https://doi.org/10.1002/jcb.27624.

Toomey, S., Eustace, A. J., Fay, J., Sheehan, K. M., Carr, A., Milewska, M., et al. (2017). Impact of somatic PI3K pathway and ERBB family mutations on pathological complete response (pCR) in HER2-positive breast cancer patients who received neoadjuvant HER2-targeted therapies. *Breast Cancer Research*. https://doi.org/10.1186/s13058-017-0883-9.

UdhayaKumar, S., Doss, C. G., & Zayed, H. (2020). Mutational landscape of K-Ras substitutions at 12th position-a systematic molecular dynamics approach. *Journal of Biomolecular Structure & Dynamics*, 1–16. https://doi.org/10.1080/07391102.2020.1830177.

Udhayakumar, S., Kumar, D., Siva, R., Doss, C., & Zayed, H. (2020). An extensive computational approach to analyze and characterize the functional mutations in the galactose-1-phosphate uridyl transferase (GALT) protein responsible for classical galactosemia. *Computers in Biology and Medicine, 117*, 103583. https://doi.org/10.1016/j.compbiomed.2019.103583.

Yuan, T. L., & Cantley, L. C. (2008). PI3K pathway alterations in cancer: Variations on a theme. *Oncogene, 27*(41), 5497–5510. https://doi.org/10.1038/onc.2008.245.

CHAPTER FIVE

Autophagy proteins and its homeostasis in cellular environment

Vrushali Guhe, Bhavnita Soni, Prajakta Ingale and Shailza Singh*
National Centre for Cell Science, NCCS Complex, Ganeshkhind, SP Pune University Campus, Pune, Maharashtra, India
*Corresponding author: E-mail: singhs@nccs.res.in

Contents

1. Introduction 74
   1.1 Autophagy- mechanism 75
      1.1.1 Induction and autophagophore formation 75
      1.1.2 Phagophore elongation and autophagosome formation 75
   1.2 Fusion, degradation and recycling 76
   1.3 Role of cytokine in autophagy 77
      1.3.1 Interplay between TNFα and IL10 for autophagy in macrophages 77
      1.3.2 Interaction between canonical and non-canonical autophagy 77
2. Methodology 79
   2.1 Reconstruction of autophagic network 79
   2.2 Phylogenetic tree construction and molecular clock analysis 79
   2.3 Reconstruction of autophagic model by mathematical modeling 81
      2.3.1 Model reconstruction and simulation 81
      2.3.2 Parameter estimation 81
      2.3.3 Sensitivity analysis 82
   2.4 Principal component analysis and model reduction 82
      2.4.1 Principal component analysis (PCA) 82
      2.4.2 Model reduction 83
   2.5 Flux analysis 83
3. Results 83
   3.1 Autophagic network analysis 83
   3.2 Phylogenetic tree construction and molecular clock analysis 84
   3.3 Mathematical modeling, principal component analysis and model reduction 84
   3.4 Flux analysis 88
4. Discussion 90
Acknowledgments 92
References 92

Advances in Protein Chemistry and Structural Biology, Volume 123
ISSN 1876-1623
https://doi.org/10.1016/bs.apcsb.2019.12.002

Abstract

Autophagy is a self-destructing mechanism of cell via lysosomal degradation, which helps to degrade/destroy hazardous substances, proteins, degenerating organelles and recycling nutrients. It plays an important role is cellular homeostasis and regulates internal environment of cell, moreover, when needed causes non-apoptotic programmed death of cell. Autophagy has been observed as one of the major factors in parasite clearance in leishmaniasis. Being an intra-cellular pathogen, the cell mediated response is the only alternative for adaptive immunity against *Leishmania* in host. Pro-inflammatory cytokines IL12 and TNFα generate Th2 response which helps in active phagocytosis of parasite whereas an anti-inflammatory cytokine like IL10 mediate parasite promotion by blocking autophagic pathways and inhibiting phagocytic actions. In the present chapter, through systems biology approach, we are trying to decipher the role of pro-inflammatory and anti-inflammatory cytokine in autophagy during leishmanial infection. TLR2/6 mediated signaling stimulated by LPG produces many pro-inflammatory cytokines like IL12, TNFα and IL6 etc. Among them TNFα, causes the activation of PI3P through a series of events, which results in activation of autophagic machinery, whereas, IL10 through ATG9 and mTOR activation, inhibits autophagy. The mathematical modeling of these pathways shows that, ATG9-PI3P act as a negative feedback loop in autophagic machinery of leishmaniasis.

1. Introduction

Autophagy is a self-destructing mechanism of cell via lysosomal degradation, which helps to degrade/destroy hazardous substances, proteins, degenerating organelles and recycling nutrients. It plays an important role is cellular homeostasis and regulates internal environment of cell, when needed causes non-apoptotic programmed death of the cell. It's usually detailed with aging and nutritional stress, but recent researches have provided support for its role in innate as well as adaptive immunity (Deretic, 2006). It is further classified into mitophagy, chaperone mediated autophagy and macroautophagy (Glick et al., 2010). Macroautophagy is found to be related to immunity of host (Deretic, 2006).

Leishmaniasis is one among the top most neglected tropical disease. It usually affects the countries with low socio-economic status, and present in three clinical forms, namely cutaneous, muco-cutaneous and visceral leishmaniasis. The present chapter focuses on the involvement of autophagy in cutaneous leishmaniasis. Literature review suggested that, the promastigote form of parasite is converted into parasitic amastigote form in macrophage through autophagy whereas host autophagic machinery is involve in parasite killing. Autophagy related protein 8 (ATG8) and ATG 12 are involved in autophagosome formation in non-canonical autophagy. An autophagy related protein 8 phosphatidylethanolamine (ATG8-PE) conjugate system is important for autophagosome formation with the help of ATG 7 and ATG3 which are E1 and E2 ligases respectively. Besides these, in recent years ATG5-ATG12 conjugate system has been emerged in *Leishmania*. ATG5 is required for both autophagosome formation and also maintenance of a fully functional mitochondrion. The mutants lacking ATG5 have increased mitochondrial mass and phospholipid content, high levels of oxidants and reduced membrane potential, all being hallmarks of a dysfunctional mitochondrion with impaired ability for energy generation. Thus, ATG5 is required for ATG8 dependent autophagy in *Leishmania major* (Williams, Smith, Cull, Mottram, & Coombs, 2012).

1.1 Autophagy- mechanism

1.1.1 Induction and autophagophore formation

The molecular machinery of autophagic processes involves several conserved autophagy (Atg) proteins called autophagy-related-gene products. Various stimuli including nutrient starvation, elimination of growth factors etc., lead to the formation of the autophagophore. This step involves two protein complexes: a complex that contains the class III PI3K/vacuolar protein sorting 34 (Vps34), Atg6/Beclin1, Atg14 and vacuolar protein sorting 15/30 (Vps15/30) and a complex that includes the serine/threonine kinase Atg1/ULK1 (human homolog of Atg1 of yeast), an important as well as an essential positive regulator of autophagosome formation (Deretic, 2006).

1.1.2 Phagophore elongation and autophagosome formation

The characteristic double-membrane autophagosome is formed in elongation step. It comprises two ubiquitin-like conjugation pathways, both catalyzed by Atg7. The first system leads to the conjugation of Atg5-Atg12, which later forms a multimeric complex with Atg16L. The Atg5-Atg12-Atg16L complex associates with the outer membrane of the extending phagophore and engages for further process.

The second system marks in the processing of light chain 3 (LC3), mammalian homolog of the yeast Atg8. Upon induction, LC3B is proteolytically sliced by Atg4 to generate LC3B—I. This molecule is activated by Atg7 and later conjugated to phosphatidylethanolamine (PE) in the membrane to produce processed LC3B-II. Processed LC3B-II is recruited onto the growing autophagophore and its integration is dependent on Atg5-Atg12 (Deretic, 2006).

Unlike Atg5-Atg12-Atg16L, LC3B-II is found on both the inner and outer surfaces of the autophagosome, where it is required for the expansion and completion of the autophagic membrane. After closure of the autophagosomal membrane, the Atg16-Atg5-Atg12 complex dissociates from the vesicle, whereas a portion of LC3B-II remains covalently bound to the membrane (Fig. 1). Therefore, LC3B-II can be used as a marker to monitor the level of autophagy in cells (Tanida, Ueno, & Kominami, 2008). Recent studies validate self-multimerization and phosphorylation of Atg9 aid membrane tethering and/or fusion (Feng et al., 2016).

1.2 Fusion, degradation and recycling

When the autophagosome formation is accomplished, LC3B-II attached to the outer membrane is cut from PE by Atg4 and released back to the cytoplasm. The fusion of the autophagosome and the lysosome is thought to require the lysosomal membrane protein LAMP-2 and the small GTPase Rab7 (Ras-related protein in brain).

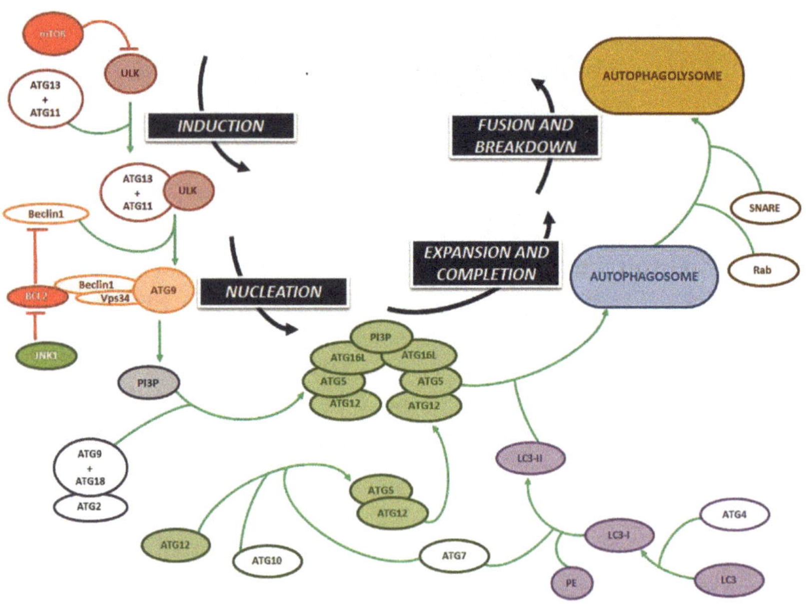

Fig. 1 Molecular mechanism of autophagy.

After fusion, a series of acid hydrolases are involved in the degradation of the confiscated cytoplasmic cargo. The small molecules resulting from the degradation, particularly amino acids are transported back to the cytosol for protein synthesis and maintenance of cellular functions under starvation conditions (Deretic, 2006).

1.3 Role of cytokine in autophagy

Autophagy has been observed as one of the major factors in parasite clearance in leishmaniasis (Franco et al., 2017). Due to intra-cellular pathogen, the cell-mediated response is the only alternative for adaptive immunity, as discussed before, via toll-like receptor (TLR) signaling.

1.3.1 Interplay between TNFα and IL10 for autophagy in macrophages

TLR2/6 mediated signaling stimulated by lipophosphoglycan (LPG) produces many pro-inflammatory cytokines like interleukin 12 (IL12), tumor necrosis factor α (TNFα) and IL6 etc (Hanada et al., 2007). In macrophages, it is found that TNFα via autocrine signaling induces autophagy and helps in parasite killing of *Toxoplasma gondii* via c-Jun NH_2-terminal kinase (JNK) phosphorylation in macrophages (Liu, Lopez Corcino, Portillo, Miao, & Subaustea, 2016). The phosphorylated JNK induces Beclin by phosphorylating Bcl2, which inhibits Beclin1 action, thus releasing Beclin1 and promoting VpS34 that recruits phosphatidylinositol 3-phosphate (PI3P) for autophagosome formation. Recruitment of PI3P is controlled by ATG9 protein, autophagy related gene protein, which in normal condition is inhibited by mechanistic target of rapamycin (mTOR) via ULK/ATG1 (Fig. 2).

IL10 has been observed to inhibit autophagy via starvation induced protein kinase B (PKB)/AKT (AKT-PI3K) pathway in macrophages via inducing mToR (Park et al., 2011). Thus, we can conclude that IL10, which is found to be encouraged by parasite interaction to host, obstructs recruitment of phosphatidylinositol 3-phosphate (PI3P) via mTOR and eventually checks autophagy for parasite survival.

1.3.2 Interaction between canonical and non-canonical autophagy

For the action of internalization of parasite, *Leishmania* evades macrophages by phagocytosis but recruitment of LC3 is reduced heavily due to the interaction with surface metalloprotease glycoprotein 63 (GP63) (Park et al., 2011).

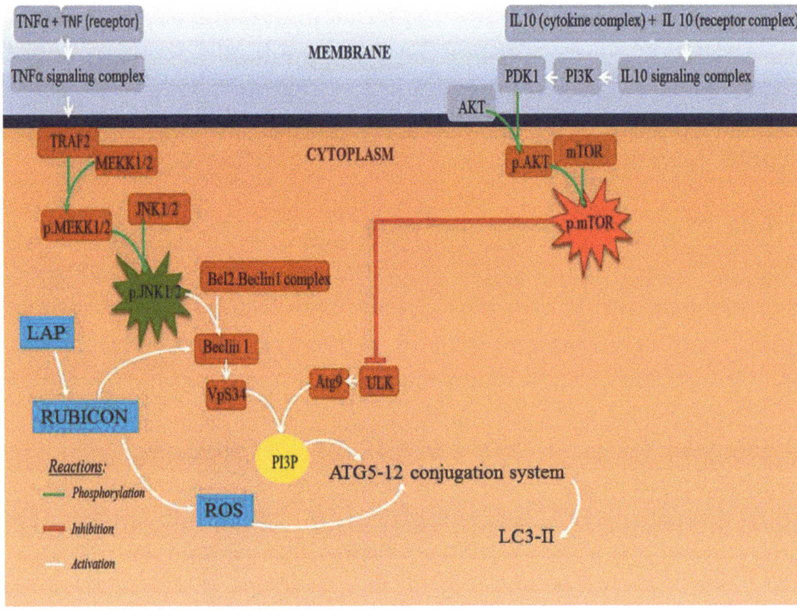

Fig. 2 Signaling pathways connecting canonical and non-canonical autophagy.

LC3 associated phagocytosis (LAP) is a kind of non-canonical autophagy which the parasite targets to manipulate host autophagic machinery in accordance. Light Chain3 (LC3) tagging on the surface of extending membrane is a crucial step and parasite alters it, thus, acquiring larger entry to host cells but persisting to survive by inactivating canonical autophagic ways. The machinery is tempered by parasite attacking or residing cells leading to less production and tags of LC3 on membrane, which in turn avoid maturation of autophagosome and fusion with lysosomes.

Henceforth, cytokine milieu present during leishmanial infection plays an important role in activation of autophagic pathway as well as determining the outcome of the disease. In past one decade, systems biology approach has emerged as one of the potent approach to churn out novel targets for therapeutic purposes (Mol, Patole, & Singh, 2014; Soni, Saha, & Singh, 2017). Keeping in mind, the role of pro and anti-inflammatory cytokines and various autophagic components, the present work focuses on further exploring autophagic machinery of host and the parasite through systems biology approach to find out important targets, which may be exploited in future to develop therapeutics for leishmaniasis.

2. Methodology

2.1 Reconstruction of autophagic network

Literature review suggested the presence of abundant databases related to autophagy (Jacomin, Gul, Sudhakar, Korcsmaros, & Nezis, 2018). Out of those, we have chosen STRING v.10.5 and Gene Ontology Consortium Databases, to retrieve protein information on the basis of their biochemical function and connectivity to the other proteins of autophagic machinery in *Homo sapiens* and *Mus musculus*. The complete autophagic network was analyzed using the network analyzer Cytoscape v.2.8.3 (Smoot, Ono, Ruscheinski, Wang, & Ideker, 2010). The total number of nodes present in the network is 100, which is a collection of autophagy related gene product proteins as well as transcription factors, cytokines and target products of autophagic machinery. From the given proteins and molecules, analysis is performed based on high values of clustering coefficient, closeness centrality, betweenness centrality, average shortest path length, outdegree and indegree. *Clustering Coefficient* is a measure of the degree to which nodes in the network tend to cluster together, thus more the clustering coefficient, denser is the network, forming a closely-knit pattern. For a particular node, high clustering coefficient specifies its highest degree of connectivity. *Closeness Centrality* measures the mean distance from a node to the other nodes, which means the more central a node, it is the closer to all other nodes. *Betweenness Centrality* measures the extent to which a node lies on paths between the other nodes. Nodes with high betweenness may have considerable influence within a network as they control information passing between others. They would also affect the network and all nodes are mostly removed because they lie on the largest number of paths taken by messages. *Average Shortest Path Length* is a concept in network analysis that defines the average number of steps along the shortest paths for all possible pairs of network nodes. It measures the efficiency of information transmitted in a network. *Indegree* and *Outdegree* measure the connection of a particular node with incoming signals and outgoing signals respectively, to execute the functions in the given network (Pavlopoulos et al., 2011).

2.2 Phylogenetic tree construction and molecular clock analysis

Phylogenetic tree construction: Coding sequences of ATGs (ATG1-32) were retrieved from NCBI databases of all the organisms available, ranging from prokaryotes to mammals. Multiple sequence alignment was performed using online multiple sequence alignment tool MultAlin. The fasta files obtained were converted to nexus format for phylogenetic tree construction. Phylogenetic tree was constructed using the software MrBayes v.3.2 (Ronquist, Huelsenbeck, Nielsen, & Bollback, 2001). MrBayes performs Bayesian inference of phylogeny using a variant of Markov chain Monte Carlo (MCMC). MrBayes uses MCMC methods to estimate the posterior distribution of model parameters. The posterior distribution is an average between knowledge about the parameters before data is observed (which is represented by the prior distribution) and the information about the parameters contained in the observed data (which is represented by the likelihood function). It provides a natural and principled way of combining prior information with data (Ronquist et al., 2001). All inferences logically follow from Bayes' theorem. MCMC, along with the other numerical methods, makes computations tractable for virtually all parametric models. The nucleotide substitution model used during the tree construction was the GTR (General time reversible) model with gamma distributed rate variation across sites and a proportion of invariable sites.

MCMC simulation was run for 1 lakh to 40 lakh generations, depending on standard deviation of split frequency of each ATG, trees were sampled every 1000 generations. Analysis was run until a standard deviation of split frequencies were below 0.01. Parameters and corresponding trees were summarized and used later for analysis. Phylogenetic tree of all the ATGs was visualized using FigTree v.1.4.3 (Rambaut, 2014).

Molecular clock analysis: Information regarding the out-groups was obtained using the phylogenetic tree construction. Only those ATGs which play utmost important role in autophagic machinery is taken for subsequent studies which includes ATG1, ATG2, ATG4, ATG5, ATG7, ATG8, ATG9, ATG10, ATG12, ATG16 and ATG18. Relaxed molecular clock was constructed using MrBayes v.3.2.1.

MCMC simulation was run for one lakh to ten lakh generations, again depending on each ATG. The model used for constructing the molecular clock was the independent gamma rates model (IGR model) (MrBayes Manual). Molecular clocks of aforementioned ATGs were constructed and visualized using the FigTree v.1.4.3.

2.3 Reconstruction of autophagic model by mathematical modeling

2.3.1 *Model reconstruction and simulation*

Signaling network reconstruction is the integration of information that describes the biochemical changes that occur in a specified network. It helps gain deeper understanding of interactions between the signaling proteins with themselves and surrounding as well as explaining the complexity of biological system at higher level. Quantitative modeling of these interactions plays an important role in understanding fundamental intra- and intercellular processes (Soni et al., 2017).

The interactions between the signaling components were entered as elemental chemical reactions in MATLAB v.7.11.1.866 SimBiology toolbox, which uses the Systems Biology Markup Language (SBML) machine language. The kinetic rate laws (mass action for association and dissociation reactions, translocation and conversion, Henri—Michaelis—Menten for phosphorylation/dephosphorylation/ubiquitination, Hills equation for gene expression) and the initial concentrations associated with the reactions were also defined in the toolbox, which were apparently applied to obtain a desired behavior.

The reconstructed network was numerically simulated using Stiff Deterministic ODE15s solver (SimBiology toolbox) which generates the first order nonlinear ODEs for each node, thus, defining the mathematical structure of the model. Complex biological systems, such as the signaling network can be viewed as networks of chemical reactions that can be analyzed mathematically using ODEs, which is the most common simulation approach used in computational systems biology. ODE helps in determining time dependent changes i.e. the time series data of the concentrations of the signaling proteins and protein complexes and thus, the dynamics associated with it. The model was exported to Copasi v.4.8.35 (Hoops et al., 2006) as a SBML file to generate the time series data.

2.3.2 *Parameter estimation*

The equations outlining the chemical transformation in the signaling flow uses variables as concentrations of protein. Often, these parameters are unknown, difficult and hard to measure. In such situation, parameter estimation is done i.e. the unknown parameters are determined indirectly using computational biology tools. The parameters for reconstructed signaling network were manually trained within a range of parameter range that was further tuned until the model was simulated to obtain a desired output with the algorithmic data applicable. This is due to the lack of homogeneity in experimental measurements (Codogno, Mehrpour, & Proikas-Cezanne, 2012).

2.3.3 Sensitivity analysis

It plays an important role in dynamic analysis of systems biology models. Sensitivity analysis is used to determine how sensitive a model is to variables changes in the model, which helps study the randomness associated with the estimated parameters and thus counter build assurance in the model (Soni et al., 2017).

It was calculated using the SimBiology toolbox, sensitivity Analysis option, which calculates the sensitivity coefficients by combining the original ODE system for a model with the auxiliary differential equations for the sensitivities. The additional equations are derivatives of the original equations with respect to the parameters. It calculates the time-dependent sensitivities of all the species with respect to species initial conditions and parameter values in the model.

In this chapter, signaling network of two different cases has been taken into consideration:

CASE:1 Interplay between TNFα and IL10 for autophagy in macrophages in leishmaniasis- In this case, two models were developed in which one was for induction of autophagy by TNFα and the second model was constructed based on inhibition of autophagy by IL10 signaling pathways. Moreover, a model for general study of LC-3 associated phagocytosis (LAP) was also reconstructed.

CASE:2 Interaction between canonical and non-canonical autophagy in leishmaniasis- In this case, three models were developed in which one explains the signaling pathway compiling both canonical and non-canonical autophagy, second model is designed that explains only canonical autophagic pathway excluding LAP and the last model is reconstructed to explain only non-canonical autophagy which is LAP excluding canonical autophagy.

2.4 Principal component analysis and model reduction

2.4.1 Principal component analysis (PCA)

Generation of a new set of variables (principal component score), called principal components, from the multivariate data sets is done using

MATLAB v.7.11.1.866. Each principal component is a linear combination of the original variables. All the principal components are orthogonal to each other, so there is no redundant information. PCA was done in MATLAB using the function i.e. (score_coefficient) = princomp (a) where, "a" is the y matrix of sensitivity coefficients of each species in the signaling network. This command returns principal component coefficient and score of the sensitivity matrix, used to identify the principal component i.e. significant individual reactions in the signaling network. The graphical representation of the derived score coefficient is visualized using SigmaPlot v.10.0 by scatter plot formation (Soni et al., 2017).

2.4.2 Model reduction

Model reduction is a systematic method for eliminating reactions that do not contribute considerably to a network output. In this study, flux analysis, sensitivity analysis and concentration at which the species shows the highest sensitivity was combined to reduce the dimensionality of the model. The concentrations were retrieved from the time series data using Copasi v.4.8.35. The graphical representation of the collected values is visualized using SigmaPlot v.10.0 by 3D-mesh plot formation (Soni et al., 2017).

2.5 Flux analysis

Flux analysis is applicable for systems that are in a pseudo-steady state. Under this condition, the differential equation system of metabolite mass balances reduces to a linear equation system, which relies solely on the known stoichiometry of the biochemical reaction network. For this system, flux analysis for each reaction is done in Copasi v.4.8.35 under the task - steady state analysis and the graphical representation is created using MS Excel v.2010 (Soni et al., 2017).

3. Results

3.1 Autophagic network analysis

From the analysis of autophagy network in human and mouse, the crucial crosstalk point molecules obtained are Beclin1, ATG9, unc-51 like autophagy activating kinase (ULK) and PI3P (Fig. 3). This provided us an insight with an understanding that these molecules play a critical role in signaling machinery of autophagy i.e. they could be used as a target to perturb the network accordingly and may help in decrease of parasite load.

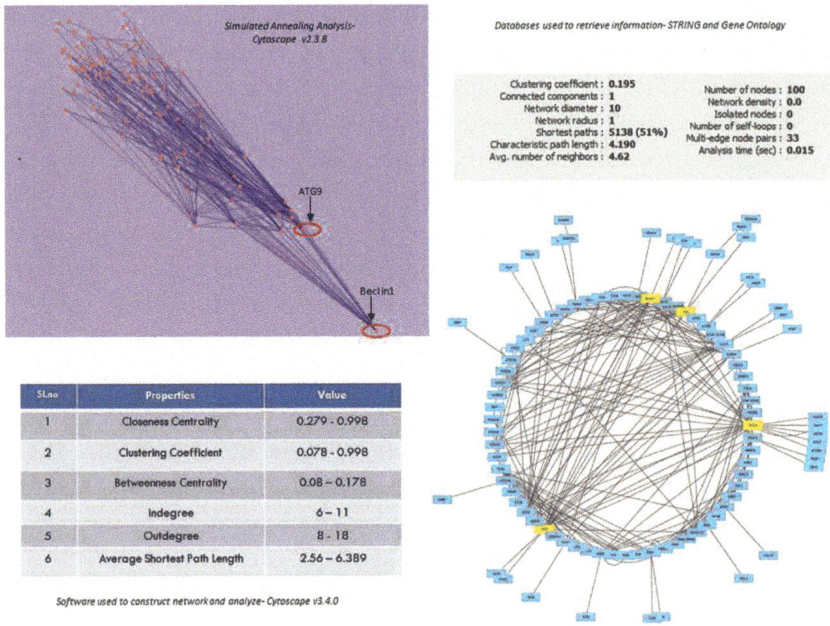

SI.no	Properties	Value
1	Closeness Centrality	0.279 - 0.998
2	Clustering Coefficient	0.078 - 0.998
3	Betweenness Centrality	0.08 – 0.178
4	Indegree	6 – 11
5	Outdegree	8 - 18
6	Average Shortest Path Length	2.56 – 6.389

Fig. 3 Autophagy network analysis and parameters used to get crucial crosstalk points of network.

3.2 Phylogenetic tree construction and molecular clock analysis

Due to the presence of core autophagic proteins i.e. autophagy related gene products in the crucial components; they are taken for phylogenetic analysis. From all the coding sequences of different autophagy related genes, few genes are further processed in molecular clock that are defined well in the process of autophagy. Moreover, from this analysis, we come to refine our result as ATG5, ATG8, ATG9, ATG12 and ATG16 are conserved among variable taxon and could further be used to construct mathematical model to analyze principal components of autophagic machinery (Fig. 4).

3.3 Mathematical modeling, principal component analysis and model reduction

With the mentioned concentration, reaction laws and parameters, models for two different set of cases were constructed.

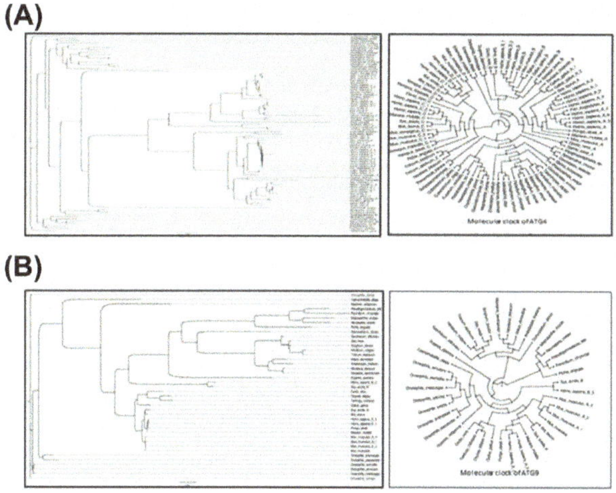

(C)

Sl.no	ATG	HPD value
1	ATG1	1.054
2	ATG2	2.059
3	ATG4	15.10
4	ATG5	0.099
5	ATG7	2.817
6	ATG8	0.606
7	ATG9	0.690
8	ATG10	0.958
9	ATG12	0.829
10	ATG16	0.595
11	ATG18	0.907

Fig. 4 Phylogenetic analysis: Molecular clock of (A) ATG 4 (B) ATG 9 (C) Table showing 95% highest posterior density (HPD) value of various ATG.

Case 1- Interplay between TNFα and IL10 for autophagy in macrophages in leishmaniasis:

A. For Induction of Autophagy:

B. For Inhibtion of Autophagy

In this case of interplay between the two cytokines, the induction (Fig. 5A) and inhibition (Fig. 5B) of autophagy is studied, where TNFα is determined to be the cause of induction and IL10 for inhibition. In the model system concerning the induction of autophagy, the principal components ATG9 and PI3P are obtained, whereas, for inhibition phosphorylated AKT in membrane and phosphorylated mTOR generated by AKT pathway are obtained. Model is then further reduced to two reactions viz., interaction between ATG9 and PI3P as well as phosphorylation of mTOR.

C. Model reconstruction for LAP (LC3 associated phagocytosis)-

In case of general study on LAP, RUN domain and cysteine-rich domain containing, Beclin 1 - interacting protein —UV radiation resistance-associated gene (Rubicon-Beclin1-UVRAG complex), Beclin1 and LC3II are produced as principal components (Fig. 5C).

Case 2- Interaction between canonical and non-canonical autophagy in leishmaniasis:

A. Model reconstruction of system having combination of canonical and non-canonical autophagy

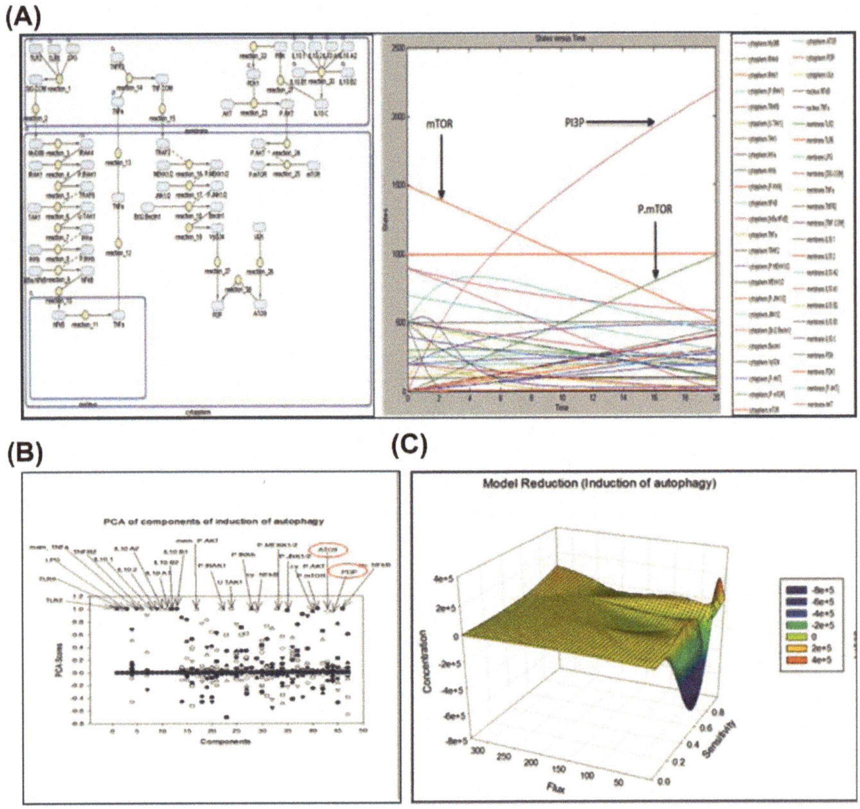

Fig. 5A Interplay between TNFα and IL10 for autophagy in macrophages in leishmaniasis: (A) Model *reconstruction* of induction of autophagy and simulated graph (B) Principal component analysis showing mTOR and PI3P as an important component (C) Graphical representation of model reduction.

B. Model reconstruction of system having canonical autophagy signaling excluding LAP

C. Model reconstruction of system having only non-canonical autophagy (LAP) excluding canonical autophagy

In the case of interaction between canonical and non-canonical autophagy in leishmaniasis (Fig. 6A, B and C), the principal components in the system obtained are p22phox, LC3 variants, ATG4, ULK and reactive

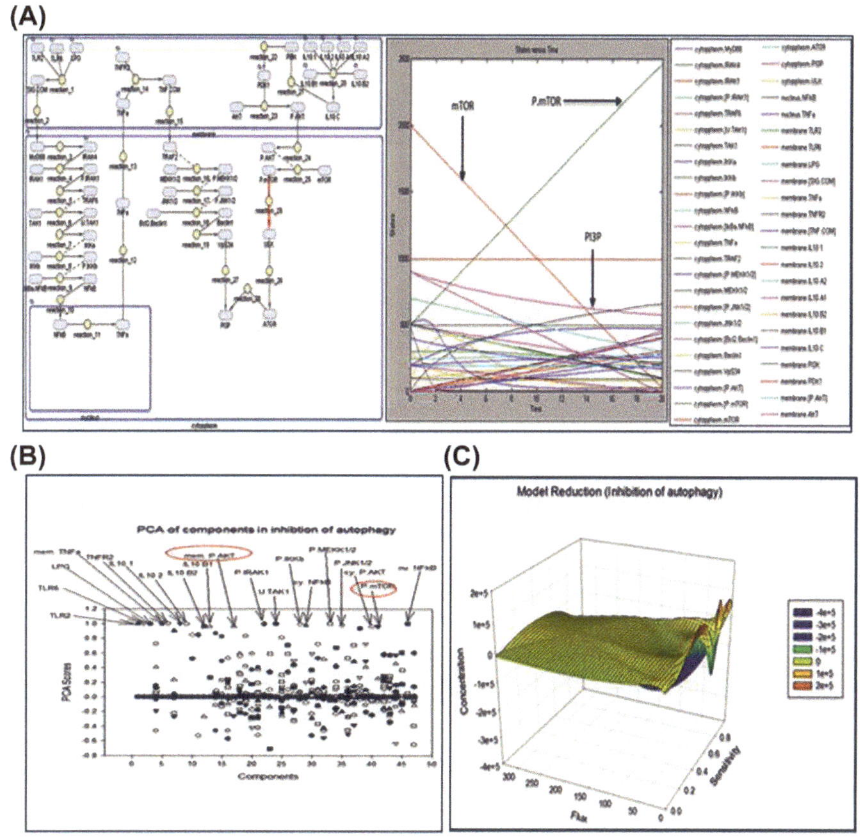

Fig. 5B Interplay between TNFα and IL10 for autophagy in macrophages in leishmaniasis: (A) Model *reconstruction* of inhibition of autophagy and simulated graph (B) Principal component analysis showing mTOR and membrane phosphorylated AKT as an important component (C) Graphical representation of model reduction.

oxygen species (ROS) (for combined study of canonical and non-canonical autophagy), Tumor necrosis factor receptor-associated factor (TRAF2), ULK, ATG4, LC3, ATG5-12-16L complex and Beclin1 (for system with only canonical autophagic machinery signaling) and p22phox, LC3 variants, ATG4 and ROS (for system with only non-canonical autophagic machinery signaling). The reduced model elucidates four important cross-talk point molecules as ATG4, ATG9, Beclin1 and Rubicon.

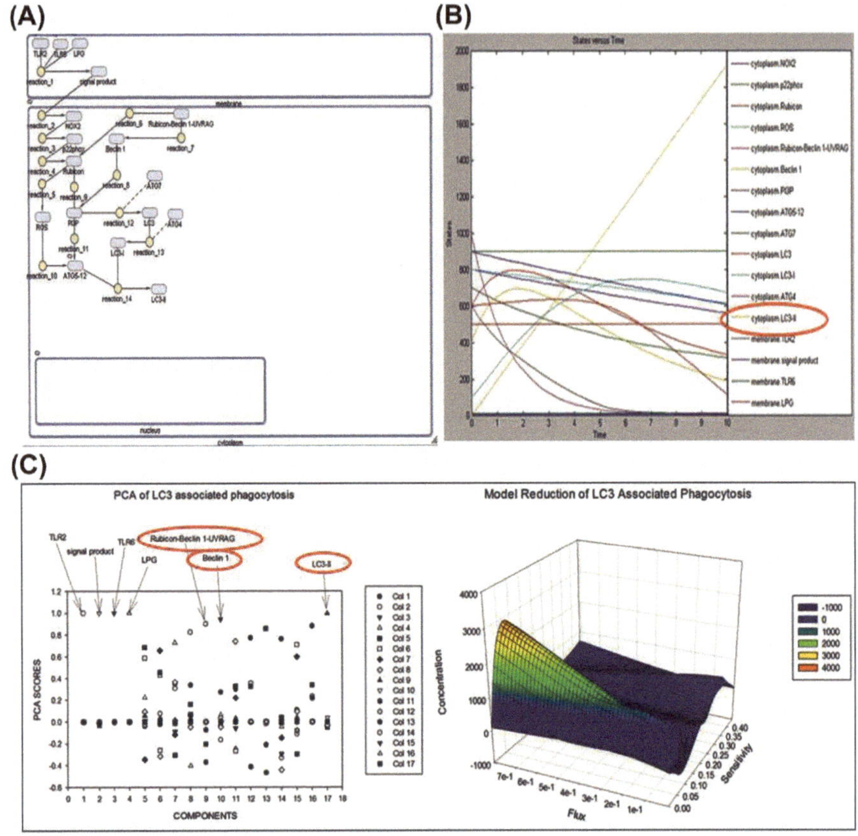

Fig. 5C (A) Model reconstruction of LC3 associated phagocytosis and its simulated graph constructed through Simbiology (B) Principal component analysis showing Beclin1, Rubicon and LC3 as an important component (C) Graphical representation of model reduction.

3.4 Flux analysis

The major reaction of the system is collection of LC3II, link of ATG9 and PI3P and disassembling of Beclin1 from the Rubicon complex in study of interplay between TNFα and IL10 governing autophagy. Whereas, in the case, which explains interaction between canonical and non-canonical autophagy in leishmaniasis, four reactions are obtained as crucial reactions that govern the whole system. They are activity of ATG9, ATG4, activity of Beclin1 and processing and accumulation of LC3 on autophagosomal membrane (Fig. 7).

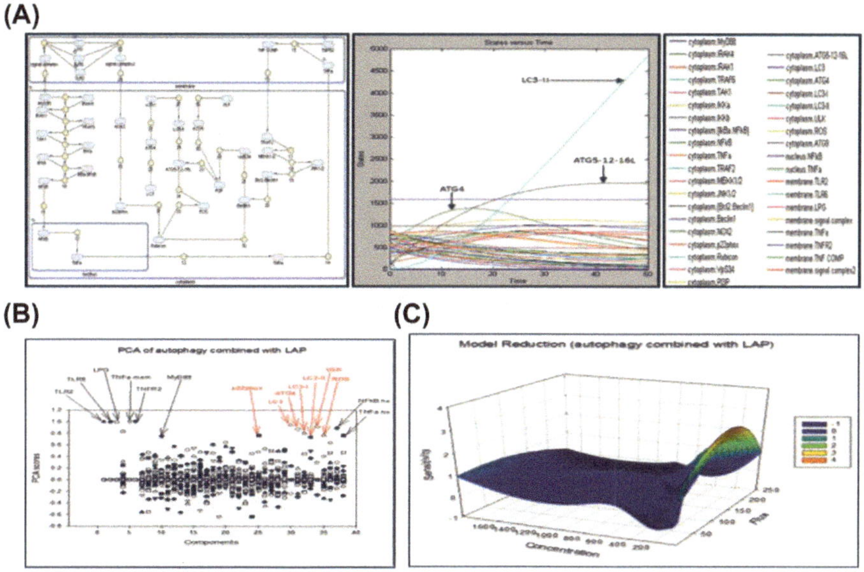

Fig. 6A Interaction between canonical and non-canonical autophagy in leishmaniasis: (A). Model reconstruction of system having combination of canonical and non-canonical autophagy and its simulated graph (B) Principal component analysis depicting ATG4 as an important component (C) Graphical representation of model reduction.

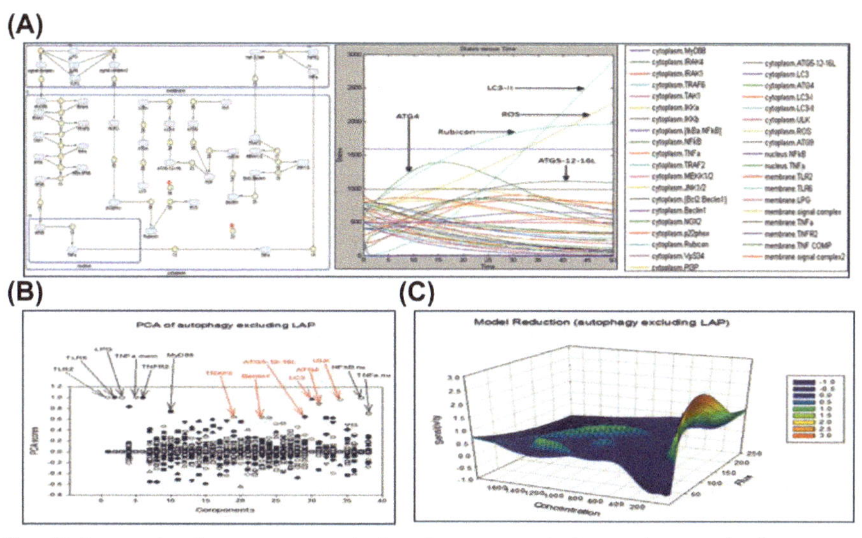

Fig. 6B Interaction between canonical and non-canonical autophagy in leishmaniasis: (A) Model reconstruction of system having canonical autophagy signaling excluding LAP and its simulated graph (B) Principal component analysis showing ATG4, Beclin1 and LC3 as an important component (C) Graphical representation of model reduction.

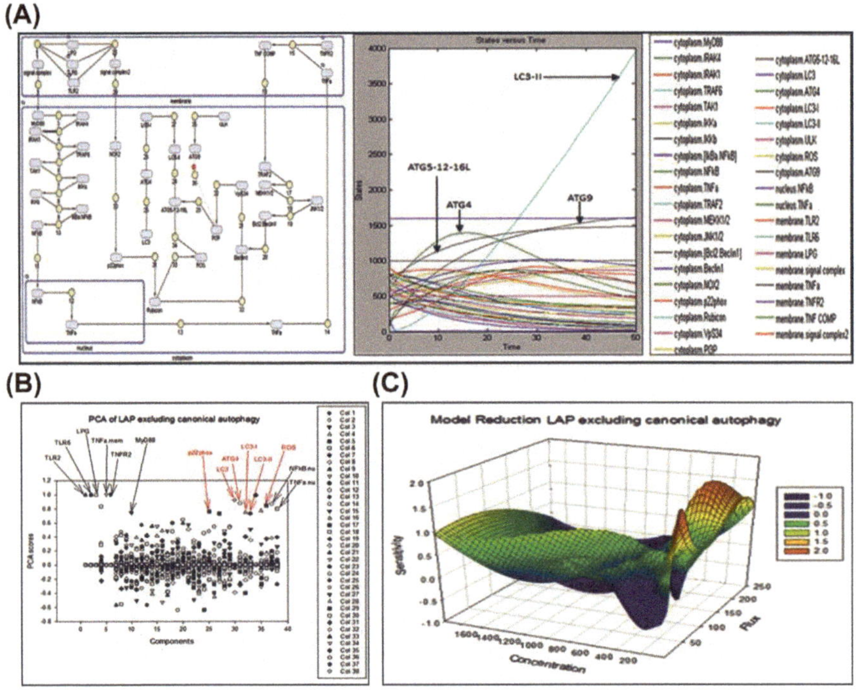

Fig. 6C Interaction between canonical and non-canonical autophagy in leishmaniasis: (A) Model reconstruction of system having only non-canonical autophagy (LAP) excluding canonical autophagy and its simulated graph (B) Principal component analysis showing ATG4, Beclin1 and LC3 as an important component (C) Graphical representation of model reduction.

4. Discussion

The in-detailed systems study of autophagic pathways involved in leishmaniasis shows that ATG9 act as a negative feedback loop with interaction to PI3P in LC3 associated phagocytosis, since there is low level of LC3-II accumulation when the non-canonical autophagic pathway is switched off. The simulated graph depicts significant change in concentration of LC3 accumulation in the system which helps PI3P accumulation in growing membrane but decreases affinity of LC3 attachment when LAP is not working in tandem. Apart from systems study, the phylogenetic study also suggests ATG9 as a relevant component in autophagic machinery, due to its conservedness observed during evolutionary processes. Study of leishmanial ATG9 may help understand the role of PI3P usage in parasite for nutrition

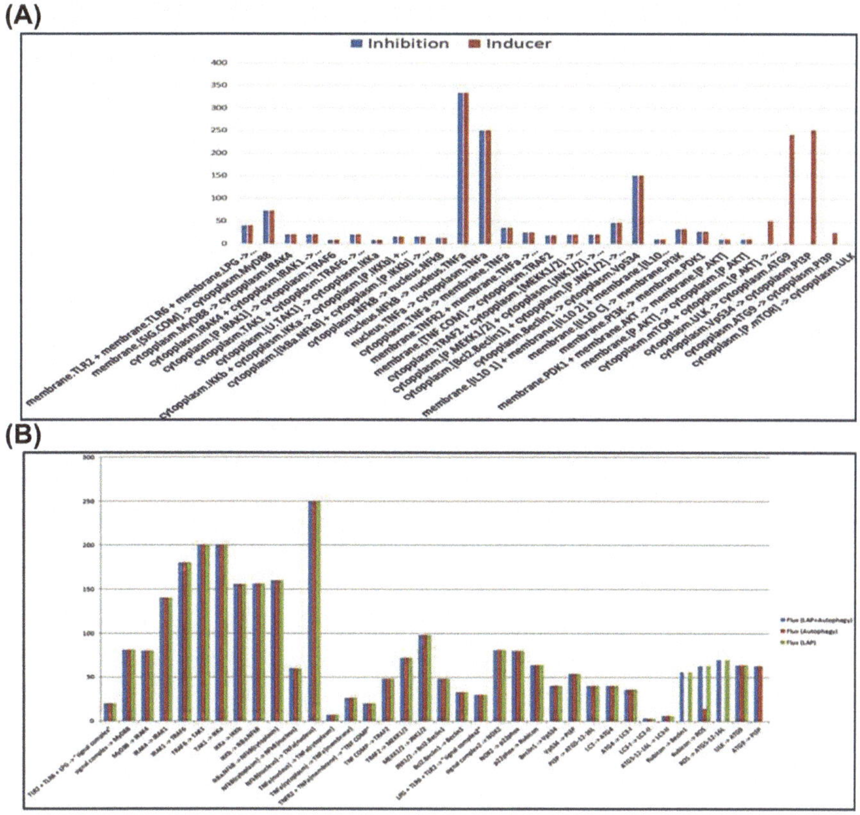

Fig. 7 Flux analysis: (A) Showing interplay between TNFα and IL10 for autophagy in macrophages in leishmaniasis (B) Showing interaction between canonical and non-canonical autophagy in leishmaniasis.

and thus may help parasite prevalence in the host. The interaction between leishmanial ATG9 and PI3P shows its similarity with the host autophagic machinery ATG9, hence it could be concluded that if the parasitic ATG9 is found to interfere with the host machinery, it could be targeted. Moreover, ATG9 is linked with the lipid accumulation for autophagosome formation and interaction with PI3P which governs membrane dynamics of the host and thus proves to be an important factor for parasite prevalence.

The other major component of autophagy reported is the ATG5-ATG12 complex (Williams et al., 2012) which is mainly involved in ATG8-PE conjugation. Post activation by ATG4, ATG8 has to be conjugated with phosphatidylethanolamine (PE) for further autophagy processes. Interestingly, it has also been observed that the conjugation of ATG8 to PE is facilitated by E3 like action of ATG5-ATG12 complex. ATG5-ATG12 complex also recruits ATG8-PE complex to the autophagosome membrane. Once ATG8-PE complex is recruited on the autophagosome membrane, ATG5-ATG12 complex dissociates, due to which ATG5-ATG12 complex is absent from fully formed autophagosomes. The amino acid residues mainly involved in ATG5-ATG12 conjugation system include, K128 of ATG5 and G185 of ATG12 (Williams et al., 2012). ATG5-ATG12 system plays a major role in the process of autophagy as it is involved in conjugation of ATG8 to PE and recruitment of ATG8-PE complex on the autophagosome membrane. Primarily, ATG8 proteins of *L.major* are selectively cleaved by ATG4 cysteine peptidases. The process involves complex interactions between amino acid residues of both ATG8 and ATG4 proteins. The amino acid residues of ATG8 actively involved in interaction with ATG4 include Leucine54 (L54), Phenylalanine 81 (F81), Phenylalanine 83 (F83), Phenylalanine (F53) and Glycine 120 (G120) (William et al., 2009). Thus, ATG4 and ATG8 conjugation is a very crucial component of the autophagy process as it marks the activation of ATG8.

The interconnected biochemical pathways shed an important knowledge governing the core cellular processes about autophagy into therapeutics. Macroautophagy plays an important role in the inflammatory conditions, and more so, when infection with *L.major* is taken into account. Small molecule/inhibitor designed against the target identified (Atg9) could help the remedial measures not getting averse both for autophagic machinery and other pathophysiological processes associated with inflammatory diseases.

Acknowledgments

The authors would like to thank the Director, National Center for Cell Science (NCCS) for supporting the Bioinformatics and High Performance Computing Facility (BHPCF) at NCCS, Pune, India. Vrushali Guhe thanks University Grants Commission (UGC), New Delhi for her Junior Research Fellowship. Bhavnita thanks her DST-INSPIRE, Senior Research fellowship. Prajakta Ingale thanks her Senior Research Fellowship from Indian Council of Medical research (ICMR), New Delhi.

Conflict of interest

The authors potentially declare no conflict of interest.

References

Codogno, P., Mehrpour, M., & Proikas-Cezanne, T. (2012). Canonical and non-canonical autophagy: Variations on a common theme of self-eating? *Nature Reviews Molecular Cell Biology, 13*, 7.

Deretic, V. (2006). Autophagy as an immune defense mechanism. *Current Opinion in Immunology, 18*, 375–382.

Feng, Y., Backues, S. K., Baba, M., Heo, J.-M., Wade Harper, J., & Klionsky, D. J. (2016). Phosphorylation of Atg9 regulates movement to the phagophore assembly site and the rate of autophagosome formation. *Autophagy, 12*, 648–658.

Franco, L. H., Fleuri, A. K. A., Pellison, N. C., Quirino, G. F. S., Horta, C. V., de Carvalho, R. V. H., et al. (2017). Autophagy downstream of endosomal Toll-like receptor signaling in macrophages is a key mechanism for resistance to *Leishmania major* infection. *Journal of Biological Chemistry, 292*, 13087–13096.

Glick, D., Barth, S., & Macleod, K. F. (2010). Autophagy: Cellular and molecular mechanisms. *The Journal of Pathology, 221*, 3–12.

Hanada, T., Noda, N. N., Satomi, Y., Ichimura, Y., Fujioka, Y., Takao, T., et al. (2007). The Atg12-Atg5 conjugate has a novel E3-like activity for protein lipidation in autophagy. *Journal of Biological Chemistry, 282*(52), 37298–37302. Epub 2007 Nov 6.

Hoops, S., Sahle, S., Gauges, R., Lee, C., Pahle, J., Simus, N., et al. (2006). COPASI–a COmplex PAthway SImulator. *Bioinformatics*, 3067–3074.

Jacomin, A. C., Gul, L., Sudhakar, P., Korcsmaros, T., & Nezis, I. P. (2018). What we learned from big data for autophagy research. *Frontiers in cell and developmental biology.* http://dx.doi.org/10.3389/fcell.2018.00092.

Liu, E., Lopez Corcino, Y., Portillo, J.-A. C., Miao, Y., & Subaustea, C. S. (2016). Identification of signaling pathways by which CD40 stimulates autophagy and antimicrobial activity against *Toxoplasma gondii* in macrophages. *Infection and Immunity, 84*, 2616–2626.

Mol, M., Patole, M. S., & Singh, S. (2014). Immune signal transduction in leishmaniasis from natural to artificial systems: Role of feedback loop insertion. *Biochimica et Biophysica Acta (BBA) – General Subjects, 1840*, 71–79.

Park, H.-J., Lee, S. J., Kim, S.-H., Han, J., Bae, J., Kim, S. J., et al. (2011). IL-10 inhibits the starvation induced autophagy in macrophages via class I phosphatidylinositol 3-kinase (PI3K) pathway. *Molecular Immunology, 48*, 720–727.

Pavlopoulos, G. A., Secrier, M., Moschopoulos, C. N., Soldatos, T. G., Kossida, S., Aerts, J., et al. (2011). Using graph theory to analyze biological networks. *BioData Mining, 4*(1), 10.

Rambaut, A. (2014). *Figtree, a graphical viewer of phylogenetic trees.*

Ronquist, F., Huelsenbeck, J. P., Nielsen, R., & Bollback, J. P. (2001). Bayesian inference of phylogeny and its impact on evolutionary biology. *Science, 294*, 2310–2314.

Smoot, M. E., Ono, K., Ruscheinski, J., Wang, P. L., & Ideker, T. (2010). Cytoscape 2.8: New features for data integration and network visualization. *Bioinformatics, 27*(3), 431–432.

Soni, B., Saha, B., & Singh, S. (2017). Systems cues governing IL6 signaling in leishmaniasis. *Cytokine, 106*, 169–175.

Tanida, I., Ueno, T., & Kominami, E. (2008). LC3 and autophagy. *Methods in Molecular Biology, 445*, 77–88.

Williams, R. A., Smith, T. K., Cull, B., Mottram, J. C., & Coombs, G. H. (2012). ATG5 is essential for ATG8-dependent autophagy and mitochondrial homeostasis in Leishmania major. *PLoS Pathogens, 8*, e1002695.

Williams, R. A., Woods, K. L., Juliano, L., Mottram, J. C., & Coombs, G. H. (2009). Characterization of unusual families of ATG8-like proteins and ATG12 in the protozoan parasite *Leishmania major. Autophagy, 5*, 159–172.

CHAPTER SIX

Expression of placental glucose transporter proteins in pregnancies complicated by fetal growth disorders

Paweł Jan Stanirowski*[,a,b], Michał Lipa[a,b], Dorota Bomba-Opoń[a] and Mirosław Wielgoś[a]

[a]1st Department of Obstetrics and Gynecology, Medical University of Warsaw, Warsaw, Poland
[b]Club 35. Polish Society of Gynecologists and Obstetricians, Warsaw, Poland
*Corresponding author: E-mail: stanirowski@gmail.com

Contents

1. Introduction 96
2. Characteristics of GLUT proteins in human placenta 104
   2.1 GLUT-1 104
   2.2 GLUT-3 107
   2.3 GLUT-4 and GLUT-12 108
   2.4 GLUT-8 110
   2.5 GLUT-9 111
3. Placental glucose transport in pregnancies complicated by fetal macrosomia 112
4. Placental glucose transport in pregnancies complicated by fetal growth restriction 118
5. Placental GLUT-mediated interventions in pregnancies complicated by fetal growth disorders 122
6. Conclusions 123
References 123

Abstract

During pregnancy fetal growth disorders, including fetal macrosomia and fetal growth restriction (FGR) are associated with numerous maternal-fetal complications, as well as due to the adverse effect of the intrauterine environment lead to an increased morbidity in adult life. Accumulating evidence suggests that occurrence of fetal macrosomia or FGR, may be associated with alterations in the transfer of nutrients across the placenta, in particular of glucose. The placental expression and activity of specific GLUT transporters are the main regulatory factors in the process of maternal-fetal glucose exchange. This review article summarizes the results of previous studies on the expression of GLUT transporters in the placenta, concentrating on human pregnancies complicated by intrauterine fetal growth disorders. Characteristics of each transporter protein

Advances in Protein Chemistry and Structural Biology, Volume 123
ISSN 1876-1623
https://doi.org/10.1016/bs.apcsb.2019.12.003

found in the placenta is presented, alterations in the location and expression of GLUT isoforms observed in individual placental compartments are described, and the factors regulating the expression of selected GLUT proteins are examined. Based on the above data, the potential function of each GLUT isoform in the maternal-fetal glucose transfer is determined. Further on, a detailed analysis of changes in the expression of glucose transporters in pregnancies complicated by fetal growth disorders is given, and significance of these modifications for the pathogenesis of fetal macrosomia and FGR is discussed. In the final part novel interventional approaches that might reduce the risk associated with abnormalities of intrauterine fetal growth through modifications of placental GLUT-mediated glucose transfer are explored.

1. Introduction

According to the epidemiological data, intrauterine fetal growth disorders, including fetal macrosomia and fetal growth restriction (FGR), are diagnosed in 7.3%–10.5% and 3%–7% of pregnant women, respectively (Boulet, Alexander, Salihu, & Pass, 2003; Li et al., 2014; Romo, Carceller, & Tobajas, 2009). Both conditions are associated with numerous maternal-fetal complications during the pregnancy period, as well as due to the adverse effect of the intrauterine environment lead to an increased morbidity in adulthood (Araujo Júnior, Peixoto, Zamarian, Elito Júnior, & Tonni, 2017; Suhag & Berghella, 2013). With regard to excessive fetal growth, researchers have identified maternal obesity and diabetes as the key risk factors (Boulet et al., 2003; Jolly, Sebire, Harris, Regan, & Robinson, 2003; Li et al., 2014). During the perinatal period, fetal macrosomia is responsible for an increased rate of Cesarean sections and birth trauma, as well as — based on the results of cohort studies — constitutes a significant risk factor for the development of metabolic syndrome in childhood and adult life (Boney, Verma, Tucker, & Vohr, 2005; Boulet et al., 2003; Jolly et al., 2003). The etiology of FGR is more complex and multifactorial. Among many other causes, placental insufficiency, fetal chromosomal abnormalities, maternal malnutrition, infections, hypertension and hypoxia lead to intrauterine fetal growth restriction (IUGR) (Suhag & Berghella, 2013). In addition, similarly to fetal overgrowth, FGR *in utero* is responsible for increased morbidity and mortality rates among fetuses and infants, as well as contributes to the development of type 2 diabetes and cardiovascular disease in adulthood (Suhag & Berghella, 2013; Osmond & Barker, 2000). The latter phenomenon, referred to as "fetal programming", seems of crucial importance with respect to the individual's health status as an adult and constitutes one of the reasons researchers are increasingly interested in the pathophysiology of abnormal fetal growth (Gluckman, Hanson, Cooper, & Thornburg, 2008).

The efficient supply of amino acids, carbohydrates and fatty acids from maternal circulation is one of the major factors determining fetal growth *in utero* (Sacks, 2004). Accumulating evidence suggests that occurrence of fetal growth disorders, such as fetal macrosomia or FGR, may be associated with alterations in the transfer of nutrients across the placenta (Jansson, Myatt, & Powell, 2009). Moreover, it has been postulated that the placenta may actually act as a nutrient sensor that adjusts fetal nutrient supply to the maternal nutrient availability (Jansson & Powell, 2006). Hence, in the conditions of compromised maternal ability to supply nutrients or oxygen to the placenta, placental function may alter, resulting in FGR. Conversely, maternal over-nutrition may lead to enhanced placental transfer and contribute to fetal macrosomia.

Among the nutrients supplied by maternal circulation, glucose is of particular importance, given its role as the principal energy substrate for the developing fetus. Simultaneously, the lack of enzymes required for gluconeogenesis in the fetal liver implicates that, transplacental transport of glucose from maternal blood is the main source of this essential substrate for fetal growth and metabolism (Kalhan & Parimi, 2000). Glucose transfer across the placenta occurs by facilitated diffusion, and the main regulatory factors in the process are the expression and activity of specific glucose transporters being members of the GLUT family. It has been confirmed that all GLUT proteins are encoded by *SLC2A* genes, and comprise of 12 transmembrane segments with a long, central linker domain and both NH_2 and COOH termini present on the cytosolic side of the membrane (Gorovits & Charron, 2003; Joost et al., 2002; Mueckler et al., 1985; Manolescu, Witkowska, Kinnaird, Cessford, & Cheeseman, 2007; Mueckler & Thorens, 2013). Regarding the role played by GLUTs in the human body, it has been demonstrated that a majority of facilitative transporters, except for GLUT-9 and GLUT-13, are responsible for sodium-independent and gradient-compliant hexose transfer. Interestingly, according to recent studies, some of GLUT proteins also participate in the transport of non-hexose substrates such as myo-inositol, urate and dehydroascorbic acid (Caulfield et al., 2008; Lee, Huang, Chang, Cheng, & Chen, 2010; Uldry et al., 2001).

Fourteen GLUT proteins have been identified so far and, based on their sequence similarity, divided into three sub-classes: class I (GLUTs 1–4 and

14); class II (GLUTs 5, 7, 9, 11) and class III (GLUTs 6, 8, 10, 12, 13) (Joost et al., 2002; Mueckler & Thorens, 2013). It should be emphasized, however, that despite being members of the same family of glycoproteins, GLUT isoforms demonstrate discrete differences in substrate specificity, transport kinetics, expression and tissue distribution. Conducted studies have revealed the presence of at least one GLUT isoform in virtually every human cell type, a predominant expression of certain transporters in the membrane (e.g. GLUT-1) or intracellular compartment (e.g. GLUT-4, GLUT-8), as well as different responses of transporters to various exo- and endogenous factors, including an increased expression of some GLUT proteins in reaction to the stimulating activity of insulin (e.g. GLUT-4, GLUT-12) (Mueckler & Thorens, 2013; Schmidt, Joost, & Schürmann, 2009; Stuart, Howell, Zhang, & Yin, 2009). It is widely accepted that these diversified characteristics of GLUT transporters result from the central role of glucose in cell metabolism, and, as a consequence, from the need to provide energy substrates in quantities adjusted to the individual cell requirements, with a minimal risk of disturbance.

The placenta constitutes a heterogeneous and multicomponent structure, in which individual tissue compartments, such as syncytio- and cytotrophoblast, villous stroma and endothelium of the vessels are characterized by varied expressions of GLUT transporters (Stanirowski, Szukiewicz, Pazura-Turowska, Sawicki, & Cendrowski, 2018). In addition, throughout the course of pregnancy, the expression of glucose transporters in the placenta changes quantitatively. Therefore, isoforms that are essential for maternal-fetal glucose exchange in the early pregnancy period (e.g. GLUT-3) play only a minor role in the perinatal period, and vice versa (Brown, Heller, Zamudio, & Illsley, 2011).

Over 30 years of studies on glucose transport has confirmed the presence of proteins GLUT-1, GLUT-3, GLUT-4, GLUT-8, GLUT-9 and GLUT-12, as well as mRNA of GLUT-10 and GLUT-11 in the placenta (Table 1) (Bibee, Illsley, & Moley, 2011; Dawson et al., 2001; Gude et al., 2003; Hauguel-De Mouzon et al., 1997; Limesand, Regnault, & Hay, 2004; Scheepers et al., 2005; Takata, Kasahara, Kasahara, Ezaki, & Hirano, 1992; Xing et al., 1998). Such a variety of isoforms implicates that maternal-fetal transport of glucose is a complex and highly regulated process. Apart from the expression and activity of glucose transporters, the maternal-fetal glucose concentration gradient, utero-placental blood flow, placental exchange surface area and glucose metabolism, as well as multiple regulatory factors, such as hypoxia, hormones (e.g. insulin, glucocorticoids), growth

Table 1 Characteristics of glucose transporter proteins in human placenta.

	GLUT class	Predominant substrates	Affinity for glucose (Km)	Placental GLUT localization	Protein expression across the gestation	Regulatory factors
GLUT -1	I	glucose, galactose, glucosamine	1–5 mM	I trim.: ST (MVM > BM) (Ericsson et al., 2005; Hahn et al., 1995; Korgun et al., 2005; Ogura et al., 2000; Tadokoro et al., 1996) CT (Ericsson et al., 2005; Hahn et al., 1995; Korgun et al., 2005; Tadokoro et al., 1996) VE (Hahn et al., 1995; Korgun et al., 2005; Tadokoro et al., 1996) EVT (Hahn et al., 1995; Korgun et al., 2005) III trim.: ST (MVM > BM) (Acosta et al., 2015; Barros et al., 1995; Bibee et al., 2011; Ericsson et al., 2005; Gao et al., 2012; Hahn et al., 1995; Hahn et al., 1999;	I-III trim.: ↑ (Ericsson et al., 2005; Sakata et al., 1995)	Hyperglycemia *in vitro*: ↓ (Hahn et al., 1998; Illsley et al., 1998) Hypoglycemia *in vitro*: ↑ (Illsley et al., 1998) IGF-I: ST (MVM ↔, BM ↑) (Baumann et al., 2014), ↑ (Jones et al., 2013) HGF: ↑ (Visiedo et al., 2015) Insulin: ST (MVM ↔, BM ↔) (James–Allan et al., 2019) Hypoxia *in vitro*: ↑ (Basak et al., 2019; Baumann et al., 2007; Hayashi et al., 2004) Hypoxia *in vivo*: ST (MVM ↔, BM ↓) (Zamudio et al., 2006, 2010) Glucocorticoids *in vitro*: ↓

(Continued)

Table 1 Characteristics of glucose transporter proteins in human placenta.—cont'd

GLUT class	Predominant substrates	Affinity for glucose (Km)	Placental GLUT localization	Protein expression across the gestation	Regulatory factors
			Hauguel–De Mouzon et al., 1997; James–Allan et al., 2019; Jansson et al., 1993; Sakata et al., 1995; Stanirowski et al., 2017; Tadokoro et al., 1996; Takata et al., 1992; Zhang et al. 2016) CT (Gao et al., 2012; Hahn et al., 1995; Sakata et al., 1995; Tadokoro et al., 1996; Takata et al., 1992) VE (Gao et al., 2012; Hahn et al., 1995; Hahn et al., 1999; Hauguel–De Mouzon et al., 1997; Tadokoro et al., 1996; Takata et al., 1992; Zhang et al., 2016), villous stroma (Zhang et al., 2016)		(Hahn et al., 1999), HPEC ↑ (Kipmen–Korgun et al., 2012) CRH *in vitro*: ↑ (Gao et al., 2012) Resistin *in vitro*: 10 ng/mL ↑, 50–100 ng/mL ↓ (Di Simone et al., 2009) HX/XO *in vitro*: ↓ (Lappas et al., 2012)

GLUT-3	I	glucose, galactose	1–2 mM	I trim.: ST MVM (Brown et al., 2011; Ericsson et al., 2005) CT (Brown et al., 2011; Ericsson et al., 2005; Hahn et al., 2001; Korgun et al., 2005; Ogura et al., 2000) VE (Hahn et al., 2001; Korgun et al., 2005) EVT (Hahn et al., 2001; Korgun et al., 2005) III trim.: ST (Gao et al., 2012; Janzen et al., 2013; Zhang et al., 2016), ST MVM (Brown et al., 2011) CT (Brown et al., 2011; Janzen et al., 2013) VE (Brown et al., 2011; Gao et al., 2012; Hahn et al., 1999; Hahn et al., 2001; Hauguel–De Mouzon et al., 1997; Zhang et al., 2016), villous stroma (Hauguel–De Mouzon et al., 1997; Zhang et al., 2016) EVT (Hahn et al., 2001; Janzen et al., 2013)	I-III trim.: ST (MVM) ↓ (Brown et al., 2011)	IGF-I: ↑ (Jones et al., 2013) Hypoxia *in vitro*: ↑ (Baumann et al., 2007) Glucocorticoids *in vitro*: ↓ (Hahn et al., 1999), HPEC ↑ (Kipmen–Korgun et al., 2012) CRH *in vitro*: ↓ (Gao et al., 2012) HX/XO *in vitro*: ↔ (Lappas et al., 2012)

(Continued)

Table 1 Characteristics of glucose transporter proteins in human placenta.—cont'd

	GLUT class	Predominant substrates	Affinity for glucose (Km)	Placental GLUT localization	Protein expression across the gestation	Regulatory factors
GLUT-4	I	glucose, glucosamine	5 mM	I trim.: ST (Ericsson et al., 2005), ST BM (James–Allan et al., 2019) III trim.: ST (Ericsson et al., 2005; Stanirowski et al., 2017; Zhang et al., 2016), ST BM (James–Allan et al., 2019), villous stroma (Xing et al., 1998)	I-III trim.: ↓ (Ericsson et al., 2005), ↑ BM (James–Allan et al., 2019)	Insulin: ST BM ↑ (James – Allan et al., 2019)
GLUT-8	III	glucose, fructose, galactose	2–2.4 mM	III trim.: ST (Janzen et al., 2018; Jones et al., 2013) CT (Janzen et al., 2018) VE (Jones et al., 2013) EVT (Janzen et al., 2018)	NA	IGF-I: ↑ (Jones et al., 2013) Hypoxia *in vitro*: ↑ (Janzen et al., 2018) Serum deprivation *in vitro*: ↓ (Janzen et al., 2018)
GLUT-9	II	urate, glucose, fructose	0.3 mM	III trim.: ST (MVM, BM) (Acosta et al., 2015; Bibee et al., 2011; Stanirowski et al., 2017) VE (Bibee et al., 2011)	NA	IGF-I: ↑ (Jones et al., 2013)

GLUT-12	III	glucose, fructose	4–5 mM	I trim.: ST (Gude et al., 2003) CT (Gude et al., 2003) EVT (Gude et al., 2003) III trim.: villous stroma (Gude et al., 2003), vascular smooth muscle cells (Gude et al., 2003)	NA	NA

BM, basal membrane; *CT*, cytotrophoblast; *CRH*, corticotropin-releasing hormone; *EVT*, extravillous trophoblast; *GLUT*, glucose transporter; trim., trimester; *HGF*, hepatocyte growth factor; *HPECs*, human placental endothelial cells; *HX/XO*, hypoxanthine/xanthine oxidase; *IGF-I*, insulin-like growth factor I; *MVM*, microvillous membrane; *NA*, not applicable; *ST*, syncytiotrophoblast; *VE*, vascular endothelium; ↑, increased; ↓, decreased; ↔, unaltered.

factors (e.g. insulin-like growth factor 1, IGF-1), cytokines (e.g. resistin) and oxidative stress are known to modulate the placental transport capacity (Table 1) (Baumann et al., 2014; Di Simone et al., 2009; Hauguel, Desmaizieres, & Challier, 1986; Hay, 2006; Illsley, Hall, & Stacey, 1987; Jones, Powell, & Jansson, 2007; Lappas, Andrikopoulos, & Permezel, 2012; Schneider, Reiber, Sager, & Malek, 2003). As a result, the analysis of GLUT expression in the placenta and its effect on intrauterine fetal growth must include numerous components that potentially affect maternal-fetal glucose exchange.

This review article summarizes the results of previous studies on the expression of GLUT transporters in the placenta, concentrating on pregnancies complicated by intrauterine fetal growth disorders. Considering the significant differences in placental structure and GLUT expression in animal models, only human pregnancies were analyzed. Characteristics of each transporter protein found in the placenta is presented, alterations in the location and expression of GLUT isoforms observed in individual placental compartments are described, and the factors regulating the expression of selected GLUT proteins are examined. Based on the above data, the potential function of each GLUT isoform in the maternal-fetal glucose transfer is determined. Further on, a detailed analysis of changes in the expression of glucose transporters in pregnancies complicated by fetal growth disorders is given, and significance of these modifications for the pathogenesis of fetal macrosomia and FGR is discussed. In the final part novel interventional approaches that might reduce the risk associated with abnormalities of intrauterine fetal growth through modifications of placental GLUT-mediated glucose transfer are explored.

2. Characteristics of GLUT proteins in human placenta

2.1 GLUT-1

The first discovered glucose transporter, GLUT-1, is a ubiquitous isoform present in the majority of human tissues and therefore considered primarily responsible for glucose transfer (Mueckler & Thorens, 2013). In the placenta, the presence of GLUT-1 has been confirmed in syncytio- and cytotrophoblast, as well as in the endothelium of the villous vessels beginning from the fifth week of gestation (Table 1) (Acosta et al., 2015; Jansson, Wennergren, & Illsley, 1993; Barros, Yudilevich, Jarvis, Beaumont, & Baldwin, 1995; Ericsson, Hamark, Powell, & Jansson, 2005; Gao et al.,

2012; Hahn et al., 1995; Hahn et al., 1999; Hauguel-De Mouzon et al., 1997; James-Allan, Arbet, Teal, Powell, & Jansson, 2019; Korgun, Celik-Ozenci, Seval, Desoye, & Demir, 2005; Ogura et al., 2000; Sakata et al., 1995; Stanirowski et al., 2017; Tadokoro et al., 1996; Takata et al., 1992; Zhang et al., 2016). Moreover, it has been demonstrated that as the pregnancy progresses, the expression of the transporter in the trophoblast increases, reaching the highest levels in the third trimester of gestation (Table 1) (Ericsson et al., 2005; Jansson et al., 1993; Sakata et al., 1995).

In the syncytiotrophoblast, being a direct barrier between the maternal and fetal circulation, expression of GLUT-1 is asymmetrical, and almost 3-fold higher in the microvillous membrane (MVM) as compared to the internally located basal membrane (BM) (Barros et al., 1995; Jansson et al., 1993). This asymmetry in the expression of GLUT-1, combined with the 7 to 9-fold greater surface of the MVM, resulting from the folding of the villi, increase the transport capacity of the maternal-facing surface by nearly twenty times as compared to adjacent to the fetal vessels BM (Teasdale & Jean-Jacques, 1985). As a consequence, a hypothesis was formulated stating that the BM may actually act as a rate-limiting step in the trans-syncytial glucose transport (Illsley, 2000). The presented hypothesis is of crucial importance, considering the fact that, given the high permeability of the villous endothelium, syncytiotrophoblast constitutes the first and main barrier for glucose transfer across the placenta (Leach & Firth, 1997). As a result, glucose transport across the syncytial BM may determine the total transplacental glucose transport. Indeed, *in vitro* studies using BeWo choriocarcinoma cells as a model of syncytiotrophoblast demonstrated that with the increasing inhibition of basal GLUT transporters, the transepithelial glucose transport became gradually down-regulated (Vardhana & Illsley, 2002). Simultaneously, no similar correlation was observed for the transporters located in the apical membrane of the trophoblast monolayer. In the view of these results, the hypothesis of a rate-limiting role of the syncytial BM seem very convincing, despite certain omissions, such as disregard of the placental glucose metabolism or glucose enzymatic transformation pathway in the syncytiotrophoblast alone. From the perspective of intrauterine fetal growth, the most important implication of the *in vitro* studies is the conclusion that every alteration in GLUT-1 expression in the syncytial BM may reduce or increase the transport of glucose to fetal circulation, and result in fetal growth disorders. For instance, a study by Jansson et al. reported a 2-fold increase in the expression of GLUT-1 and glucose uptake in the syncytial BM between 16-22 and 27—30 weeks of gestation, corresponding to an increasing demand for energy substrates and intensification of fetal growth in the second half of pregnancy (Jansson et al., 1993). In addition, a recently published study proved that GLUT-1 expression and glucose uptake in the placental BM positively correlate with fetal birth-weight among healthy non-diabetic women (Borges et al., 2019).

Analyzing the relationship between the transplacental glucose transport and fetal growth disorders, one must consider results of the *in vitro* studies evaluating the effect of different concentrations of glucose, insulin, IGF-1 and hypoxia on GLUT-1 expression in trophoblast cells (Table 1). Hyperglycemia with concurrent hyperinsulinemia is a recognized condition of diabetic women and their macrosomic offspring, whereas severe growth-restricted fetuses often present with hypoglycemia and hypoxia (Economides & Nicolaides, 1989; Marconi et al., 1996; Taricco et al., 2009; Wang et al., 2018; Westgate et al., 2006). Experimentally, it was demonstrated that both extreme hypo- and hyperglycemia (i.e. 0 or >20–25 mmol/L) result in alterations of GLUT-1 expression, causing an increase or a reduction in the transporter expression and glucose uptake, respectively (Hahn, Barth, Weiss, Mosgoeller, & Desoye, 1998; Illsley, Sellers, & Wright, 1998). Conversely, glucose concentrations of 1–12 mmol/L, commonly observed in physiologic or well-controlled diabetic pregnancies do not affect GLUT-1 expression, despite causing a gradual reduction of the glucose uptake into cultured trophoblast cells (Illsley et al., 1998). This latter discrepancy indicates the possibility of participation of other GLUT isoforms in glucose transport across the syncytium, as well as post-translational modification of GLUT-1 affecting its transport capacity.

Results of previous studies regarding the effect of insulin on GLUT-1 expression and glucose transfer in the placenta are inconclusive. In cultured ED27 and HTR8/SVneo cell lines corresponding to the first trimester trophoblast, incubation with insulin led to increased GLUT-1 mRNA expression and glucose uptake (Basak, Vilasagaram, Naidu, & Duttaroy, 2019; Gordon, Zimmerman, Landon, Gabbe, & Kniss, 1995). Conversely, in a recent study by James-Allan et al., insulin had no effect on the expression of the GLUT-1 protein in the MVM and BM derived from early and late pregnancy placental explants (James-Allan et al., 2019).

Numerous studies point to the role of factors such as IGF-1 in the regulation of fetoplacental growth, and one of the suggested mechanisms of action may involve stimulation of transepithelial glucose transfer due to increased expression of GLUT-1 transporter in syncytial BM (Baumann et al., 2014; Borges et al., 2019; Hiden, Glitzner, Hartmann, & Desoye,

2009; Luo et al., 2012). According to recent study, hepatocyte growth factor (HGF) may also be responsible for an increase in GLUT-1 expression and glucose uptake in placental explants in the population of obese women (Visiedo et al., 2015).

With regard to hypoxia, it was established that in first trimester placental explants, cultured BeWo choriocarcinoma or HTR8/SVneo cells reduced oxygen concentration increases the expression of GLUT-1 and GLUT-3 transporters via the hypoxia-inducible factor-1α (HIF-1α) mediated pathway, and stimulates glucose uptake (Basak et al., 2019; Baumann, Zamudio, & Illsley, 2007; Hayashi et al., 2004). On the contrary, among high-altitude pregnant women exposed to low oxygen concentrations, a significant decrease in GLUT-1 expression in the syncytial BM was noted (Zamudio et al., 2006, 2010). Due to observed differences between the *in vitro* and *in vivo* hypoxia models, it is highly probable that in the latter group, decreased GLUT-1 expression is attributable to numerous modulating factors (e.g. decreased levels of fetal insulin), and not merely to the reduced oxygen concentration (Zamudio et al., 2010). Importantly, in compliance with the rate-limiting role of the syncytial BM, the observed decreased expression of GLUT-1 may explain the reduction in the transplacental glucose transfer and fetal birth-weight among women exposed to chronic hypoxia during pregnancy (Zamudio et al., 2006).

2.2 GLUT-3

Another transporter found in the placenta, GLUT-3, demonstrates a distinctive high affinity to glucose ($K_m = \sim 1-2$ mM), which enables the effective transfer of energy substrates into the cells, even in conditions of significant glucose deprivation (Manolescu, Witkowska et al., 2007; Mueckler & Thorens, 2013). Studies consecutively confirmed the presence of GLUT-3 protein in the vascular endothelium, extravillous trophoblast, cytotrophoblast and syncytiotrophoblast of placental villi (Table 1) (Brown et al., 2011; Ericsson et al., 2005; Gao et al., 2012; Hahn et al., 1999; Hahn et al., 2001; Hauguel-De Mouzon et al., 1997; Janzen et al., 2013; Korgun et al., 2005; Ogura et al., 2000; Zhang et al., 2016). Moreover, it was revealed that contrary to GLUT-1, the expression of GLUT-3 in syncytium is limited to the MVM, and it demonstrates a decreasing trend in the course of pregnancy i.e. in the second and third trimester constitutes 48% and 34% of the expression observed in the first weeks of pregnancy, respectively (Brown et al., 2011; Ericsson et al., 2005; Hahn et al., 2001). Furthermore, in a study by Brown et al., the authors using BeWo choriocarcinoma cells as an example of the first trimester trophoblast proved that GLUT-3 is responsible for nearly 50% of the total cellular glucose uptake (Brown et al., 2011).

Based on these findings, it was suggested that the GLUT-3 protein may be of importance for cells with high energy demand and intensive metabolism, such as proliferating trophoblast cells in early pregnancy. During this period of fetal development, and possibly as early as embryonic implantation, the transporter could secure energy supply in the setting of insufficiently developed uterine-placental circulation, i.e. in conditions of reduced glucose and oxygen concentrations (Brown et al., 2011; Ericsson et al., 2005; Korgun et al., 2005). This assumption has been partially confirmed by *in vitro* studies that demonstrated a stimulating effect of hypoxia on GLUT-3 expression in first trimester trophoblast cells (Baumann et al., 2007). Correspondingly, the progressing maturation of intervillous circulation in the course of pregnancy, resulting in an increased supply of glucose and oxygen, reduces the importance of GLUT-3, and in the perinatal period GLUT-1 becomes primarily responsible for maternal-fetal glucose transfer.

2.3 GLUT-4 and GLUT-12

In 1998, Xing et al. were first to discover the presence of the GLUT-4 transporter in the proximity of insulin receptors in placental villous stromal cells (Xing et al., 1998). Presented finding was of particular importance, considering the fact that transporter expression and activity are sensitive to insulin stimulation, which makes GLUT-4 the main insulin-responsive isoform in human tissues (Mueckler & Thorens, 2013). Equally important were the results of further research, which confirmed the expression of GLUT-4 in placental syncytiotrophoblast in the first and third trimester of pregnancy (Table 1) (Ericsson et al., 2005; James-Allan et al., 2019; Stanirowski et al., 2017; Zhang et al., 2016). It was also revealed that, contrary to GLUT-1 and GLUT-3 transporters, primarily expressed in the trophoblast cell membranes, GLUT-4 is mainly located in the intracellular compartment (Ericsson et al., 2005; Stanirowski et al., 2017). Considering that only transporters present in the cell membrane are capable of active glucose uptake, the existence of stimuli responsible for the effective subcellular redistribution of GLUT-4 appears evident. In the skeletal muscles and adipose tissue, insulin stimulates translocation of GLUT-4 toward the cell membrane, which increases glucose uptake and reduces its blood concentration, thus allowing the maintenance of systemic glucose homeostasis (Bryant, Govers, & James, 2002; Smith, Charron, Shah, Lodish, & Jarett, 1991; Somwar et al., 2001).

According to a recently published study by James-Allan et al., a similar effect of insulin stimulation can be observed in the placental trophoblast (James-Allan et al., 2019). The authors demonstrated that incubation of the first and third trimester placental villous explants with insulin increased GLUT-4 expression in the BM of the syncytiotrophoblast. Furthermore, similarly to GLUT-1, the expression of GLUT-4 in the syncytial BM was significantly higher at term as compared to early pregnancy, and it was accompanied by the increased expression of the insulin receptors in the apical and basal membranes of the trophoblast. The above findings are somewhat revolutionary, as they contradict the previously formulated hypotheses on insulin-dependent and GLUT-4-mediated glucose transfer in the human placenta. Until recently, it was suggested that the expression of insulin receptors in the syncytiotrophoblast gradually decreases with the progression of pregnancy, and in the third trimester, their density is highest in the vascular endothelium, pericytes and smooth muscle cells of the villi (Desoye et al., 1997). Moreover, a study by Ericsson et al. revealed that, similarly to insulin receptors, expression of GLUT-4 in placental homogenates decreases in the course of pregnancy (Ericsson et al., 2005). Considering the role of the syncytiotrophoblast in the maternal-fetal glucose exchange, this spatio-temporal change in insulin receptors and GLUT-4 expression between the first trimester and term implies that maternal insulin may effectively affect the transplacental glucose transport via GLUT-4, primarily in early pregnancy. In accordance with the presented hypothesis, most studies regarding transplacental glucose transport have demonstrated a stimulating effect of insulin on glucose transport in the first trimester of pregnancy, and the absence or only temporary effect in the placenta at term (Challier, Hauguel, & Desmaizieres, 1986; Ericsson et al., 2005; Urbach, Mor, Ronen, & Brandes, 1989; Osmond, King, Brennecke, & Gude, 2001). In addition, it was reported that the expression of the GLUT-12 isoform, presenting characteristics similar to GLUT-4, also decreases in the trophoblast during pregnancy (Gude et al., 2003). In the only study published so far, the authors demonstrated that the dominant site of GLUT-12 expression in the first trimester of pregnancy is the cytosol of the syncytio- and cytotrophoblast, whereas at term the transporter primarily localizes in the stromal and vascular smooth muscle cells of the villi (Table 1). As the changes in transporter distribution correspond to alterations in the location of insulin receptors, GLUT-12 alongside GLUT-4 might participate in insulin-dependent glucose uptake in early pregnancy (Gude et al., 2003). Regarding the above discrepancies between study results, it appears that the potential cause of the recorded differences in the expression of GLUT-4 and insulin receptors is using antibodies of varied specificity, as well as conducting experiments in different fractions of the placenta, i.e. tissue homogenates or isolated plasma membranes (Desoye et al., 1997; Ericsson et al., 2005; James-Allan et al., 2019).

To sum up, it is plausible that transplacental glucose transport increases under stimulation by maternal insulin, and GLUT-4/GLUT-12 transporters are among the factors directly engaged in the process. As a result of insulin activity and the increased supply of energy substrates, the trajectory of fetal growth may alter toward macrosomia as early as in the first weeks of gestation, which indicates the key importance of this pregnancy period in programming of fetal development. Contrarily, in term pregnancy, the localization of insulin receptors and GLUT-4/GLUT-12 proteins in the vicinity of placental vessels implies that the insulin from the fetal circulation might be primarily responsible for glucose uptake by the endothelial, stromal and smooth muscle cells of the villi, including storage of the excess glucose as glycogen (Acevedo, Marquez, Rojas, & Bravo, 2005; Desoye et al., 1997; Gude et al., 2003; Hiden et al., 2009; Xing et al., 1998). New light has been shed on the role of insulin-dependent transfer of glucose in term placenta by the results of recent studies that reveal an increased expression of insulin receptors and GLUT-4 protein in the apical and basal membranes of the trophoblast along with the progression of pregnancy. The above finding, combined with the stimulatory effect of the hormone on GLUT-4 expression in the syncytial BM — a rate-limiting step in the transplacental glucose transfer — suggests that maternal insulin can effectively increase glucose flux across the placenta, in both early and late pregnancy (Brunette, Lajeunesse, Leclerc, & Lafond, 1990).

2.4 GLUT-8

GLUT-8 belongs to class III GLUT transporters, and is characterized by high-affinity transport activity for glucose, fructose and galactose (Mueckler & Thorens, 2013; Schmidt et al., 2009). The transporter's specificity involves its exclusive presence in the intracellular compartment, particularly in the endoplasmic reticulum, lysosomes and late endosomes (Augustin, Riley, & Moley, 2005; Janzen et al., 2018; Schmidt et al., 2009). As a consequence of this localization, some authors suggest that, similarly to GLUT-4, a stimulus is required for the effective recruitment of GLUT-8 to the plasma membrane (Limesand et al., 2004). For example, in mouse blastocysts, insulin stimulates translocation of GLUT-8 toward the cell membrane, increasing the glucose uptake (Carayannopoulos et al., 2000). Nevertheless, as no similar effect of insulin was observed in other cells, most researchers postulate the exclusive participation of GLUT-8 in the exchange of hexoses and their derivatives between intracellular organelles (Augustin et al. 2005; Schmidt et al., 2009).

Although the presence of the transporter protein in the human placenta was initially confirmed in homogenates, and then in the syncytiotrophoblast, cytotrophoblast, extravillous trophoblast and the endothelium of villous vessels in a term pregnancy, the significance of GLUT-8 mediated hexose transfer for intrauterine fetal development remains unclear (Table 1) (Janzen et al., 2018; Jones, Crombleholme, & Habli, 2013; Limesand et al., 2004). Experimentally, it has been shown that serum deprivation reduced transporter expression, whereas hypoxia up-regulated GLUT-8 expression in HTR8/SVneo and primary trophoblast cells (Janzen et al., 2018). As a result, the first evidence is available for the correlation between altered GLUT-8 expression and placental insufficiency, and therefore possibly IUGR.

2.5 GLUT-9

The last transporter whose presence in the placenta has been confirmed, GLUT-9, is the only member of class II GLUT proteins, and occurs in two splice variants: GLUT-9a and GLUT-9b, which are characterized by different membrane targeting and localization in human tissues (Mueckler & Thorens, 2013). In the term placenta, positive GLUT-9 expression has been observed in the membranes and cytosol of the syncytiotrophoblast and in the endothelium of villous vessels (Table 1) (Acosta et al., 2015; Bibee et al., 2011; Stanirowski et al., 2017). Moreover, regarding both splice variants, Bibee et al. demonstrated that the dominant form in the syncytial BM is GLUT-9a, whereas in the MVM, there is an enrichment of GLUT-9b (Bibee et al., 2011). Initially, studies revealed that GLUT-9, similarly to all class II GLUT isoforms (apart from GLUT-5) participates in the facilitative transfer of glucose and fructose; however, according to recent publications, uric acid is the primary substrate for the transporter, and the mechanism of transfer is electrogenic (Caulfield et al., 2008; Manolescu, Augustin, Moley, & Cheeseman, 2007; Witkowska et al., 2012). Interestingly, studies performed in Xenopus oocytes suggested a significant role of GLUT-9 in urate efflux from the cell, which might indicate a protective function of the transporter in conditions of elevated intracellular urate concentrations (Witkowska et al., 2012).

Considering the not fully understood effect of uric acid on the development of the fetal-placental unit and the dominant role of the GLUT-1 and GLUT-3 isoforms in transplacental glucose transfer, participation of GLUT-9 in the transport of fructose represents an interesting issue. Despite the scientifically proven role of fructose in the pathogenesis of increased adiposity and metabolic disorders, previous studies, conducted using animal models, did not provide an unambiguous explanation of the effect of fructose on intrauterine fetal development (Jürgens et al., 2005; Tappy, 2018). On the one hand, Vickers et al. demonstrated a significant increase of birth-weight in the offspring of rats receiving 20% of their daily caloric intake from fructose, on the other hand Asghar et al. observed presence of growth-restricted fetuses in mice fed with a high-fructose diet (Asghar et al., 2016; Vickers, Clayton, Yap, & Sloboda, 2011). In addition, in the latter study, the authors pointed to fructose as a substrate for the synthesis of placental uric acid, whose excess may result in oxidative stress and placental insufficiency (Asghar et al., 2016).

3. Placental glucose transport in pregnancies complicated by fetal macrosomia

Isolated fetal macrosomia, not associated with maternal obesity/diabetes, constitutes a rare entity, therefore only two studies have assessed the placental expression of GLUT proteins regardless of the presence of both maternal co-morbidities (Table 2) (Kainulainen, Jarvinen, & Heinonen, 1997; Reza-López et al., 2014). A study by Kainulainen et al. did not find any differences in the expression of the GLUT-3 and GLUT-4 transporters between the placentas from physiological pregnancies and pregnancies complicated by fetal macrosomia (Kainulainen et al., 1997). As a result, the authors concluded that excessive fetal growth is probably due to a higher weight of the placenta in the macrosomic group, and the resulting higher total number of glucose transporters rather than their density. Similarly, the second study reported the absence of changes in GLUT-1 and GLUT-3 expression in the placentas of macrosomic fetuses (Reza-López et al., 2014). However, it was noted that a greater abundance of GLUT-3 was associated with a 5-fold higher risk of a fetus being overgrowth.

As mentioned in the introduction, both obesity and diabetes are known risk factors of fetal macrosomia (Boulet et al., 2003; Jolly et al., 2003; Li et al., 2014). Previous studies in obese patients who delivered macrosomic fetuses demonstrated a significant reduction or increase of GLUT-1

Table 2 Placental expression of glucose transporter proteins in human pregnancies complicated by fetal growth disorders.

	Fetal macrosomia	Fetal growth restriction
GLUT -1	Idiopathic: ↔ (Reza–López et al. 2014) Obesity: ST (MVM ↓, BM ↑) (James–Allan et al., 2019) Diabetes: GDM: ST (MVM ↔, BM ↔) (Jansson et al., 2001); ↑ (Yao et al., 2017) GDMG1: ST (MVM ↔, BM ↑) (Gaither et al., 1999); ↔ (Colomiere et al., 2009; Stanirowski et al., 2017; Zhang et al., 2016) GDMG2: ST (MVM ↔, BM ↑) (Borges et al., 2019; Gaither et al., 1999); ↑ (Colomiere et al., 2009); ↔ (Stanirowski et al., 2017) PGDM: ST (MVM ↔, BM ↑) (Borges et al., 2019; Gaither et al., 1999; Jansson et al., 1999); ↑ (Stanirowski et al., 2017)	Idiopathic: ST (MVM ↔, BM ↔) (Jansson et al., 1993, 2002); ↔ (Janzen et al., 2013) PM with IVS: ST (MVM ↔, BM ↓) (Chandrasiri et al., 2014) High-altitude exposure: ST (MVM ↔, BM ↓) (Zamudio et al., 2006, 2010) PE: ST (MVM ↓, BM ↔) (Lüscher et al., 2017)
GLUT-3	Idiopathic: ↔ (Kainulainen et al., 1997; Reza–López et al. 2014) Diabetes: GDMG1: ↓ (Zhang et al., 2016) PGDM: ↔ (Kainulainen et al., 1997)	Idiopathic: ↔ (Kainulainen et al., 1997); ↑ on the maternal side of the placenta (Janzen et al., 2013)
GLUT-4	Idiopathic: ↔ (Kainulainen et al., 1997) Obesity: ST BM ↓ (James–Allan et al., 2019) Diabetes: GDMG1: ↔ (Colomiere et al., 2009; Stanirowski et al., 2017); ↓ (Zhang et al., 2016) GDMG2: ↓ (Colomiere et al., 2009); ↑ (Stanirowski et al., 2017) PGDM: ↔ (Kainulainen et al., 1997; Xing et al., 1998); ↑ (Stanirowski et al., 2017)	Idiopathic: ↔ (Janzen et al., 2013; Kainulainen et al., 1997)

(Continued)

Table 2 Placental expression of glucose transporter proteins in human pregnancies complicated by fetal growth disorders.—cont'd

	Fetal macrosomia	Fetal growth restriction
GLUT-8	NA	Idiopathic: ↑ on the maternal side of the placenta (Janzen et al., 2018)
GLUT-9	Diabetes: GDMG1: GLUT-9a ST (MVM ↔, BM ↑) (Bibee et al., 2011); GLUT-9b ST (MVM ↔, BM ↔) (Bibee et al., 2011); GLUT-9 ↔ (Stanirowski et al., 2017) GDMG2: GLUT-9a ST (MVM ↔, BM ↑) (Bibee et al., 2011); GLUT-9b ST (MVM ↑, BM ↑) (Bibee et al., 2011); GLUT-9 ↑ (Stanirowski et al., 2017) PGDM: GLUT-9a ST (MVM ↔, BM ↑) (Bibee et al., 2011); GLUT-9b ST (MVM ↔, ↑; BM ↔, ↑) (Bibee et al., 2011); GLUT-9 ↑ (Stanirowski et al., 2017)	NA
GLUT-12	NA	NA

BM, basal membrane; *GLUT*, glucose transporter; *GDM*, gestational diabetes mellitus; *GDMG1*, diet-controlled gestational diabetes mellitus; *GDMG2*, insulin-controlled gestational diabetes mellitus; *IVS*, intervillositis; *MVM*, microvillous membrane; *NA*, not applicable; *PE*, pre-eclampsia; *PGDM*, pre-gestational diabetes mellitus; *PM*, placental malaria; *ST*, syncytiotrophoblast; ↑, increased; ↓, decreased; ↔, unaltered.

expression in the MVM and BM of the syncytiotrophoblast, respectively (Table 2) (James-Allan et al., 2019). Moreover, in the group of women whose body mass index exceeded 25 kg/m^2, the expression of GLUT-1 in the BM and the fetal birth-weight remained positively correlated (Acosta et al., 2015). The absence of a similar relationship with regard to the expression of GLUT-1 in the MVM, and GLUT-9 in the MVM and BM of the trophoblast partially confirms the role of GLUT-1 as the predominant glucose transporter in the human placenta, as well as the hypothesis that the BM acts as a rate-limiting step in the transplacental glucose transfer (Acosta et al., 2015). Contrary to GLUT-1, the expression of GLUT-4 protein in the syncytial BM was considerably lower in obese mothers of macrosomic fetuses as compared to non-obese women (James-Allan et al., 2019).

Alterations in the placental expression of GLUT proteins in diabetes-complicated pregnancies have already been extensively reviewed elsewhere (Stanirowski et al., 2018). In brief, it has been confirmed that in pre-gestational diabetes mellitus (PGDM), the expression of GLUT-1 and GLUT-9 transporters in the trophoblast of term placenta is significantly increased (Table 2) (Bibee et al., 2011; Borges et al., 2019; Gaither, Quraishi, & Illsley, 1999; Jansson, Wennergren, & Powell, 1999; Stanirowski et al., 2017). In addition, two studies reported a positive correlation between placental expression of both glucose transporters and birth-weight of the fetus (Jansson et al., 1999; Stanirowski et al., 2019). As a result, in pregnancies complicated by PGDM, the increased expression of GLUT-1 and GLUT-9 transporters may lead to an excessive transfer of hexoses and the intensification of fetal growth.

With regard to pregnancies complicated by gestational diabetes mellitus (GDM), more discrepancies in study results were observed, mainly due to the use of different methodologies, diagnostic criteria of diabetes, methods of treatment, and optimal levels of glycemic control. Among the patients with GDM treated exclusively with diet (GDMG1), representing the mildest form of glucose intolerance in pregnancy, most studies did not report any changes in the placental expression of the GLUT-1, GLUT-4 or GLUT-9 transporters (Table 2) (Bibee et al., 2011; Colomiere, Permezel, Riley, Desoye, & Lappas, 2009; Jansson, Ekstrand, Wennergren, & Powell, 2001; Stanirowski et al., 2017; Zhang et al., 2016). Only in the studies by Gaither et al. and Yao et al. the authors observed increased expression of GLUT-1, while Zhang et al. found a significant decrease in GLUT-3 and GLUT-4 proteins expression in the placenta (Gaither et al., 1999; Yao et al., 2017; Zhang et al., 2016). In patients with GDM and more severe insulin resistance, requiring a combined treatment with diet and insulin (GDMG2), the expression of GLUT-1 remained unchanged or increased; the expression of GLUT-4 increased or decreased, and the expression of GLUT-9 was only increased (Table 2) (Bibee et al., 2011; Borges et al., 2019; Colomiere et al., 2009; Gaither et al., 1999; Jansson et al., 2001; Stanirowski et al., 2017). Furthermore, similarly to PGDM, in patients with GDM, the expression of the GLUT-1 and GLUT-4 transporters in the placenta correlated positively with the fetal birth-weight (Stanirowski et al., 2019; Yao et al., 2017).

Following one of the hypotheses, diabetic macrosomia takes its origin from the increased glucose concentrations in maternal circulation, occurring before the diagnosis of diabetes, or being result of suboptimal glycemic control during pregnancy, and causing fetal hyperglycemia (Illsley, 2000). The latter may subsequently contribute to increased secretion of fetal insulin and IGF-1, responsible for the changes in the expression of glucose transporters, such as GLUT-1 and/or GLUT-4, in the BM of the trophoblast, and accelerated fetal growth (Borges et al., 2019; James-Allan et al., 2019). In accordance with the presented sequence of events, disturbed glucose homeostasis, more often observed in the patients requiring insulin treatment (GDMG2+PGDM), causes more pronounced changes in the placental GLUT expression, resulting in increased rates of macrosomic fetuses (Stanirowski et al., 2019). On the other hand, increased glucose concentrations in the umbilical vein and arteries, as well as the presence of fetal macrosomia, are frequently observed even in well-controlled GDMG1, GDMG2 and PGDM pregnancies (Gaither et al., 1999; Jansson et al., 2001; Stanirowski et al., 2017; Taricco et al., 2009; Yao et al., 2017). Alongside the alterations in the expression and activity of other nutrient transporters (e.g. amino acids, fatty acids), the increased placental exchange surface area, decreased placental glucose utilization rate, hormonal stimulation (e.g. by insulin) as well as the inability to achieve truly optimal glucose control in diabetic patients could all contribute to such unexpected macrosomia occurrence (Jansson, Ekstrand, Björn, Wennergren, & Powell, 2002; Magnusson, Waterman, Wennergren, Jansson, & Powell, 2004; Stanirowski et al., 2019). According to the hypothesis proposed by Desoye et al. this "unexplained" fetal overgrowth in diabetic pregnancies could also be the result of reduced placental capacity to store the excess fetal glucose as glycogen, which in turn leads to long-term fetal hyperglycemia (Desoye, Korgun, Ghaffari-Tabrizi, & Hahn, 2002).

Finally, the role of GLUT-4 and insulin-dependent glucose transfer in the pathogenesis of fetal macrosomia in pregnancies complicated by diabetes still remains controversial. Studies evaluating GLUT-4 expression in PGDM patients found no differences, or an increased expression of the transporter in term placentas (Table 2) (Kainulainen et al., 1997; Stanirowski et al., 2017; Xing et al., 1998). On the other hand in pregnancies with concurrent GDM, the introduction of insulin therapy resulted in an increased or reduced expression of GLUT-4 protein (Colomiere et al., 2009; Stanirowski et al., 2017). Considering the fact that the isoform is expressed predominantly in cytosol, using of different methods for the assessment of GLUT-4 expression in individual trophoblast compartments seems to be the probable cause of the discrepancies between the results, and difficulties with their interpretation. According to the outcomes of most studies, the expression of both insulin receptors and GLUT-4 is highest in the first trimester trophoblast, which may explain the increase in insulin-dependent glucose uptake in this period of pregnancy (Basak et al., 2019; Desoye et al., 1997; Ericsson et al., 2005; Gordon et al., 1995). Consequently, in pregnancies complicated by PGDM, early therapy with exogenous insulin combined with transient hyperglycemic episodes may result, as early as in the first weeks of pregnancy, in elevated GLUT-4 expression in the syncytial BM and increased glucose transfer to fetal circulation, leading to increased incidence of fetal macrosomia (James-Allan et al., 2019; Stanirowski et al., 2019). In GDMG2 patients, the introduction of insulin therapy at the end of the second, or the beginning of the third trimester of pregnancy relatively reduces the period of intensified placental uptake and transfer of glucose determined by hormonal stimulation, which results in a lower percentage of macrosomic fetuses (Stanirowski et al., 2019).

Summing up, alterations in the expression of the placental glucose transporters may constitute one of the main metabolic derangements leading to fetal overgrowth. Indeed, in pregnancies complicated by maternal obesity and/or diabetes, the expression of certain GLUT proteins is increased and positively correlated with birth-weight of the fetus (Acosta et al., 2015; Jansson et al., 1999; Stanirowski et al., 2019). In the most extensively studied population of diabetic women, the key factors determining placental GLUT expression apparently include optimal glycemic control, which reduces the frequency of hyperglycemic episodes, and the necessity to introduce insulin therapy, especially in the first trimester of pregnancy. At the same time, detailed research of the processes responsible for the occurrence of fetal overgrowth in the cases unrelated to maternal diabetes or obesity, and in the GDM/PGDM patients with optimal glycemic control, is required. Moreover, the absence or limited number of studies exploring the expression of transporters other than GLUT-1 prevents a full picture of the changes in the placental glucose transport in pregnancies complicated by fetal macrosomia from being obtained.

4. Placental glucose transport in pregnancies complicated by fetal growth restriction

FGR is an example of growth disorder with complex etiology, resulting from a number of causative factors. Frequent causes of FGR include chromosomal abnormalities of the fetus, placental insufficiency due to arterial hypertension and vasculopathy in the mother, malnutrition, infections, hypoxia, as well as effects of toxins and drugs (Suhag & Berghella, 2013). This diversity in the group of affected fetuses makes the proper selection of study populations and interpretation of results difficult. Moreover, as a consequence of the use of outdated FGR definitions, not allowing for the appropriate selection of small for gestational age (SGA) fetuses, some studies did not discriminate between the pathologically growth restricted and constitutionally low birth-weight fetuses. According to the most recent Delphi consensus, certain biometric and/or ultrasonographic criteria must be met to diagnose restricted growth in a fetus or neonate during the antenatal or postnatal period, respectively (Beune et al., 2018; Gordijn et al., 2016).

The analysis of glucose concentrations in umbilical vessels demonstrated that some growth restricted fetuses, in particular those severely compromised, are hypoglycemic as compared to appropriate for gestational age (AGA) fetuses (Economides & Nicolaides, 1989; Marconi et al., 1996). The above finding inspired studies on the role of transplacental glucose transport in the pathogenesis of FGR. In one of the first publications by Jansson et al. the authors did not demonstrate differences in GLUT-1 expression between the MVM and BM of placentas from pregnancies complicated by IUGR (defined as fetal birth-weight 2 SD below population mean) and those from physiological pregnancies (Table 2) (Jansson et al., 1993; Jansson, Ylvén, Wennergren, & Powell, 2002). Similarly, a study by Kainulainen et al. did not find any differences in the placental expression of GLUT-3 and GLUT-4 transporters (Kainulainen et al., 1997). As a result, it was posited that in the placentas from pregnancies complicated by idiopathic FGR, the expression of GLUT proteins did not alter, and the observed fetal hypoglycemia resulted from reduced utero-placental blood flow and maternal supply combined with high placental and fetal glucose utilization rates (Challis, Pfarrer, Ritchie, Koren, & Adamson, 2000; Economides & Nicolaides, 1989; Limesand et al., 2004; Hay, 2006). According to some authors' suggestions, the latter process, in conjunction with unchanged maternal glucose concentrations observed in the majority of FGR-complicated pregnancies, may be an adaptive mechanism that would enhance glucose flux across the placenta due to an increased maternal-fetal glucose concentration gradient (Hay, 2006; Marconi 1996). Nonetheless, one must take into consideration that the populations included in the above-mentioned studies were not homogeneous, with some patients presenting with concomitant hypertension or diabetes, and that the majority of fetuses were not severely compromised, as assessed by Doppler ultrasound (Jansson et al., 1993, 2002). These limitations were partially eliminated in further studies by Janzen et al. (Janzen et al., 2013, 2018). In the group of idiopathic IUGR fetuses, the authors demonstrated a significantly increased expression of GLUT-3 and GLUT-8 transporters on the maternal side of the placenta, and unaltered GLUT-1 and GLUT-4 protein expression (Table 2). Moreover, similarly to previous *in vitro* studies, the increased expression of both proteins was associated with the increased expression of HIF-1α, which indicates the presence of compensatory mechanisms in case of hypoxia-associated placental insufficiency (Baumann et al., 2007; Janzen et al., 2013, 2018). With regard to GLUT-3, the increased expression of the transporter in the apical membrane of the trophoblast, may be responsible for a higher glucose transfer to the placenta and fetal circulation. On the other hand, the previously mentioned frequent occurrence of hypoglycemia among growth restricted fetuses suggests that the principal goal of up-regulated transporter expression may be the provision of energy substrates exclusively to cover the demands of the intensified placental metabolism (Challis et al., 2000). With regard to GLUT-8, due to its cytoplasmic localization, it is suggested that the transporter may participate in glucose supply to intracellular organelles in the setting of increased oxidative stress (Janzen et al., 2018). It should be noted, however, that despite a more careful selection of patients, in both studies by Janzen et al. the study groups were not entirely homogeneous, as some of the neonates classified as asymmetric IUGR did not meet the criteria of growth restriction according to the recommended biometric and ultrasound data (Beune et al., 2018).

On the contrary, recently conducted studies involving more meticulously selected groups of patients revealed significant differences in GLUT expression depending on the etiology of the FGR. For instance, in placental malaria associated with intervillositis and among high-altitude pregnant women exposed to low oxygen concentrations, decreased GLUT-1 expression in the syncytial BM was observed (Table 2) (Chandrasiri et al., 2014; Zamudio et al., 2006, 2010). Supporting previous data, in both studies, GLUT-1 expression in the BM of the trophoblast was positively correlated with the birth-weight of the fetus. On the other hand, in patients with early pre-eclampsia, the GLUT-1 expression and glucose uptake were significantly reduced in the MVM of the syncytium, and remained unaltered in the BM when compared to normotensive women (Lüscher et al., 2017). Although an apical membrane is not considered as a rate limiting step, it is plausible that the observed decrease in GLUT-1 expression could contribute to a reduced intrasyncytial glucose concentration and, as a consequence, to a decreased trans-syncytial concentration gradient. Nevertheless, as the study did not analyze the presence of growth restriction in fetuses, the effect of the observed changes in GLUT-1 expression on the occurrence of IUGR in pregnancies complicated by pre-eclampsia requires further research.

The last group of studies that should be mentioned in the context of the etiopathogenesis of FGR are those conducted *in vitro* and in animal models (Table 1) (Winterhager & Gellhaus, 2017). It was demonstrated experimentally that glucocorticoid administration associated with lower FBW led to the reduced expression of GLUT-1 and GLUT-3 proteins in human trophoblast cell cultures and in the placentas of rats intraperitoneally injected with triamcinolone in a dose-dependent manner (Hahn et al., 1999). On the contrary, in a study by Langdown et al. subcutaneous dexamethasone administration to pregnant rats reduced FBW, resulting in an increased placental expression of GLUT-1 and GLUT-3 proteins (Langdown & Sugden, 2001). Also, in cultured human placental endothelial cells, the incubation with triamcinolone increased GLUT-1 and GLUT-3 protein expression, indicating the presence of tissue-specific compensatory mechanisms in the case of excessive glucocorticoids exposure (Kipmen-Korgun et al., 2012). Therefore, it seems that, although the use of steroids in pregnant women may cause FGR, the observed alterations in GLUT expression in the placenta depend on the type of medication, its dose and the analyzed tissue compartment. In the rat model of maternal hypothyroxinemia, FGR was accompanied by decreased GLUT-1 and increased GLUT-3 placental expression (Pickard et al., 1999). The opposite effect was observed with regard to placental corticotropin-releasing hormone (CRH), whose elevated concentrations are found in pregnancies complicated by IUGR, and which stimulated the expression of GLUT-1, and reduced the expression of GLUT-3 in human trophoblast cells *in vitro* (Gao et al., 2012). Undernutrition is another established cause of FGR, which effect on the placental expression of glucose transporters was analyzed with the use of an animal model. In rodents maternal undernutrition in late pregnancy caused a reduction in FBW, as well as a decrease in placental GLUT-3 expression without affecting the levels of GLUT-1 (Ganguly, Collis, & Devaskar, 2012; Lesage et al., 2002). On the other hand, maternal nutrient restriction led to down-regulation of GLUT-1 protein expression in the placentas of non-human primates (Kavitha et al., 2014). Finally, in models of placental insufficiency caused by uterine artery ligation in mice and rats, or following the exposure of pregnant ewes to elevated temperatures, a decreased expression of GLUT-1 and/or GLUT-8 transporters was noted (Das, Sadiq, Soares, Hay, & Devaskar, 1998; Jones et al., 2013; Limesand et al., 2004; Nüsken et al., 2016).

To conclude, the presented study results indicate that glucose transfer in the placenta is altered in pregnancies complicated by FGR, being closely determined by etiological factors. Obviously, due to differences in the placental structure and GLUT expression between species, it is not possible to extrapolate the data obtained from animal studies directly to humans. Nonetheless, it was revealed experimentally that factors such as placental insufficiency, maternal undernutrition or abnormal hormone secretion result in restricted fetal growth *in utero,* and one of the possible mechanisms is through alterations in the expression of GLUT transporters in the placenta. Contrary to increased GLUT expression in pregnancies complicated by fetal macrosomia, the placentas of FGR fetuses generally demonstrate lower expression of glucose transporter proteins. Nevertheless, it remains to be elucidated whether the observed alterations in GLUT expression indeed contribute to IUGR, and if so, then their role in the pathogenesis of this pregnancy complication compared to other nutrients should be investigated. Furthermore, the potential compensatory mechanisms in the placentas of growth restricted fetuses are still insufficiently examined. For example, can a reduced glucose uptake resulting from down-regulated expression of a single transporter be compensated by an increased expression of another GLUT isoform? To answer these questions further research into placental expression of the whole spectrum of GLUT transporters in animal models, as well as in properly selected and homogeneous groups of patients with IUGR is necessary.

5. Placental GLUT-mediated interventions in pregnancies complicated by fetal growth disorders

The placenta represents an active interface between the maternal and fetal circulations, and by responding to different perturbations in the maternal environment, plays an important role in the regulation of fetal growth. Furthermore, considering the fetal programming of adult disease, alterations in the placental structure and function may be conducive to metabolic and cardiovascular diseases in future generations. It is therefore plausible that through the development of novel interventional strategies aiming to revert or alleviate deteriorated placental function, such adverse consequences as fetal growth disorders and related diseases in adulthood might be prevented.

The number of published studies evaluating possible therapeutic interventions in pregnancies complicated by fetal growth disorders via the GLUT-mediated pathway is scarce. For obvious reasons these studies are conducted exclusively *in vitro* or in animal models. With regard to FGR, a relationship has been found between the maternal IGF-1 concentration and fetal birth-weight, and it was demonstrated that delivery of the protein into the amniotic fluid or directly into the placenta through adenoviral transfection improves the fetal growth rate (Eremia, de Boo, Bloomfield, Oliver, & Harding, 2007; Keswani et al., 2015; Luo et al., 2012). According to some authors, one of the possible mechanisms of IGF-1 action may be through increased transplacental glucose transfer following the increased expression of GLUT transporters in the trophoblast. Indeed, Jones et al. found that in murine placentas after uterine artery branch ligation, adenoviral-mediated IGF-1 treatment up-regulated the previously reduced GLUT-8 expression (Jones et al., 2013). Moreover, in the same study, the authors observed an increased expression of GLUT-1, GLUT-3, GLUT-8 and GLUT-9 proteins in the BeWo choriocarcinoma cell line following adenoviral-IGF-1 transfection. Similarly, Baumann et al. demonstrated increased expression of GLUT-1 in the BM of placental explants incubated in the presence of IGF-1, associated with increased transepithelial glucose transfer (Baumann et al., 2014).

Collectively, despite their limited number, the above studies suggest a possible use of exogenous modulators of transplacental glucose transfer in pregnancies complicated by fetal growth pathologies. Importantly, such interventions may not only reduce the risk of complications during the pregnancy period, but also prevent the development of metabolic and cardiovascular diseases in adult life.

6. Conclusions

To conclude, intrauterine fetal growth disorders are frequent pregnancy complications, contributing to increased morbidity and mortality rates, from the perinatal period to adult age. As presented in this paper, the pathogenesis of disorders such as fetal macrosomia and FGR may be determined to a certain degree by changes in the expression of GLUT proteins in the placenta, affecting the transplacental glucose transport to fetal circulation. With few exceptions, including cases of idiopathic fetal overgrowth and IUGR, it can be stated that in pregnancies complicated by fetal macrosomia, the expression of GLUT transporters in the placenta is increased, resulting in excessive transfer of energy substrates, and intensified fetal growth. On the contrary, in pregnancies complicated by FGR, the placental expression of GLUT and glucose transport are reduced, limiting nutrient availability for the fetus. An emerging number of studies in animal and human models published in recent years has provided valuable information regarding placental glucose transport in the case of fetal growth pathology, as well as questioned some of the prevailing views, e.g. regarding the effect of insulin on the glucose transfer in term placenta. Evidence was also demonstrated for the presence of compensatory mechanisms in the setting of an adverse intrauterine environment due to placental insufficiency, and the first interventions were implemented to modify the GLUT-mediated glucose transfer in the placenta. Nonetheless, further investigations are necessary to fully elucidate the interrelations between certain etiological factors and alterations in placental GLUT expression as well as to uncover the responsible molecular mechanisms in pregnancies complicated by fetal growth disorders.

References

Acevedo, G. C., Marquez, J. L., Rojas, S., & Bravo, I. (2005). Insulin and nitric oxide stimulates glucose transport in human placenta. *Life Sciences, 76*, 2643–2653.

Acosta, O., Ramirez, V. I., Lager, S., Gaccioli, F., Dudley, D. J., Powell, T. L., et al. (2015). Increased glucose and placental GLUT-1 in large infants of obese non diabetic mothers. *American Journal of Obstetrics and Gynecology, 212*, 227.e1-7.

Araujo Júnior, E., Peixoto, A. B., Zamarian, A. C., Elito Júnior, J., Tonni, G., & Macrosomia. (2017). Best practice and research. *Clinical Obstetrics & Gynaecology, 38*, 83–96.

Asghar, Z. A., Thompson, A., Chi, M., Cusumano, A., Scheaffer, S., Al-Hammadi, N., et al. (2016). Maternal fructose drives placental uric acid production leading to adverse fetal outcomes. *Scientific Reports, 6*, 25091.

Augustin, R., Riley, J., & Moley, K. H. (2005). GLUT8 contains a [DE]XXXL[LI] sorting motif and localizes to a late endosomal/lysosomal compartment. *Traffic, 6*(12), 1196–1212.

Barros, L. F., Yudilevich, D. L., Jarvis, S. M., Beaumont, N., & Baldwin, S. A. (1995). Quantitation and immunolocalization of glucose transporters in the human placenta. *Placenta, 16*, 623–633.

Basak, S., Vilasagaram, S., Naidu, K., & Duttaroy, A. K. (2019). Insulin-dependent, glucose transporter 1 mediated glucose uptake and tube formation in the human placental first trimester trophoblast cells. *Molecular and Cellular Biochemistry, 451*(1–2), 91–106.

Baumann, M. U., Schneider, H., Malek, A., Palta, V., Surbek, D. V., Sager, R., et al. (2014). Regulation of human trophoblast GLUT1 glucose transporter by insulin-like growth factor I (IGF-I). *PLoS One, 9*(8), e106037.

Baumann, M. U., Zamudio, S., & Illsley, N. P. (2007). Hypoxic upregulation of glucose transporters in BeWo choriocarcinoma cells is mediated by hypoxia-inducible factor-1. *American Journal of Physiology – Cell Physiology, 293*(1), C477–C485.

Beune, I. M., Bloomfield, F. H., Ganzevoort, W., Embleton, N. D., Rozance, P. J., van Wassenaer-Leemhuis, A. G., et al. (2018). Consensus based definition of growth restriction in the newborn. *The Journal of Pediatrics, 196*, 71–76.e1.

Bibee, K. P., Illsley, N. P., & Moley, K. H. (2011). Asymetric syncytial expression of GLUT9 splice variants in human term placenta and alterations in diabetic pregnancies. *Reproductive Sciences, 18*(1), 20–27.

Boney, C. M., Verma, A., Tucker, R., & Vohr, B. R. (2005). Metabolic syndrome in childhood: Association with birth weight, maternal obesity, and gestational diabetes mellitus. *Pediatrics, 115*(3), e290–e296.

Borges, M. H., Pullockaran, J., Catalano, P. M., Baumann, M. U., Zamudio, S., & Illsley, N. P. (2019). Human placental GLUT1 glucose transporter expression and the fetal insulin-like growth factor axis in pregnancies complicated by diabetes. *Biochimica et Biophysica Acta – Molecular Basis of Disease, 1865*(9), 2411–2419.

Boulet, S. L., Alexander, G. R., Salihu, H. M., & Pass, M. (2003). Macrosomic births in the United States: Determinants, outcomes, and proposed grades of risk. *American Journal of Obstetrics and Gynecology, 188*(5), 1372–1378.

Brown, K., Heller, D. S., Zamudio, S., & Illsley, N. (2011). Glucose transporter 3 (GLUT3) protein expression in human placenta across gestation. *Placenta, 32*, 1041–1049.

Brunette, M. G., Lajeunesse, D., Leclerc, M., & Lafond, J. (1990). Effect of insulin on D-glucose transport by human placental brush border membranes. *Molecular and Cellular Endocrinology, 69*(1), 59–68.

Bryant, N. J., Govers, R., & James, D. E. (2002). Regulated transport of the glucose transporter GLUT4. *Nature Reviews Molecular Cell Biology, 3*(4), 267–277.

Carayannopoulos, M. O., Chi, M. M., Cui, Y., Pingsterhaus, J. M., McKnight, R. A., Mueckler, M., et al. (2000). GLUT8 is a glucose transporter responsible for insulin-stimulated glucose uptake in the blastocyst. *Proceedings of the National Academy of Sciences of the United States of America, 97*(13), 7313–7318.

Caulfield, M. J., Munroe, P. B., O'Neill, D., Witkowska, K., Charchar, F. J., Doblado, M., et al. (2008). SLC2A9 is a high-capacity urate transporter in humans. *PLoS Medicine, 5*(10), e197.

Challier, J. C., Hauguel, S., & Desmaizieres, V. (1986). Effect of insulin on glucose uptake and metabolism in the human placenta. *Journal of Clinical Endocrinology & Metabolism, 62*(5), 803–807.

Challis, D. E., Pfarrer, C. D., Ritchie, J. W., Koren, G., & Adamson, S. L. (2000). Glucose metabolism is elevated and vascular resistance and maternofetal transfer is normal in perfused placental cotyledons from severely growth-restricted fetuses. *Pediatric Research, 47*(3), 309–315.
Chandrasiri, U. P., Chua, C. L., Umbers, A. J., Chaluluka, E., Glazier, J. D., Rogerson, S. J., et al. (2014). Insight into the pathogenesis of fetal growth restriction in placental malaria: Decreased placental glucose transporter isoform 1 expression. *Journal of Infectious Diseases, 209*(10), 1663–1667.
Colomiere, M., Permezel, M., Riley, C., Desoye, G., & Lappas, M. (2009). Defective insulin signaling in placenta from pregnancies complicated by gestational diabetes mellitus. *European Journal of Endocrinology, 160*, 567–578.
Das, U. G., Sadiq, H. F., Soares, M. J., Hay, W. W., Jr., & Devaskar, S. U. (1998). Time-dependent physiological regulation of rodent and ovine placental glucose transporter (GLUT-1) protein. *American Journal of Physiology, 274*(2), R339–R347.
Dawson, P. A., Mychaleckyj, J. C., Fossey, S. C., Mihic, S. J., Craddock, A. L., & Bowden, D. W. (2001). Sequence and functional analysis of GLUT10: A glucose transporter in the type 2 diabetes-linked region of chromosome 20q12-13.1. *Molecular Genetics and Metabolism, 74*, 186–199.
Desoye, G., Hartmann, M., Jones, C. J. P., Wolf, H., Kohnen, G., Kosanke, G., et al. (1997). Location of insulin receptors in the placenta and its progenitor tissues. *Microscopy Research and Technique, 38*, 63–75.
Desoye, G., Korgun, E. T., Ghaffari-Tabrizi, N., & Hahn, T. (2002). Is fetal macrosomia in adequately controlled diabetic women the result of a placental defect? — A hypothesis. *Journal of Maternal-Fetal and Neonatal Medicine, 11*(4), 258–261.
Di Simone, N., Di Nicuolo, F., Marzioni, D., Castellucci, M., Sanguinetti, M., D'lppolito, S., et al. (2009). Resistin modulates glucose uptake and glucose transporter-1 (GLUT-1) expression in trophoblast cells. *Journal of Cellular and Molecular Medicine, 13*(2), 388–397.
Economides, D. L., & Nicolaides, K. H. (1989). Blood glucose and oxygen tension levels in small-for-gestational-age fetuses. *American Journal of Obstetrics and Gynecology, 160*(2), 385–389.
Eremia, S. C., de Boo, H. A., Bloomfield, F. H., Oliver, M. H., & Harding, J. E. (2007). Fetal and amniotic insulin-like growth factor-I supplements improve growth rate in intrauterine growth restriction fetal sheep. *Endocrinology, 148*(6), 2963–2972.
Ericsson, A., Hamark, B., Powell, T. L., & Jansson, T. (2005). Glucose transporter isoform 4 is expressed in the syncytiotrophoblast of first trimester human placenta. *Human Reproduction, 20*(2), 521–530.
Gaither, K., Quraishi, A. N., & Illsley, N. P. (1999). Diabetes alters the expression and activity of the human placental GLUT 1 glucose transporter. *Journal of Clinical Endocrinology & Metabolism, 84*, 695–701.
Ganguly, A., Collis, L., & Devaskar, S. U. (2012). Placental glucose and amino acid transport in calorie-restricted wild-type and Glut3 null heterozygous mice. *Endocrinology, 153*(8), 3995–4007.
Gao, L., Lv, C., Xu, C., Li, Y., Cui, X., Gu, H., et al. (2012). Differential regulation of glucose transporters mediated by CRH receptor type 1 and type 2 in human placental trophoblasts. *Endocrinology, 153*(3), 1464–1471.
Gluckman, P. D., Hanson, M. A., Cooper, C., & Thornburg, K. L. (2008). Effect of in utero and early-life conditions on adult health and disease. *New England Journal of Medicine, 359*(1), 61–73.
Gordijn, S. J., Beune, I. M., Thilaganathan, B., Papageorghiou, A., Baschat, A. A., Baker, P. N., et al. (2016). Consensus definition of fetal growth restriction: A Delphi procedure. *Ultrasound in Obstetrics and Gynecology, 48*(3), 333–339.

Gordon, M. C., Zimmerman, P. D., Landon, M. B., Gabbe, S. G., & Kniss, D. A. (1995). Insulin and glucose modulate glucose transporter messenger ribonucleic acid expression and glucose uptake in trophoblasts isolated from first-trimester chorionic villi. *American Journal of Obstetrics and Gynecology, 173*(4), 1089–1097.

Gorovits, N., & Charron, M. J. (2003). What we know about facilitative glucose transporters. *Biochemistry and Molecular Biology Education, 31*(3), 163–172.

Gude, N. M., Stevenson, J. L., Rogers, S., Best, J. D., Kalionis, B., Huisman, M. A., et al. (2003). GLUT12 expression in human placenta in first trimester and term. *Placenta, 24*(5), 566–570.

Hahn, T., Barth, S., Graf, R., Engelmann, M., Beslagic, D., Reul, J. M. H. M., et al. (1999). Placental glucose transporter expression is regulated by glucocorticoids. *Journal of Clinical Endocrinology & Metabolism, 84*(4), 1445–1452.

Hahn, T., Barth, S., Weiss, U., Mosgoeller, W., & Desoye, G. (1998). Sustained hyperglycemia downregulates the GLUT-1 glucose transport system of cultured human placental trophoblast: A mechanism to protect fetal development? *The FASEB Journal, 12*, 1221–1231.

Hahn, D., Blaschitz, A., Korgun, E. T., Lang, I., Desoye, G., Skofitsch, G., et al. (2001). From maternal glucose to fetal glycogen: Expression of key regulators in the human placenta. *Molecular Human Reproduction,* 7(12), 1173–1178.

Hahn, T., Hartmann, M., Blaschitz, A., Skofitsch, G., Graf, R., Dohr, G., et al. (1995). Localisation of the high affinity facilitative glucose transporter protein GLUT 1 in the placenta of human, marmoset monkey (*Callithrix jacchus*) and rat at different developmental stages. *Cell and Tissue Research, 280*, 49–57.

Hauguel-De Mouzon, S., Challier, J. C., Kacemi, A., Cauzac, M., Malek, A., & Girard, J. (1997). The GLUT3 glucose transporter isoform is differentially expressed within human placental cell types. *Journal of Clinical Endocrinology & Metabolism, 82*, 2689–2694.

Hauguel, S., Desmaizieres, V., & Challier, J. C. (1986). Glucose uptake, utilization and transfer by the human placenta as functions of maternal glucose concentration. *Pediatric Research, 20*(3), 269–273.

Hayashi, M., Sakata, M., Takeda, T., Yamamoto, T., Okamoto, Y., Sawada, K., et al. (2004). Induction of glucose transporter 1 expression through hypoxia-inducible factor 1alpha under hypoxic conditions in trophoblast-derived cells. *Journal of Endocrinology, 183*(1), 145–154.

Hay, W. W., Jr. (2006). Placental-fetal glucose exchange and fetal glucose metabolism. *Transactions of the American Clinical and Climatological Association, 117*, 321–339.

Hiden, U., Glitzner, E., Hartmann, M., & Desoye, G. (2009). Insulin and the IGF system in the human placenta of normal and diabetic pregnancies. *Journal of Anatomy, 215*(1), 60–68.

Illsley, N. P. (2000). Placental glucose transport in diabetic pregnancy. *Clinical Obstetrics and Gynecology, 43*, 116–126.

Illsley, N., Hall, S., & Stacey, T. E. (1987). The modulation of glucose transfer across the human placenta by intervillous flow rates: An in vitro perfusion study. *Trophoblast Research, 2*, 535–544.

Illsley, N. P., Sellers, M. C., & Wright, R. L. (1998). Glycemic regulation of glucose transporter expression and activity in the human placenta. *Placenta, 19*, 517–524.

James-Allan, L. B., Arbet, J., Teal, S. B., Powell, T. L., & Jansson, T. (2019). Insulin stimulates GLUT4 trafficking to the syncytiotrophoblast basal plasma membrane in the human placenta. *Journal of Clinical Endocrinology & Metabolism, 104*(9), 4225–4238.

Jansson, T., Ekstrand, Y., Björn, C., Wennergren, M., & Powell, T. L. (2002). Alterations in the activity of placental amino acid transporters in pregnancies complicated by diabetes. *Diabetes, 51*(7), 2214–2219.

Jansson, T., Ekstrand, Y., Wennergren, M., & Powell, T. L. (2001). Placental glucose transport in gestational diabetes mellitus. *American Journal of Obstetrics and Gynecology, 184*, 111–116.

Jansson, T., Myatt, L., & Powell, T. L. (2009). The role of trophoblast nutrient and ion transporters in the development of pregnancy complications and adult disease. *Current Vascular Pharmacology,* 7(4), 521–533.

Jansson, T., & Powell, T. L. (2006). IFPA 2005 award in placentology lecture. Human placental transport in altered fetal growth: Does the placenta function as a nutrient sensor? – A review. *Placenta, 27*(Suppl. A), S91–S97.

Jansson, T., Wennergren, M., & Illsley, N. P. (1993). Glucose transporter expression and distribution in the human placenta throughout gestation and in intrauterine growth retardation. *Journal of Clinical Endocrinology & Metabolism, 77*, 1554–1562.

Jansson, T., Wennergren, M., & Powell, T. L. (1999). Placental glucose transport and GLUT 1 expression in insulin dependent diabetes. *American Journal of Obstetrics and Gynecology, 180*, 163–168.

Jansson, T., Ylvén, K., Wennergren, M., & Powell, T. L. (2002). Glucose transport and system A activity in syncytiotrophoblast microvillous and basal plasma membranes in intrauterine growth restriction. *Placenta, 23*(5), 392–399.

Janzen, C., Lei, M. Y., Cho, J., Sullivan, P., Shin, B. C., & Devaskar, S. U. (2013). Placental glucose transporter 3 (GLUT3) is up-regulated in human pregnancies complicated by late-onset intrauterine growth restriction. *Placenta, 34*(11), 1072–1078.

Janzen, C., Lei, M. Y. Y., Jeong, I. S. D., Ganguly, A., Sullivan, P., Paharkova, V., et al. (2018). Humanin (HN) and glucose transporter 8 (GLUT8) in pregnancies complicated by intrauterine growth restriction. *PLoS One, 13*(3), e0193583.

Jolly, M. C., Sebire, N. J., Harris, J. P., Regan, L., & Robinson, S. (2003). Risk factors for macrosomia and its clinical consequences: A study of 350,311 pregnancies. *European Journal of Obstetrics & Gynecology and Reproductive Biology, 111*(1), 9–14.

Jones, H. N., Crombleholme, T., & Habli, M. (2013). Adenoviral-mediated placental gene transfer of IGF-1 corrects placental insufficiency via enhanced placental glucose transport mechanisms. *PLoS One, 8*(9), e74632.

Jones, H. N., Powell, T. L., & Jansson, T. (2007). Regulation of placental nutrient transport – A review. *Placenta, 28*, 763–774.

Joost, H. G., Bell, G. I., Best, J. D., Birnbaum, M. J., Charron, M. J., Chen, Y. T., et al. (2002). Nomenclature of the GLUT/SLC2A family of sugar/polyol transport facilitators. *American Journal of Physiology – Endocrinology and Metabolism, 282*(4), E974–E976.

Jürgens, H., Haass, W., Castañeda, T. R., Schürmann, A., Koebnick, C., Dombrowski, F., et al. (2005). Consuming fructose-sweetened beverages increases body adiposity in mice. *Obesity Research, 13*(7), 1146–1156.

Kainulainen, H., Jarvinen, T., & Heinonen, P. K. (1997). Placental glucose transporters in fetal intrauterine growth retardation and macrosomia. *Gynecologic and Obstetric Investigation, 44*, 89–92.

Kalhan, S., & Parimi, P. (2000). Gluconeogenesis in the fetus and neonate. *Seminars in Perinatology, 24*(2), 94–106.

Kavitha, J. V., Rosario, F. J., Nijland, M. J., McDonald, T. J., Wu, G., Kanai, Y., et al. (2014). Down-regulation of placental mTOR, insulin/IGF-I signaling, and nutrient transporters in response to maternal nutrient restriction in the baboon. *The FASEB Journal, 28*(3), 1294–1305.

Keswani, S. G., Balaji, S., Katz, A. B., King, A., Omar, K., Habli, M., et al. (2015). Intraplacental gene therapy with Ad-IGF-1 corrects naturally occurring rabbit model of intrauterine growth restriction. *Human Gene Therapy, 26*(3), 172–182.

Kipmen-Korgun, D., Ozmen, A., Unek, G., Simsek, M., Demir, R., & Korgun, E. T. (2012). Triamcinolone up-regulates GLUT 1 and GLUT 3 expression in cultured human placental endothelial cells. *Cell Biochemistry and Function, 30*(1), 47–53.

Korgun, E. T., Celik-Ozenci, C., Seval, Y., Desoye, G., & Demir, R. (2005). Do glucose transporters have other roles in addition to placental glucose transport during early pregnancy? *Histochemistry and Cell Biology, 123*, 621–629.

Langdown, M. L., & Sugden, M. C. (2001). Enhanced placental GLUT1 and GLUT3 expression in dexamethasone-induced fetal growth retardation. *Molecular and Cellular Endocrinology, 185*(1-2), 109–117.

Lappas, M., Andrikopoulos, S., & Permezel, M. (2012). Hypoxanthine-xanthine oxidase down-regulates GLUT1 transcription via SIRT1 resulting in decreased glucose uptake in human placenta. *Journal of Endocrinology, 213*(1), 49–57.

Leach, L., & Firth, J. A. (1997). Structure and permeability of human placental microvasculature. *Microscopy Research and Technique, 38*(1-2), 137–144.

Lee, Y. C., Huang, H. Y., Chang, C. J., Cheng, C. H., & Chen, Y. T. (2010). Mitochondrial GLUT10 facilitates dehydroascorbic acid import and protects cells against oxidative stress: Mechanistic insight into arterial tortuosity syndrome. *Human Molecular Genetics, 19*(19), 3721–3733.

Lesage, J., Hahn, D., Léonhardt, M., Blondeau, B., Bréant, B., & Dupouy, J. P. (2002). Maternal undernutrition during late gestation-induced intrauterine growth restriction in the rat is associated with impaired placental GLUT3 expression, but does not correlate with endogenous corticosterone levels. *Journal of Endocrinology, 174*(1), 37–43.

Li, G., Kong, L., Li, Z., Zhang, L., Fan, L., Zou, L., et al. (2014). Prevalence of macrosomia and its risk factors in China: A multicentre survey based on birth data involving 101,723 singleton term infants. *Paediatric and Perinatal Epidemiology, 28*(4), 345–350.

Limesand, S. W., Regnault, T. R. H., & Hay, W. W. (2004). Characterization of glucose transporter 8 (GLUT 8) in the ovine placenta of normal and growth restricted fetuses. *Placenta, 25*, 70–77.

Luo, Z. C., Nuyt, A. M., Delvin, E., Audibert, F., Girard, I., Shatenstein, B., et al. (2012). Maternal and fetal IGF-I and IGF-II levels, fetal growth, and gestational diabetes. *Journal of Clinical Endocrinology & Metabolism, 97*(5), 1720–1728.

Lüscher, B. P., Marini, C., Joerger-Messerli, M. S., Huang, X., Hediger, M. A., Albrecht, C., et al. (2017). Placental glucose transporter (GLUT)-1 is down-regulated in preeclampsia. *Placenta, 55*, 94–99.

Magnusson, A. L., Waterman, I. J., Wennergren, M., Jansson, T., & Powell, T. L. (2004). Triglyceride hydrolase activities and expression of fatty acid binding proteins in the human placenta in pregnancies complicated by intrauterine growth restriction and diabetes. *Journal of Clinical Endocrinology & Metabolism, 89*(9), 4607–4614.

Manolescu, A. R., Augustin, R., Moley, K., & Cheeseman, C. (2007a). A highly conserved hydrophobic motif in the exofacial vestibule of fructose transporting SLC2A proteins acts as a critical determinant of their substrate selectivity. *Molecular Membrane Biology, 24*(5-6), 455–463.

Manolescu, A. R., Witkowska, K., Kinnaird, A., Cessford, T., & Cheeseman, C. (2007). Facilitated hexose transporters: New perspectives on form and function. *Physiology, 22*, 234–240.

Marconi, A. M., Paolini, C., Buscaglia, M., Zerbe, G., Battaglia, F. C., & Pardi, G. (1996). The impact of gestational age and fetal growth on the maternal-fetal glucose concentration difference. *Obstetrics & Gynecology, 87*(6), 937–942.

Mueckler, M., Caruso, C., Baldwin, S. A., Panico, M., Blench, I., Morris, H. R., et al. (1985). Sequence and structure of a human glucose transporter. *Science, 229*, 941–945.

Mueckler, M., & Thorens, B. (2013). The SLC2 (GLUT) family of membrane transporters. *Molecular Aspects of Medicine, 34*, 121–138.

Nüsken, E., Gellhaus, A., Kühnel, E., Swoboda, I., Wohlfarth, M., Vohlen, C., et al. (2016). Increased rat placental fatty acid, but decreased amino acid and glucose transporters potentially modify intrauterine programming. *Journal of Cellular Biochemistry, 117*(7), 1594–1603.

Ogura, K., Sakata, M., Okamoto, Y., Yasui, Y., Tadokoro, C., Yoshimoto, Y., et al. (2000). 8-bromo-cyclicAMP stimulates glucose transporter-1 expression in a human choriocarcinoma cell line. *Journal of Endocrinology, 164*(2), 171–178.

Osmond, C., & Barker, D. J. (2000). Fetal, infant, and childhood growth are predictors of coronary heart disease, diabetes, and hypertension in adult men and women. *Environmental Health Perspectives, 108*(Suppl. 3), 545–553.

Osmond, D. T. D., King, R. G., Brennecke, S. P., & Gude, N. M. (2001). Placental glucose transport and utilisation is altered at term in insulin-treated, gestational-diabetic patients. *Diabetologia, 44*, 1133–1139.

Pickard, M. R., Sinha, A. K., Ogilvie, L. M., Leonard, A. J., Edwards, P. R., & Ekins, R. P. (1999). Maternal hypothyroxinemia influences glucose transporter expression in fetal brain and placenta. *Journal of Endocrinology, 163*(3), 385–394.

Reza-López, S. A., Sánchez-Ramírez, B. E., Gutiérrez-Torres, D., Rodríguez-Salinas, M., Ortega-Valerio, M. G., Chávez-Corral, D. V., et al. (2014). Expression of glucose transporters in placenta from macrosomic infants. *Global Advanced Research Journal of Medicine and Medical Sciences, 3*(10), 315–321.

Romo, A., Carceller, R., & Tobajas, J. (2009). Intrauterine growth retardation (IUGR): Epidemiology and etiology. *Pediatric Endocrinology Reviews, 6*(Suppl. 3), 332–336.

Sacks, D. A. (2004). Determinants of fetal growth. *Current Diabetes Reports, 4*(4), 281–287.

Sakata, M., Kurachi, H., Imai, T., Tadokoro, C., Yamaguchi, M., Yoshimoto, Y., et al. (1995). Increase in human placental glucose transporter-1 during pregnancy. *European Journal of Endocrinology, 132*, 206–212.

Scheepers, A., Schmidt, S., Manolescu, A., Cheeseman, C. I., Bell, A., Zahn, C., et al. (2005). Characterization of the human SLC2A11 (GLUT11) gene: Alternative promoter usage, function, expression, and subcellular distribution of three isoforms, and lack of mouse orthologue. *Molecular Membrane Biology, 22*(4), 339–351.

Schmidt, S., Joost, H. G., & Schürmann, A. (2009). GLUT8, the enigmatic intracellular hexose transporter. *American Journal of Physiology – Endocrinology and Metabolism, 296*, E614–E618.

Schneider, H., Reiber, W., Sager, R., & Malek, A. (2003). Asymmetrical transport of glucose across the in vitro perfused human placenta. *Placenta, 24*(1), 27–33.

Smith, R. M., Charron, M. J., Shah, N., Lodish, H. F., & Jarett, L. (1991). Immunoelectron microscopic demonstration of insulin-stimulated translocation of glucose transporters to the plasma membrane of isolated rat adipocytes and masking of the carboxyl-terminal epitope of intracellular GLUT4. *Proceedings of the National Academy of Sciences of the United States of America, 88*(15), 6893–6897.

Somwar, R., Kim, D. Y., Sweeney, G., Huang, C., Niu, W., Lador, C., et al. (2001). GLUT4 translocation precedes the stimulation of glucose uptake by insulin in muscle cells: Potential activation of GLUT4 via p38 mitogen-activated protein kinase. *Biochemical Journal, 359*(Pt 3), 639–649.

Stanirowski, P. J., Szukiewicz, D., Pazura-Turowska, M., Sawicki, W., & Cendrowski, K. (2018). Placental expression of glucose transporter proteins in pregnancies complicated by gestational and pregestational diabetes mellitus. *Canadian Journal of Diabetes, 42*(2), 209–217.

Stanirowski, P. J., Szukiewicz, D., Pyzlak, M., Abdalla, N., Sawicki, W., & Cendrowski, K. (2017). Impact of pre-gestational and gestational diabetes mellitus on the expression of glucose transporters GLUT-1, GLUT-4 and GLUT-9 in human term placenta. *Endocrine, 55*, 799–808.

Stanirowski, P. J., Szukiewicz, D., Pyzlak, M., Abdalla, N., Sawicki, W., & Cendrowski, K. (2019). Analysis of correlations between the placental expression of glucose transporters GLUT-1, GLUT-4 and GLUT-9 and selected maternal and fetal parameters in pregnancies complicated by diabetes mellitus. *Journal of Maternal-Fetal and Neonatal Medicine, 32*(4), 650–659.

Stuart, C. A., Howell, M. E., Zhang, Y., & Yin, D. (2009). Insulin-stimulated translocation of glucose transporter (GLUT) 12 parallels that of GLUT4 in normal muscle. *Journal of Clinical Endocrinology & Metabolism, 94*(9), 3535–3542.

Suhag, A., & Berghella, V. (2013). Intrauterine growth restriction (IUGR): Etiology and diagnosis. *Current Obstetrics and Gynecology Reports, 2*, 102–111.

Tadokoro, C., Yoshimoto, Y., Sakata, M., Fujimiya, H., Kurachil, E., Adachi, E., et al. (1996). Localization of human placental glucose transporter 1 during pregnancy. *Histology & Histopathology, 11*, 673–681.

Takata, K., Kasahara, T., Kasahara, M., Ezaki, O., & Hirano, H. (1992). Localization of erythrocyte/HepG2-type glucose transporter (GLUT-1) in human placental villi. *Cell and Tissue Research, 267*, 407–412.

Tappy, L. (2018). Fructose metabolism and noncommunicable diseases: Recent findings and new research perspectives. *Current Opinion in Clinical Nutrition and Metabolic Care, 21*(3), 214–222.

Taricco, E., Radaelli, T., Rossi, G., Nobile de Santis, M. S., Bulfamante, G. P., Avagliano, L., et al. (2009). Effects of gestational diabetes on fetal oxygen and glucose levels in vivo. *BJOG: An International Journal of Obstetrics and Gynaecology, 116*(13), 1729–1735.

Teasdale, F., & Jean-Jacques, G. (1985). Morphometric evaluation of the microvillous surface enlargement factor in the human placenta from mid-gestation to term. *Placenta, 6*(5), 375–381.

Uldry, M., Ibberson, M., Horisberger, J. D., Chatton, J. Y., Riederer, B. M., & Thorens, B. (2001). Identification of a mammalian H(+)-myo-inositol symporter expressed predominantly in the brain. *The EMBO Journal, 20*(16), 4467–4477.

Urbach, J., Mor, L., Ronen, N., & Brandes, J. M. (1989). Does insulin affect placental glucose metabolism and transfer? *American Journal of Obstetrics and Gynecology, 161*, 953–959.

Vardhana, P. A., & Illsley, N. P. (2002). Transepithelial glucose transport and metabolism in BeWo choriocarcinoma cells. *Placenta, 23*(8-9), 653–660.

Vickers, M. H., Clayton, Z. E., Yap, C., & Sloboda, D. M. (2011). Maternal fructose intake during pregnancy and lactation alters placental growth and leads to sex-specific changes in fetal and neonatal endocrine function. *Endocrinology, 152*(4), 1378–1387.

Visiedo, F., Bugatto, F., Carrasco-Fernández, C., Sáez-Benito, A., Mateos, R. M., Cózar-Castellano, I., et al. (2015). Hepatocyte growth factor is elevated in amniotic fluid from obese women and regulates placental glucose and fatty acid metabolism. *Placenta, 36*(4), 381–388.

Wang, X., Yang, T., Miao, J., Liu, H., Wu, K., Guo, J., et al. (2018). Correlation between maternal and fetal insulin resistance in pregnant women with gestational diabetes mellitus. *Clinical Laboratory, 64*(6), 945–953.

Westgate, J. A., Lindsay, R. S., Beattie, J., Pattison, N. S., Gamble, G., Mildenhall, L. F., et al. (2006). Hyperinsulinemia in cord blood in mothers with type 2 diabetes and gestational diabetes mellitus in New Zealand. *Diabetes Care, 29*(6), 1345–1350.

Winterhager, E., & Gellhaus, A. (2017). Transplacental nutrient transport mechanisms of intrauterine growth restriction in rodent models and humans. *Frontiers in Physiology, 8*, 951.

Witkowska, K., Smith, K. M., Yao, S. Y., Ng, A. M., O'Neill, D., Karpinski, E., et al. (2012). Human SLC2A9a and SLC2A9b isoforms mediate electrogenic transport of urate with different characteristics in the presence of hexoses. *American Journal of Physiology – Renal Physiology, 303*(4), F527–F539.

Xing, A. Y., Challier, J. C., Lepercq, J., Cauzac, M., Charron, M. J., Girard, J., et al. (1998). Unexpected expression of glucose transporter 4 in villous stromal cells of human placenta. *Journal of Clinical Endocrinology & Metabolism, 83*, 4097–4101.

Yao, G., Zhang, Y., Wang, D., Yang, R., Sang, H., Han, L., et al. (2017). GDM-induced macrosomia is reversed by Cav-1 via AMPK-mediated fatty acid transport and GLUT1-mediated glucose transport in placenta. *PLoS One, 12*(1), e0170490.

Zamudio, S., Baumann, M. U., & Illsley, N. P. (2006). Effects of chronic hypoxia in vivo on the expression of human placental glucose transporters. *Placenta, 27*(1), 49–55.

Zamudio, S., Torricos, T., Fik, E., Oyala, M., Echalar, L., Pullockaran, J., et al. (2010). Hypoglycemia and the origin of hypoxia-induced reduction in human fetal growth. *PLoS One, 5*(1), e8551.

Zhang, B., Jin, Z., Sun, L., Zheng, Y., Jiang, J., Feng, C., et al. (2016). Expression and correlation of sex hormone-binding globulin and insulin signal transduction and glucose transporter proteins in gestational diabetes mellitus placental tissue. *Diabetes Research and Clinical Practice, 119*, 106–117.

CHAPTER SEVEN

The role of transferrins and iron-related proteins in brain iron transport: applications to neurological diseases

Camila Campos-Escamilla*
Programa de Doctorado en Ciencias Biomédicas, Universidad Nacional Autónoma de México. Unidad de Posgrado, Ciudad Universitaria, Ciudad de México, Mexico
*Corresponding author: E-mail: camila.cescamilla@gmail.com

Contents

1. Introduction 134
2. Brain iron transport 135
3. Iron and neurodegeneration 138
   3.1 Oxidative stress and ferroptosis 139
   3.2 Damage to the blood-brain barrier (BBB) 142
   3.3 Alzheimer's disease (AD) 142
   3.4 Parkinson's disease (PD) 143
   3.5 Huntington's disease (HD) 144
   3.6 Multiple sclerosis (MS) 145
   3.7 Sanfilippo syndrome 145
4. Iron transport related proteins 146
   4.1 Transferrin receptors 1 and 2 (TfR1 and TfR2) 146
   4.2 Divalent metal transporter 1 (DMT1) 149
   4.3 Ferroportin 1 (Fpn1) 149
   4.4 Ferritin 150
   4.5 Hephaestin (Heph) 150
   4.6 Ceruloplasmin (Cp) 151
5. Transferrins 151
   5.1 Human serum transferrin (Tf) 152
6. Perspectives on the use of transferrin as a drug delivery system 154
7. Concluding remarks 156
Acknowledgments 156
References 156

Advances in Protein Chemistry and Structural Biology, Volume 123
ISSN 1876-1623
https://doi.org/10.1016/bs.apcsb.2020.09.002

Abstract

Iron transport in the central nervous system (CNS) is a highly regulated process in which several important proteins participate to ensure this important metal reaches its sites of action. However, iron accumulation has been shown to be a common factor in different neurological disorders such as Alzheimer's disease, Parkinson's disease, Huntington's disease, Multiple Sclerosis, and Sanfilippo syndrome. This review is divided into four parts. The first part describes brain iron transport in homeostasis, mentioning the main proteins involved, whereas the second part contrasts the consequences of iron dysregulation, elaborating on its role in the aforementioned neurodegenerative diseases. The third part details the functions of the main proteins involved in brain iron homeostasis and their role in neurodegeneration. In the fourth part, in order to highlight the importance of transport proteins, the focus is set on human serum transferrin, the main iron transport protein. This final part describes perspectives about the mechanisms and chemical properties of human transferrin for the development of potential targeted drug delivery systems across the blood-brain barrier (BBB) or enhancers for the treatment of neurological diseases.

1. Introduction

Neurological diseases are a topic of major interest given the complexity of their nature. Particularly, neurodegenerative diseases are taking the spotlight because as the elderly population increases, the incidence of age-related maladies increases as well (Acevedo, Masaldan, Opazo, & Bush, 2019; Hodgson et al., 2020; Licher et al., 2019; McWhirter, Ritchie, Stone, & Carson, 2020; Sfera, Bullock, Price, Inderias, & Osorio, 2018). Nowadays, the search for treatments for neurodegenerative diseases remains a challenge, as the knowledge about their causes is still rather unknown, and some of them are still considered to be part of the natural course of aging; such is the case of Alzheimer's disease (Chang, 2019; Mills, Dong, Wang, & Xu, 2010; Muhoberac & Vidal, 2019;Wang et al., 2020). However, neurodegeneration is not exclusive of the geriatric population, since it is also observed in Sanfilippo Syndrome, a disorder that affects children (Heon-Roberts, Nguyen, & Pshezhetsky, 2020). Therefore, it is important to analyze the panorama of disease in order to understand the potential source of the problem. Several studies have been reviewed over the last decade to shed light on the causes and factors that lead to neurodegeneration, and thus to find the means to deal with this health issue. These studies have shown an important relationship between iron and neurological alterations, particularly neurodegeneration (Acevedo et al., 2019; Ashraf, Clark, & So, 2018;

Chang, 2019; Rouault, 2013; Roy Sarkar & Dutta, 2019; Wang et al., 2020). It is known that the lack of sufficient iron leads to deficient overall brain performance at any stage of life (Hare, Ayton, Bush, & Lei, 2013; Scassellati, Bonvicini, Benussi, Ghidoni, & Squitti, 2020); however, increasing evidence shows that iron overload becomes more common in people of older age and is linked to neurodegenerative diseases (Buijs et al., 2017; Picca et al., 2019; Uranga & Salvador, 2018; Wawer, Jennings, & Fairweather-Tait, 2018). On the other hand, a close look to the biomolecules that contribute to the regulation of some of the main biochemical processes, such as transferrin and other iron transport-related proteins, which become affected when disease strikes, is also a good starting point (Acevedo et al., 2019; Chang, 2019; Guan, Wang, Lu, & Zhao, 2020; Ingrassia, Garavaglia, & Memo, 2019; Muhoberac & Vidal, 2019; Prakash, Dhaliwal, Kumar, & Majeed, 2017; Singh, Pandey, Vishwakarma, & Modi, 2019; Uranga & Salvador, 2018; Wang et al., 2020). In the following sections, we will discuss the main mechanisms of brain iron transport, the role of iron in neurodegeneration, the main iron transport-related proteins, and some perspectives on the use of transport proteins, namely human serum transferrin, for the treatment of neurological diseases.

2. Brain iron transport

Iron is an important metal involved in various physiological processes, such as ATP generation and DNA replication (Chang, 2019; Mills et al., 2010; Qian & Ke, 2019). The brain is the organ with the highest energy demand; thus, the mitochondrial electron transport chain must keep up with ATP production. In order to achieve this, iron must be present in sufficient proportion to maintain its function as a cofactor for mitochondrial enzymes such as ferredoxins and cytochromes. Particularly, iron is essential for a variety of neurological processes (McCarthy & Kosman, 2015; Rouault, 2013). In neuronal cells, iron is necessary for dopaminergic neurotransmitter synthesis and myelination of axons. Iron is involved in the synthesis and signaling of dopamine, noradrenaline, adrenaline and serotonin, as it acts as a cofactor for proteins such as phenylalanine hydroxylase, tyrosine hydroxylase, and tryptophan hydroxylase (Hare et al., 2013). Myelination is an essential process for protection and conductivity-optimization in axons in the central nervous system (CNS), which takes an important role when pathogenesis appears. In the case of iron deficiency, myelin components are decreased; consequently, myelination impairment slows neuronal conduction and reflex response (Hare et al., 2013; Möller et al., 2019; Nasrabady, Rizvi, Goldman, & Brickman, 2018).

Iron transport in the brain is effectuated by several pathways; namely, transferrin-dependent iron transport, non-transferrin bound iron (NTBI) mobilization, uptake and export by and from neurons, oligodendrocytes, astrocytes, and microglia (Hohnholt & Dringen, 2013; Roy Sarkar & Dutta, 2019). Although there are other molecular processes involved in iron uptake in each cell type, the Tf/TfR1/DMT1 iron uptake mechanism, which involves transferrin (Tf), transferrin receptor 1 (TfR1), and divalent metal transporter 1 (DMT1), is the main one present in neurons, astrocytes, and microglia. In the case of oligodendrocytes, this mechanism has only been found in early stages, so the main iron source in mature oligodendrocytes is provided by ferritin (Möller et al., 2019; Todorich, Zhang, Slagle-Webb, Seaman, & Connor, 2008). On the other hand, cellular iron export is mainly mediated by ferroportin 1 (Fpn1), in conjunction with ceruloplasmin (Cp)—a ferroxidase that converts ferrous iron to ferric iron— in the case of neurons and astrocytes, or with hephaestin (Heph) in the case of neurons, as well, oligodendrocytes, and microglia (Bonaccorsi di Patti et al., 2018; Jiang et al., 2015; Qian & Ke, 2019; Sokolov et al., 2017, 2017; White et al., 2012). The mechanisms of iron transport are illustrated in Fig. 1, and more detailed information about each of the aforementioned proteins involved in iron transport will be presented in section 4.

As much as iron is needed for several functions in the brain, this important organ requires a very strict control on its intake. This control is possible thanks to two main structures: the blood-brain barrier (BBB), and the blood-cerebrospinal fluid barrier (BCSFB) (Duck, Simpson, & Connor, 2017; McCarthy & Kosman, 2015; Roy Sarkar & Dutta, 2019; Simpson et al., 2015). The BBB consists of several tightly bound brain vascular endothelial cells (BVECs) that limit the flux of material between the blood and the brain (Mills et al., 2010). Regulation of iron influx by BVECs involves controlling receptor expression, internalization of Tf-TfR complexes, and acidification of cell endosomes (Qian & Ke, 2019; Roy Sarkar & Dutta, 2019). Astrocytes are perhaps the most important part of the BBB in terms of iron metabolism, since they're the first cells to encounter iron ions and compounds after these have traversed the endothelial cells of the BBB (see Fig. 1). Also, astrocytes may play a protective role toward neurons, since they can prevent the saturation of iron into the synaptic cleft by taking the NTBI into themselves (Hohnholt & Dringen, 2013). On the other

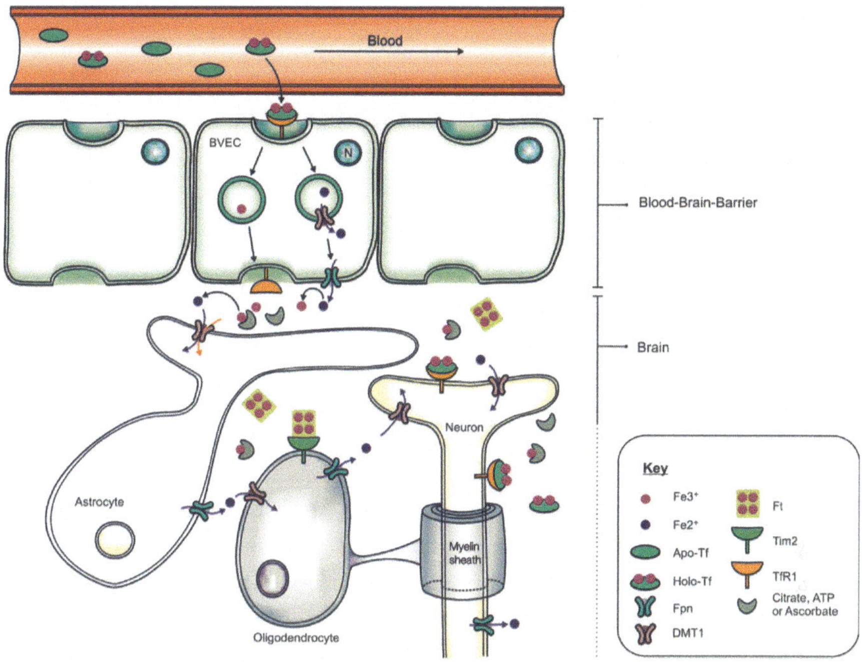

Fig. 1 Schematic representation of iron transport in the central nervous system. *Apo-Tf*, iron-free transferrin; *Holo-Tf*, iron-loaded transferrin; *Fpn*, ferroportin; *DMT1*, divalent metal transporter 1; *Ft*, ferritin; *Tim2*, T-cell immunoglobulin and mucin domain-containing protein-2 (ferritin receptor); *TfR1*, transferrin receptor 1. *Reprinted with permission from Ashraf, A., Clark, M., & So, P.-W. (2018). The aging of iron man.* Frontiers in Aging Neuroscience, 10*, 65. https://doi.org/10.3389/fnagi.2018.00065 (Fig. 2), Copyright 2018 licensed under* CC BY 4.0.

hand, the BCSFB is composed of epithelial cells from the choroid plexus. This cerebral region is characterized by fenestrated capillaries through which iron-bound transferrin is able to cross. Also, the choroid plexus cells synthesize the important iron-related proteins previously described, such as Tf itself, TfR, DMT1, and Fpn1. However, it is suggested that the BCSFB mainly contributes to iron regulation through its removal through iron export mechanisms and clearance via interstitial fluid convective flow (Qian & Ke, 2019; Roy Sarkar & Dutta, 2019). Other than Tf dependent transport, NTBI, which is usually bound to citrate, ascorbate, or ATP, has an important role as it is the main source of iron for astrocytes, oligodendrocytes, and microglia (DeGregorio-Rocasolano, Martí-Sistac, & Gasull, 2019; Knutson, 2019).

Iron regulation is also effectuated by iron regulatory proteins (IRPs) at the post-transcriptional level; in case of iron insufficiency, IRPs bind specific iron responsive elements (IREs) at post-transcriptional level, in the 3' untranslated region of transferrin receptor mRNA (Joppe, Roser, Maass, & Lingor, 2019). This protects the TfR transcript from being degraded by nucleases and henceforth iron uptake is favored. The opposite occurs once iron reaches higher levels (Möller et al., 2019; Testi, Boffi, & Montemiglio, 2019).

3. Iron and neurodegeneration

Iron overload can lead to a variety of detrimental consequences, such as oxidative stress, cell death, and neurodegeneration (Fig. 2) (Chang, 2019; Han et al., 2020; Healy, McMahon, & FitzGerald, 2017; Ingrassia et al., 2019; Mills et al., 2010; Sfera et al., 2018). In addition, iron has the ability to induce the aggregation of some intrinsically disordered proteins, as it may interact with susceptible residues and thus promote the destabilization and unfolding of the usual suspects of some of the most known neurodegenerative diseases (Dekens, De Deyn, Sap, Eisel, & Naudé, 2020; Joppe et al., 2019; Liu, Fan, Yang, Wang, & Guo, 2018; Muhoberac & Vidal, 2019). As if this weren't enough, the aggregation process itself may promote the

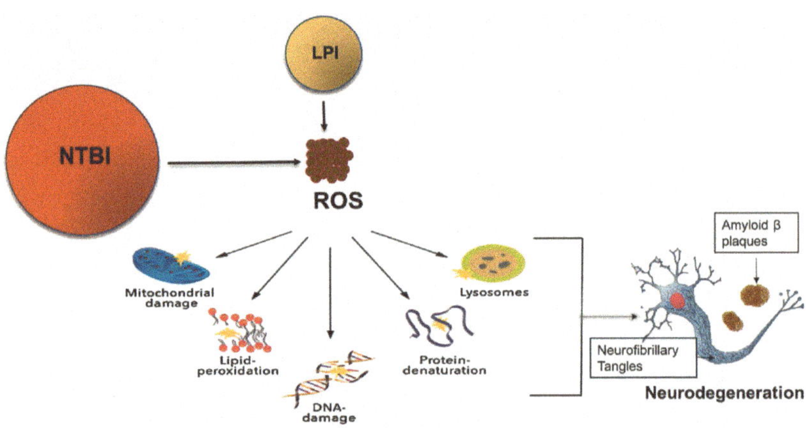

Fig. 2 Illustration of the intracellular damage caused by reactive oxygen species. *LPI*, labile plasma iron; *NTBI*, non-transferrin-bound iron; *ROS*, reactive oxygen species. *Reprinted with permission from Bagwe-Parab, S., & Kaur, G. (2020). Molecular targets and therapeutic interventions for iron induced neurodegeneration.* Brain Research Bulletin, 156, 1–9. *https://doi.org/10.1016/j.brainresbull.2019.12.011 (Fig. 1), Copyright 2019 Elsevier Inc.*

formation of new iron binding sites that become centers of reactive oxygen species (ROS) production (Muhoberac & Vidal, 2019). A thorough compilation of studies regarding the impact of iron in the neurodegenerative process has been made by (Zanella et al., 2020). Table 1 summarizes the relationship between iron and several neurodegenerative diseases that will be described in the following subsections.

3.1 Oxidative stress and ferroptosis

As a product of the high energetic consumption by the brain, a great amount of ROS are generated, which contribute to a condition known as oxidative stress (Chen et al., 2020). This condition is caused by the generation of free radicals by the disruption of the mitochondrial respiratory chain. Iron is able to donate electrons to oxygen, which is useful for redox balance but in case of overload, ROS such as hydroxyl radicals can be produced via the Fenton reaction (Eq. 1) when iron is in its Fe(II) form (Chang, 2019).

$$Fe^{2+} + H_2O_2 \rightarrow Fe^{3+} + \cdot OH + HO^- \quad (1)$$

Cell death may occur when pathways such as caspase-3-dependent apoptosis are triggered by iron accumulation, with subsequent damage to the mitochondrial membrane potential. However, a "new" type of regulatory cell death has been suggested to be exclusively activated in iron disorders: ferroptosis. The main events that take place in this particular type of cell death include the formation of ROS, lipid peroxidation, iron accumulation, inflammation, and redox homeostasis impairment (Han et al., 2020; Li et al., 2020; Mills et al., 2010; Muhoberac & Vidal, 2019). Several pathways for ferroptosis have been proposed, all of them eventually converging into lipid peroxidation. This is explained in the simplest terms by the aforementioned Fenton reaction, in which a ferrous-excessive ambient promotes the formation of ROS that cause damage to the lipids that form the cell membrane (Li et al., 2020). The main regulator in ferroptosis is glutathione peroxidase 4 (GPX4), an enzyme that repairs lipid peroxides by reducing them. However, iron overload is one of the causes of loss of GPX4 enzymatic activity and disruption of the synthesis of its precursor, glutathione (GSH), the main endogenous antioxidant molecule. The damage of proteins such as Ca^{2+}-ATPase, glutamate transporter, Na^+/K^+-ATPase, and N-methyl-D-aspartate (NMDA) receptor caused by intracellular iron accumulation is also a feature of ferroptosis (Hare et al., 2013). Even though there haven't been defined yet which are the specific markers associated to this process, the

Table 1 Summary of neurodegenerative diseases and their relationship with iron and iron transport proteins.

	Relationship with iron	Involved iron transport proteins	Main affected brain regions/components	References
Alzheimer's disease (AD)	Iron accumulation Iron-induced aggregation of amyloid-β protein Ferroptosis	Transferrin Ceruloplasmin Ferritin	Hippocampus Parietal cortex Motor cortex	(Ayton et al., 2018, 2017; Chang, 2019; Dekens et al., 2020; Diouf, 2020; Guan et al., 2020; Joppe et al., 2019; Liu et al., 2018; Roy Sarkar & Dutta, 2019; Sabbir, 2018; Sfera et al., 2018; Taniguchi et al., 2008; Uranga & Salvador, 2018)
Parkinson's disease (PD)	Iron accumulation Restless leg syndrome Fatigue	Transferrin receptor 1 Divalent metal transporter 1 Ferroportin 1 Transferrin	*Substantia nigra* (neurons and glia)	(Abeyawardhane & Lucas, 2019; Cerri et al., 2019; Chang, 2019; Leung et al., 2020; Piao et al., 2017; Qian & Ke, 2019; Roy Sarkar & Dutta, 2019; Sun et al., 2018; Vellieux & d'Ortho, 2020; Yu et al., 2018; Zuo et al., 2016)
Huntington's disease (HD)	Excessive intracellular iron Ferroptosis		Oligodendrocytes	(Agrawal et al., 2018; Roy Sarkar & Dutta, 2019; Sánchez-Castañeda et al., 2015; van Bergen et al., 2016)

Multiple sclerosis (MS)	Iron accumulation Demyelination		Gray matter Myelin Oligodendrocytes Astrocytes	(Hametner et al., 2013; Ropele et al., 2017; Roy Sarkar & Dutta, 2019; Siotto et al., 2019; Stankiewicz et al., 2014)
Sanfilippo syndrome	Iron accumulation	Ferritin	Microglia	(Fedele, 2015; Heon-Roberts et al., 2020; Hodroge et al., 2016; Kumari, 2018; Puy et al., 2018)

current research in this subject allows to summarize the key players of ferroptosis into reactive iron (Fe^{2+}) accumulation, lipid peroxidation, and a defective lipid peroxide repair system (i.e. GPX4) (Han et al., 2020). This particular cell death mechanism has been determined to be one of the main factors contributing to the development of neurodegeneration (Han et al., 2020; Knutson, 2019).

3.2 Damage to the blood-brain barrier (BBB)

It has been shown that systemic iron overload can cause damage to this structure if it reaches the brain parenchyma. For instance, after ischemic events such as strokes, elevated levels of Tf and TfR expression have been observed in cerebral regions such as the hippocampus (DeGregorio-Rocasolano et al., 2019).

A way of determining if the BBB has been damaged is by the detection of iron in the brain as a reflection of an increased iron intake (McCarthy & Kosman, 2015). Since iron has unpaired electrons at its farthest shell, it responds to external magnetic fields, a phenomenon known as magnetic susceptibility. This is useful for detecting iron compounds in the brain *in vivo* through non-invasive techniques such as Magnetic Resonance Imaging (MRI) and quantitative susceptibility mapping (QSM) (Acosta-Cabronero, Betts, Cardenas-Blanco, Yang, & Nestor, 2016; Möller et al., 2019; Ropele, Enzinger, & Fazekas, 2017; Rouault, 2013).

3.3 Alzheimer's disease (AD)

Alzheimer's disease is characterized by the deposition of neurofibrillary tangles and amyloid plaques formed by the aggregation of amyloid-β protein (Aβ) in cerebral regions such as the hippocampus, the parietal cortex, and the motor cortex (Sabbir, 2018). Around these, iron deposits have also been found, indicating that iron associates with the metal-binding sites of Aβ, catalyzing its aggregation. A study made by QSM showed this association between iron increase and Aβ deposition in the brain. Higher QSM levels measured in the hippocampus and the temporal and frontal lobes were related to low cognitive performance in subjects with Aβ pathology (Ayton et al., 2017). Intra-neuronal iron accumulation appears to be also favored by the presence of Aβ *per se*, since destabilization of Fpn and blockage of iron export channels have been observed (Sfera et al., 2018). Furthermore, the oxidative stress caused by iron accumulation contributes to the production of Aβ (Uranga & Salvador, 2018). All in all, the aggregation of Aβ protein causes the most neuronal damage, since these oligomers block the synaptic communication when they accumulate in the interneuronal junctions (Chang, 2019). Notwithstanding, iron seems to be a key factor around it. A supporting finding is that ferritin expression increases with age and particularly in Alzheimer's disease, hence promoting iron accumulation in the brain and progression in AD symptoms (Ayton, Diouf, & Bush, 2018; Roy Sarkar & Dutta, 2019).

Recent studies have determined that transferrin levels in plasma (Guan et al., 2020) and ceruloplasmin (Diouf, 2020) and ferritin (Ayton et al., 2018) concentrations in cerebrospinal fluid are all related to cognitive decline in AD patients, supporting the current notion about iron metabolism's role in AD pathogenesis. Furthermore, transferrin had already been proposed as a biomarker for identifying AD, as it was found that in this pathologic condition, transferrin glycosylations were altered and thus lost binding activity (Taniguchi et al., 2008).

3.4 Parkinson's disease (PD)

In Parkinson's disease, iron accumulates in a cerebral region known as the *substantia nigra*, including neurons and glia from that region, as evidenced by higher echogenicity during MRI studies (Yu et al., 2018). Studies in mice have related this accumulation to a dysregulation on the intake and elimination of iron, mainly due to increased activity of TfR1 and DMT1, augmented expression of lactoferrin receptor on neurons, and reduced expression of Fpn1 (Qian & Ke, 2019; Roy Sarkar & Dutta, 2019). Further, some studies suggest that in Parkinson's disease, iron import is augmented uncontrollably by oxidation of iron-sulfur clusters, promoting the accumulation of transferrin and thus of iron in mitochondria (Cerri, Milanese, & Mastroberardino, 2019).

A comparative study (Zuo et al., 2016) with geriatric patients with PD evaluated, among other factors, the levels of iron and its related proteins and their participation in one of the symptoms observed in most cases of PD: fatigue. Fatigue can be manifested as mental fatigue, physical fatigue, or both. In this study, samples taken from cerebrospinal fluid (CSF) of PD patients with mental fatigue showed significantly higher levels of iron and transferrin, in contrast with control samples. Also, samples taken from blood serum had decreased transferrin levels, inferring that most of it had reached the brain due to the lack of restriction from a damaged BBB. These observations suggested that an impaired iron regulation might be a potential mechanism of PD with mental fatigue (Zuo et al., 2016).

Some PD patients also manifest a condition called Restless Leg Syndrome (RLS), where abnormalities in iron distribution have also been observed (Vellieux & d'Ortho, 2020). A study made by (Piao et al., 2017) revealed that iron levels measured in CSF and peripheral blood in PD-RLS patients were found to be deficient, compared to PD patients without RLS. Paradoxically, Tf levels in CSF were found to be significantly increased; therefore, a possible explanation is that the iron carried by said Tf is not reaching the sites in the brain related to PD-RLS. An observation that supports this is the improvement in RLS patients who underwent iron supplementation treatments (Leung, Singh, McWilliams, Stockler, & Ipsiroglu, 2020).

Dopamine (DA) is the main neurotransmitter involved in Parkinson's disease; its deficiency is usually suggested to be one of the main players in this neurodegenerative condition. Iron, being a cofactor of tyrosine hydroxylase, contributes to the regulation of DA synthesis; hence, a deficit in iron concentration in the brain leads to a lessened production of DA (Piao et al., 2017). However, DA interaction with reactive iron (i.e. ferrous iron) results in the formation of protein-modifying quinones and in the aggravation of oxidative stress (Sun, Pham, Hare, & Waite, 2018), an event also triggered, although through different pathways, by the formation of H_2O_2 after DA's enzymatic hydrolysis by monoamine oxidase (Abeyawardhane & Lucas, 2019; Chang, 2019).

3.5 Huntington's disease (HD)

Energy metabolism is principally affected in Huntington's disease from its early stages, as evidenced by MRI studies where brain iron metabolism is modified. Post-mortem studies confirm the role of iron overload, as an excess of intracellular iron has been found in HD patients' brains, compared to similarly aged controls (Roy Sarkar & Dutta, 2019). Some studies have shown decreased levels of iron concentration in the progress of HD, possibly due to a redistribution of brain iron caused by the increased density of oligodendrocytes that occurs as a response to compensate myelin and neuronal loss (van Bergen et al., 2016). A previous study suggests similarities to this, as it was observed that in preliminary stages of HD, iron is usually found to be in excess, but as the disease evolves, there is an apparent decrease; however, this phenomenon may indicate that the iron is being redistributed (Sánchez-Castañeda et al., 2015). Other findings have found increased mitochondrial iron uptake in HD patients. However, this iron is not utilized, but accumulated in a labile form which in turn may be the responsible for mitochondrial function disruption. This is also supported by a higher rate of oxidative stress evidenced by lipid peroxidation and ferroptosis in HD mice and human brains with iron overload (Agrawal, Fox, Thyagarajan, & Fox, 2018). All in all, the specific role of iron in HD pathogenesis is yet unclear, but definitely present, mostly at early stages.

3.6 Multiple sclerosis (MS)

Neurodegeneration in multiple sclerosis is usually attributed to inflammation, accompanied by other features such as demyelination and axonal damage (Stankiewicz, Neema, & Ceccarelli, 2014). In cases of MS patients with low disability, oxidative stress has been evidenced as a potential marker and therapeutic target (Siotto et al., 2019). These factors all share a relationship with iron management. MRI studies have revealed elevated iron concentration in multiple sclerosis patients' brains *in vivo*, particularly in gray matter and in the vicinity of MS lesions. These findings may provide an explanation to myelin loss and mitochondrial malfunction associated to this neurological condition (Ropele et al., 2017; Roy Sarkar & Dutta, 2019). Some authors have suggested that inflammatory cytokines may promote iron liberation from oligodendrocytes and astrocytes during the neuroinflammatory state associated to MS. This idea is supported by the findings, in histological and MRI analyses, of free iron in the extracellular space following the destruction of oligodendrocytes and myelin in active MS lesions (Hametner et al., 2013).

3.7 Sanfilippo syndrome

Contrarily to most neurodegenerative diseases, Sanfilippo syndrome affects the pediatric population and its etiology is genetic, as it shows an autosomal recessive inheritance pattern (Fedele, 2015; Heon-Roberts et al., 2020; Kumari, 2018). However, brain iron accumulation is a common factor with other "adult" neurological diseases (Hodroge et al., 2016; Puy et al., 2018). This pediatric neurodegenerative disease is characterized by the accumulation of abnormal heparan sulfate oligosaccharides (HSOs) in several tissues including the CNS (Fedele, 2015; Hodroge et al., 2016), which alters the child's overall development, causes cognitive decline, and is usually fatal within the first twenty years of life (Heon-Roberts et al., 2020; Kumari, 2018). The accumulation of HSOs increases the production of several pro-inflammatory molecules in microglia, as well as the expression of hepcidin, a peptide responsible for downregulating iron, as evidenced in a mice model study (Puy et al., 2018). This study also found abnormalities in ferritin expression possibly due to other iron dysregulation pathways yet to be clarified.

4. Iron transport related proteins

In this section, we briefly describe some of the main proteins involved in the iron transport in the CNS (Fig. 3), their role in homeostasis and in disease. Table 2 summarizes them in terms of location and function.

4.1 Transferrin receptors 1 and 2 (TfR1 and TfR2)

Also known as Cluster of Differentiation 71 (CD71), the TfR1 is an essential homodimeric protein for cellular iron uptake. It is constituted by a cytoplasmic domain, a transmembrane domain, and an extracellular domain, where it binds ferric iron-loaded transferrin (Tf) (Testi et al., 2019). Binding of Tf to TfR1 involves a conformational change in TfR1, specifically at its dimer interface (Eckenroth, Steere, Chasteen, Everse, & Mason, 2011).

It is thanks to this receptor that transferrin is able to traverse cells, as both proteins form a complex that is internalized in a process known as receptor-mediated endocytosis, which takes place via a clathrin dependent pathway.

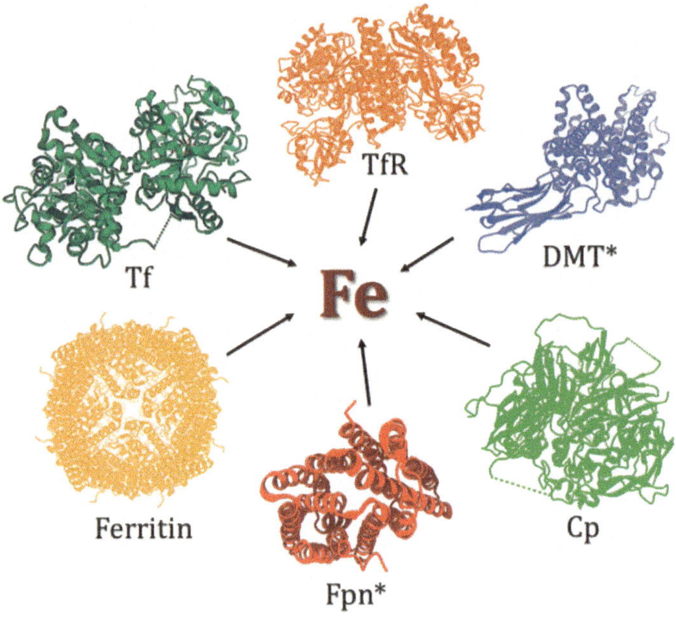

Fig. 3 Illustrative representation of the main proteins related to iron transport in the CNS. *Tf*, human serum transferrin (PDB: 4 × 1B); *TfR*, transferrin receptor ectodomain (PDB: 1CX8); Ferritin (PDB: 5N27); *Cp*, ceruloplasmin (PDB: 4ENZ); *Fpn*, ferroportin (PDB: 5AYM); *DMT*, divalent metal transporter (PDB: 4WGW). *Proteins from non-human sources.

Table 2 Summary of the main iron transport proteins, their location and function in human brain iron homeostasis.

	Location	Function	References
Transferrin (Tf)	Plasma (bloodstream)	Iron transport in blood circulation	(Fernandes et al., 2020; Kawabata, 2019; Möller et al., 2019)
Lactoferrin (LTf)	Glandular epithelial cells	Iron binding from Tf during inflammation	(Bonaccorsi di Patti et al., 2018; García-Montoya et al., 2012; Lambert, 2012; Liu et al., 2018; Page, 2019)
Melanotransferrin (MTf)	Tumor cells (melanomas) Cerebrospinal fluid (in AD)	Iron binding for tumor cells	(Ashraf et al., 2019; Chang, 2019; Lambert, 2012)
Transferrin receptor (TfR)	Cell Surface (endothelial cells of the BBB)	Receptor for cellular iron uptake Recognition of Tf on the cell surface	(Eckenroth et al., 2011; Lambert, 2012; Mills et al., 2010; Testi et al., 2019)
Divalent Metal Transporter 1 (DMT1)	Early endocitic compartments (astrocytes)	Metal transporter Uptake of iron from circulating Tf Internalization of NTBI in endosomes	(Bonaccorsi di Patti et al., 2018; Hohnholt & Dringen, 2013; Ingrassia et al., 2019; Knutson, 2019; Mills et al., 2010; Qian & Ke, 2019)
Ferroportin 1 (Fpn1)	Abluminal membrane of the brain (neurons and astrocytes)	Iron export	(Chang, 2019; Hare et al., 2013; Mills et al., 2010; Qian & Ke, 2019; Roy Sarkar & Dutta, 2019)
Ferritin	Neurons Microglia Oligodendrocytes	Iron storage	(Chang, 2019; Möller et al., 2019; Muhoberac & Vidal, 2019; Rouault, 2013; Todorich et al., 2008)

(Continued)

Table 2 Summary of the main iron transport proteins, their location and function in human brain iron homeostasis.—cont'd

	Location	Function	References
Hephaestin (Heph)	Cytoplasm (astrocytes) Epithelial cells (choroid plexus)	Ferroxidase; oxidizes ferrous iron to ferric iron in the intracellular milieu. Aids iron export by Fpn1.	(Dlouhy et al., 2019; Jiang et al., 2015; Rouault, 2013; Yanase et al., 2020)
Ceruloplasmin (Cp)	Cerebrospinal fluid	Ferroxidase in the extracellular milieu. Removes iron from Fpn1. Binds to LTf during inflammation.	(Bonaccorsi di Patti et al., 2018; Diouf, 2020; Jiang et al., 2015; Sokolov et al., 2017; White et al., 2012)

TfR1 is expressed on the surface of virtually every cell type, mostly in highly differentiated and rapidly dividing cells, immature erythroid cells and placental tissue; but also in endothelial cells of the blood-brain barrier, hepatocytes, reticulocytes and hepatocytes (Lambert, 2012; Mills et al., 2010; Testi et al., 2019). TfR2 is a paralog to TfR1, which means that it develops a different function although it originates from a duplicated gene. In contrast to TfR1, TfR2 is a transmembrane protein expressed only in hepatic, duodenal, and erythroid cells. Also, an important difference is that TfR2 doesn't have many regulative mechanisms, as its transcript doesn't include iron responsive elements (IREs) and therefore there is no downregulation of TfR2 in the event of iron overload (Lambert, 2012).

4.2 Divalent metal transporter 1 (DMT1)

Localized in the early endocytic compartments, DMT1 is the main transmembrane transporter for metal ions such as Fe^{2+}, Mn^{2+}, Co^{2+}, and Cd^{2+}. Overall, its main functions in iron metabolism involve the uptake of iron from dietary ingestion, as it is expressed in the apical membrane of enterocytes in the proximal small intestine (Bonaccorsi di Patti et al., 2018), and in developing erythroid cells of the bone marrow, assimilating iron that comes from circulating transferrin (Knutson, 2019). In most mammalian cells, DMT1 is responsible for releasing iron obtained from the Tf-TfR1 cycle, once it is reduced from Fe^{3+} to Fe^{2+} in the endosome, to the cytosol (Mills et al., 2010). DMT1 is expressed in astrocytes, thus playing part in the uptake of ferrous iron (Hohnholt & Dringen, 2013). It has also been observed that DMT1 is able to transport another form of circulating iron known as Non-Transferrin Bound Iron (NTBI), which consists of iron being bound to anions such as citrate or ascorbate, or to other proteins such as albumin, ferritin, or lactoferrin (Qian & Ke, 2019). When Tf transport is impaired or absent, DMT1 becomes the main mediator of iron by internalizing NTBI in the endosome (Knutson, 2019). In relation to neurodegeneration, some findings have shown an increase in the expression of DMT1; however, this event is not enough to cause neurodegeneration by itself. On the other hand, DMT1 increase and iron accumulation have been found in the *substantia nigra* of PD patients, indicating a potential role in the pathogenesis of PD (Ingrassia et al., 2019).

4.3 Ferroportin 1 (Fpn1)

Fpn1 is the protein responsible for iron export across the abluminal membrane (i.e. the brain side of the brain interstitial membrane that separates it from the bloodstream) (Mills et al., 2010; Qian & Ke, 2019). This transmembrane protein is expressed in neurons and astrocytes, and it is assisted by the ferroxidase hephaestin, which converts the divalent iron into ferric iron, so that iron efflux is accomplished (Chang, 2019; Hare et al., 2013). Iron accumulation has also been proposed to be due to the activation of N-metyl-D-aspartate (NMDA) receptor, related partly to a reduced Fpn1 expression (Qian & Ke, 2019; Roy Sarkar & Dutta, 2019).

4.4 Ferritin

A much structurally larger molecule, Ferritin is the main iron storage protein, conformed by two types of subunits, H type (heavy) and L type (light), which co-assemble into a supramolecular spherical-shaped protein (Chang, 2019). In the CNS, the expression of these subunits varies, as the H type is mainly expressed in neurons, the L type in microglia, and a combination of both in oligodendrocytes (Möller et al., 2019). Oligodendrocytes need a great amount of iron for myelin synthesis, which explains the significantly higher expression of ferritin and its receptor, T-cell immunoglobulin and mucin domain-containing protein 2 (Tim-2) (Todorich et al., 2008). Ferritin is able to store up to 4,500 iron atoms in a non-reactive state, taking the excess of ferrous iron from the labile iron pool (LIP) to avoid the generation of reactive oxygen species (ROS), the aggregation of proteins, and the formation of iron deposits, which may all cause cellular damage. Additionally, it also serves as an arsenal of iron for when it is needed for physiological functions. The mechanisms by which iron is released from ferritin consist on degradation of this protein, either by lysosomes or by proteasomes (Muhoberac & Vidal, 2019; Rouault, 2013).

4.5 Hephaestin (Heph)

Heph is a ferroxidase expressed in the cytoplasm of human brain astrocytes and in the epithelial cells of the choroid plexus (Yanase et al., 2020). It is the main enzyme responsible for oxidizing iron from its ferrous form (Fe^{2+}) to its ferric, non-toxic form (Fe^{3+}) and thus allowing the binding of iron to Tf (Rouault, 2013). The importance of this protein resides in its function, as it works hand to hand with Fpn1 to regulate iron export (Dlouhy, Bailey, Steimle, Parker, & Kosman, 2019). A study using Heph knockout mice demonstrated its essential role in brain iron homeostasis, considering its interaction with other iron regulatory proteins, as the lack of this iron export-related protein led to iron accumulation in several brain regions, and a consequential deficiency in other cell types (Jiang et al., 2015).

4.6 Ceruloplasmin (Cp)

This protein has been proposed as a potential marker for neurodegeneration due to its increased concentration in cerebrospinal fluid in patients with amyloid pathology and mild cognitive decline (Diouf, 2020). Ceruloplasmin is an important regulator of iron toxicity due to its function as a ferroxidase in the extracellular milieu, since it removes iron from ferroportin after iron export to prevent its saturation and degradation (Bonaccorsi di Patti et al., 2018). Some studies have shown that it works in a slightly different but interrelated manner with the aforementioned hephaestin (Jiang et al., 2015; White et al., 2012). Additionally, it is a key player during inflammation, a relevant factor of several diseases, during which it readily binds to lactoferrin (Sokolov et al., 2017), an iron-binding protein involved in the immune response which we will review later on.

5. Transferrins

Transferrins are described separately from the rest of the iron-transport proteins reviewed so far because we want to highlight the relevance of human serum transferrin (Tf), the main iron transport protein, as a potential drug delivery system, as it will be further discussed in the following section.

Transferrins are a family of glycoproteins found in several organisms whose main function is to transport iron for its utilization in different cells (Lambert, 2012). Lactoferrin (LTf) is mainly expressed by glandular epithelial cells and neutrophils, and it is found in bodily secretions such as milk, saliva, and tears (Page, 2019). It plays an important role in the immune response toward infection, as it binds to bacterial lipopolysaccharides to induce peroxidation and trigger cell lysis (García-Montoya, Cendón, Arévalo-Gallegos, & Rascón-Cruz, 2012; Page, 2019) and to iron to prevent its utilization by invading pathogens (Bonaccorsi di Patti et al., 2018). In fact, lactoferrin's affinity to iron is 300-fold higher than transferrin's, especially in acidic conditions, so that iron from circulating Tf is transferred to LTf during inflammation (Lambert, 2012; Page, 2019). Interestingly, studies have observed that LTf plays a neuroprotective role in neurodegenerative diseases, particularly in AD, through several mechanisms such as metal chelation, regulation of inflammatory processes, and cell autophagy, making LTf a promising molecule of study for targeting neurodegeneration (Liu et al., 2018).

Another member of the transferrin family is melanotransferrin (MTf), primarily found in human tumor cells known as melanomas, although expression of this protein has also been found in normal tissues such as salivary glands, sweat glands, liver, intestine, and umbilical cord (Chang, 2019). A recent study has implied MTf as a potential marker for AD given its altered distribution in cerebrospinal fluid in patients with mild cognitive decline, a preliminary stage of this disease (Ashraf, Alepuz Guillen, Aljuhani, Hubens, & So, 2019). In MTf, iron binding only occurs in its N-lobe, it is neither recognized by TfR1 nor TfR2, and its function under normal conditions remains unclear. However, it has been proposed that its expression on the surface of melanoma cells may allow their crossing through the blood brain barrier to unchain metastasis (Lambert, 2012).

5.1 Human serum transferrin (Tf)

Also known as serotransferrin, transferrin is the main iron transport protein involved in iron homeostasis, as it travels through the bloodstream picking up free iron in its Fe(III) form and delivers it to cells through its binding to the TfR1 (Kawabata, 2019; Möller et al., 2019). It is thanks to transferrin's structural and functional properties that it has become an attractive candidate for novel targeted drug-delivery systems (Fernandes et al., 2020). Transferrin is composed of two lobes, the amino-terminal (N) and the carboxyl-terminal (C), each divided into two subdomains: N1 and N2, and C1 and C2. Both lobes are connected by a bridge formed by residues 332 through 338, at the N1 and C1 subdomains. A ribbon representation of transferrin's overall structure is presented in Fig. 4. These lobes are structurally homologous, but slightly different in terms of their function. For instance, it has been observed that the N-lobe is mainly focused on carrying iron, whereas the C-lobe is majorly involved in recognition by the TfR (Eckenroth et al., 2011; Yang, Zhang, Wang, Hao, & Sun, 2012).

Tf binds two ferric ions to each of its metal-binding sites with very high affinity, though reversibly (Chang, 2019). Some studies have demonstrated its ability to bind different metals other than iron, indicating a certain structural and coordination-bond flexibility in its metal-binding site (Adam, Trumm, Smith, MacGillivray, & Panak, 2018; Bewley et al., 1999; Yang et al., 2012). The metal-binding and release mechanisms of transferrin have been broadly described throughout the last few decades (Adams et al., 2003; Hirose, 2000; Jeffrey et al., 1998; Kawabata, 2019; H. Li & Qian, 2002; MacGillivray et al., 2000, 1998; Wally et al., 2006; Wang et al., 2015). Briefly, a conformational change in which both domains

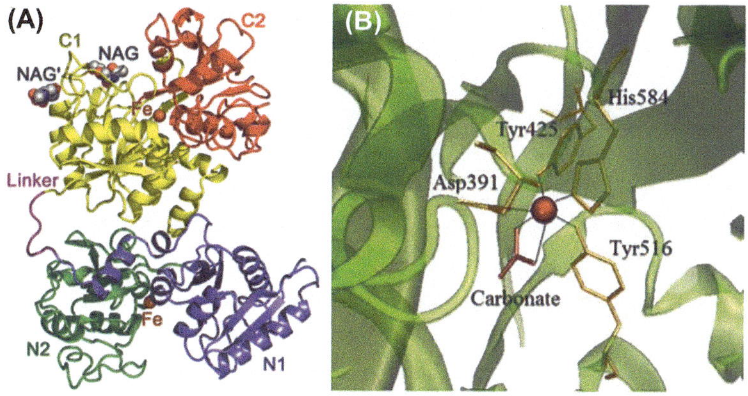

Fig. 4 Crystallographic structure of human serum transferrin. (A) Overall structure of Tf with iron bound to both of its iron-binding sites. Its subdomains are represented as N1 lobe in blue, N2 lobe in green, C1 lobe in yellow, and C2 lobe in red, and the linker peptide is shown in purple. Iron ions are represented as brown spheres and N-acetylglucosamine moieties (NAG and NAG′) are shown in sphere models. (B) Transferrin metal binding site. Metal ion is represented in a red sphere. *(A) Reprinted with permission from Springer Nature Service Center GmbH: Springer Nature, Yang, N., Zhang, H., Wang, M., Hao, Q., & Sun, H. (2012). Iron and bismuth bound human serum transferrin reveals a partially-opened conformation in the N-lobe.* Scientific Reports, 2*(1), 999. https://doi.org/10.1038/srep00999 (Fig. 1A), Copyright 2012. (B) Reprinted with permission from Duca, G., Anghel, L., & Erhan, R. V. (2018). Structural aspects of lactoferrin and serum transferrin observed by ftir spectroscopy.* Chemistry Journal of Moldova, 13*(1), 111–116. https://doi.org/10.19261/cjm.2018.482 (Fig. 1D) by Copyright 2018 licensed under CC BY 4.0.*

that conform the N lobe, identified as N1 and N2, go through a rigid body rotation, implies the burial of a region and a simultaneous exposure of another that allows solvent accessibility (Jeffrey et al., 1998).

Structural studies suggest that there are three basic conformational stages in the tertiary structure of transferrin: fully-closed, partially-opened, and fully-opened. These conformations are governed by interactions between the protein and its environment and with the formation and dissociation of the coordination bonds between the metal and the transferrin's metal-binding pocket residues. Knowledge of the partially-opened conformation of the N-lobe is of special relevance given that it has allowed to describe the mechanism by which metals bind to transferrin. A comparison made between the isolated N-lobe of serotransferrin and the isolated N-lobe of lactoferrin revealed a highly conserved nature in the mechanism that involves the conformational change between the open and closed state, indicating that ligand binding in the different members of the transferrin family follows practically the same mechanism (Jeffrey et al., 1998).

Iron release from the transferrin molecule has been proposed to be activated by a "dilysine-trigger" conformed by Lys206 and Lys296. In the iron-loaded protein, these lysines are bound by a hydrogen bond; when the pH decreases, Lys206 is protonated, leading to a destabilization of the closed conformation by repulsive forces (Jeffrey et al., 1998; Yang et al., 2012). Nonetheless, it has been observed that this "dilysine trigger" is only activated in the N-lobe, whereas in the C-lobe, the mechanism of iron release is mediated by the binding of His349 to the TfR1 and the protonation by the acid environment. When iron release is completed, a series of events takes place to ensure the stabilization of the apo-Tf molecule within the Tf-TfR1 complex before its release to blood circulation. This stabilization is made possible by the protonation of His349, as well as of His318, which favor cation-π interactions with residues Phe760 and Trp641 of the TfR1, respectively. Therefore, throughout the Tf-TfR1 complex's passage within the endosome, Tf remains bound to the TfR1 by its C1 subdomain, and the acidification that causes iron to be released is the same that allows the stabilization of apo-Tf within the complex with its receptor (Eckenroth et al., 2011).

6. Perspectives on the use of transferrin as a drug delivery system

The majority of the strategies developed so far for targeting the CNS have proven a low to null success rate in clinical trials by several reasons; namely, the strategies *per se*, as they are usually designed to avoid or disrupt the BBB, or because they used ineffective vehicles altogether. These strategies are thoroughly described by (Pardridge, 2020).

Transport proteins are thus attractive potential drug delivery systems, as they generally have well-described transport mechanisms, they're usually found in the bloodstream, and they have some degree of flexibility that allows them to bind ligands. Particularly, transferrin possesses promising physicochemical properties to propose it as such a system (Fernandes et al., 2020). Within the transferrin family, lactoferrin has also shown promising observations regarding neuroprotection, thus making it a molecule worth studying further (Liu et al., 2018). On other viewpoints, some of the promising strategies for traversing the BBB include the reengineering of molecules to create molecular BBB Trojan horses, which are defined as monoclonal antibodies against endogenous BBB receptors such as the TfR1 or the insulin receptor (Pardridge, 2017). Taking the principle of a molecular Trojan horse, which basically mimics an endogenous transport mechanism, transferrin might be the horse altogether, carrying a drug that does not cross the BBB by itself.

A study revealed that the metal-binding site of transferrin can be somewhat "molded" by modifying the anions that complete the coordination sphere; for instance, the introduction of a malonate anion instead of a carbonate anion widens the metal-binding pocket, since malonate allows for larger angles and thus more space for molecules to bind. Albeit having a wider binding pocket, the C-lobe remains closed with minor conformational changes, something essential for transferrin to be recognized by its receptor (Wang et al., 2015). This knowledge is useful for evaluating how much room there could be for other molecules, such as drugs, to bind to the Tf metal-binding site.

A recent approach toward AD treatment was the evaluation of the interaction between human transferrin and rivastigmine tartrate, an existing Alzheimer's drug already used in the therapeutics of this disease (Shamsi et al., 2019). The findings of such study in terms of the binding sites and mechanisms may be useful for future studies using other drugs and focusing on transferrin as the delivery system.

Iron chelation has become an important strategy for targeting neurodegeneration due to iron overload and some of its derivations, such as ferroptosis (Bayır et al., 2020; Han et al., 2020). Within the main iron chelators that have been tested is deferoxamine, which could help to regulate the load of intracellular iron (Joppe et al., 2019; Liu et al., 2018). In a similar manner, deferiprone has been tested and shown positive results to alleviate some of the early motor symptoms of PD. It must be noted that most of the trials developed in this area have yielded promising results in animal models, but only mild effects in clinical trials (Han et al., 2020). Herein, transferrin could be a means of getting a more efficient delivery of iron chelators.

Other approaches involving transferrin have opted for using it as the recognition molecule, as an extension of another drug delivery system. For instance, a study by (Pinheiro et al., 2020) proposes lipid nanoparticles loaded with quercetin, a neuroprotective antioxidant flavonoid, functionalized with transferrin for a more efficient delivery. This functionalization favors the neuroprotective effect by reducing amyloid-beta aggregation, the main hallmark of Alzheimer's disease, implying that transferrin works synergistically with other drug delivery systems (Pinheiro et al., 2020). This is not the first time that transferrin has been used as a tag for recognition of the

TfR, so this strategy is also promising and supports the valuable role of transferrin in novel drug delivery systems development for targeting the CNS.

In contexts other than neurodegeneration, potential targets for transferrin-mediated drug delivery are cancer cells; because of their uncontrolled proliferation rate, they require excessive quantities of iron and thus they express up to a hundred times more TfR than normal cells (during metastases) (Testi et al., 2019).

7. Concluding remarks

The studies reviewed so far have confirmed the close relationship between iron, its transport proteins, and the so far observed causal factors of several neurodegenerative diseases. In addition, several deleterious processes, such as oxidative stress, are also linked with iron misplacement, such as the relatively recently described ferroptosis. Therefore, iron transport proteins must be paid close attention, as they may be potential therapeutic targets for neurological diseases.

The search for an adequate drug delivery system for the central nervous system is still in research, but the knowledge we possess so far will allow to propose new pathways to reach this target effectively. Several attempts are being made and drugs are being evaluated to overcome the challenge of treatment beyond the blood-brain barrier. In this review, an emphasis has been made on the relevance of human serum transferrin, the main iron transport protein, as a promising molecule for targeted drug delivery to the CNS.

Acknowledgments

The author would like to acknowledge Prof. Dr. Abel Moreno from the Instituto de Química, Universidad Nacional Autónoma de México, for the technical supervision throughout the elaboration of this chapter.

References

Abeyawardhane, D. L., & Lucas, H. R. (2019). Iron redox Chemistry and implications in the Parkinson's disease brain. *Oxidative Medicine and Cellular Longevity, 2019*, 1–11. https://doi.org/10.1155/2019/4609702.

Acevedo, K., Masaldan, S., Opazo, C. M., & Bush, A. I. (2019). Redox active metals in neurodegenerative diseases. *JBIC Journal of Biological Inorganic Chemistry, 24*(8), 1141–1157. https://doi.org/10.1007/s00775-019-01731-9.

Acosta-Cabronero, J., Betts, M. J., Cardenas-Blanco, A., Yang, S., & Nestor, P. J. (2016). In vivo MRI mapping of brain iron deposition across the adult lifespan. *Journal of Neuroscience, 36*(2), 364–374. https://doi.org/10.1523/JNEUROSCI.1907-15.2016.

Adams, T. E., Mason, A. B., He, Q.-Y., Halbrooks, P. J., Briggs, S. K., Smith, V. C., et al. (2003). The position of arginine 124 controls the rate of iron release from the N-lobe of human serum transferrin: A structural study. *Journal of Biological Chemistry, 278*(8), 6027–6033. https://doi.org/10.1074/jbc.M210349200.

Adam, N., Trumm, M., Smith, V. C., MacGillivray, R. T. A., & Panak, P. J. (2018). Incorporation of transuranium elements: Coordination of Cm(iii) to human serum transferrin. *Dalton Transactions, 47*(41), 14612–14620. https://doi.org/10.1039/C8DT02915F.

Agrawal, S., Fox, J., Thyagarajan, B., & Fox, J. H. (2018). Brain mitochondrial iron accumulates in Huntington's disease, mediates mitochondrial dysfunction, and can be removed pharmacologically. *Free Radical Biology and Medicine, 120*, 317–329. https://doi.org/10.1016/j.freeradbiomed.2018.04.002.

Ashraf, A., Alepuz Guillen, J. A., Aljuhani, M., Hubens, C., & So, P.-W. (2019). Low cerebrospinal fluid levels of melanotransferrin are associated with conversion of mild cognitively impaired subjects to Alzheimer's disease. *Frontiers in Neuroscience, 13*, 181. https://doi.org/10.3389/fnins.2019.00181.

Ashraf, A., Clark, M., & So, P.-W. (2018). The aging of iron man. *Frontiers in Aging Neuroscience, 10*, 65. https://doi.org/10.3389/fnagi.2018.00065.

Ayton, S., Diouf, I., & Bush, A. I. (2018). Evidence that iron accelerates Alzheimer's pathology: A CSF biomarker study. *Journal of Neurology, Neurosurgery & Psychiatry, 89*(5), 456–460. https://doi.org/10.1136/jnnp-2017-316551.

Ayton, S., Fazlollahi, A., Bourgeat, P., Raniga, P., Ng, A., Lim, Y. Y., et al. (2017). Cerebral quantitative susceptibility mapping predicts amyloid-b-related cognitive decline. *Brain, 140*, 2112–2119. https://doi.org/10.1093/brain/awx137.

Bagwe-Parab, S., & Kaur, G. (2020). Molecular targets and therapeutic interventions for iron induced neurodegeneration. *Brain Research Bulletin, 156*, 1–9. https://doi.org/10.1016/j.brainresbull.2019.12.011.

Bayır, H., Anthonymuthu, T. S., Tyurina, Y. Y., Patel, S. J., Amoscato, A. A., Lamade, A. M., et al. (2020). Achieving life through death: Redox Biology of lipid peroxidation in ferroptosis. *Cell Chemical Biology, 27*(4), 387–408. https://doi.org/10.1016/j.chembiol.2020.03.014.

van Bergen, J. M. G., Hua, J., Unschuld, P. G., Lim, I. A. L., Jones, C. K., Margolis, R. L., et al. (2016). Quantitative susceptibility mapping suggests altered brain iron in premanifest Huntington disease. *American Journal of Neuroradiology, 37*(5), 789–796. https://doi.org/10.3174/ajnr.A4617.

Bewley, M. C., Tam, B. M., Grewal, J., He, S., Shewry, S., Murphy, M. E. P., et al. (1999). X-ray crystallography and mass spectroscopy reveal that the N-lobe of human transferrin expressed in *Pichia pastoris* is folded correctly but is glycosylated on serine-32 [†,‡]. *Biochemistry, 38*(8), 2535–2541. https://doi.org/10.1021/bi9824543.

Bonaccorsi di Patti, M. C., Cutone, A., Polticelli, F., Rosa, L., Lepanto, M. S., Valenti, P., et al. (2018). The ferroportin-ceruloplasmin system and the mammalian iron homeostasis machine: Regulatory pathways and the role of lactoferrin. *Biometals, 31*(3), 399–414. https://doi.org/10.1007/s10534-018-0087-5.

Buijs, M., Doan, N. T., van Rooden, S., Versluis, M. J., van Lew, B., Milles, J., et al. (2017). In vivo assessment of iron content of the cerebral cortex in healthy aging using 7-Tesla T2*-weighted phase imaging. *Neurobiology of Aging, 53*, 20–26. https://doi.org/10.1016/j.neurobiolaging.2016.09.005.

Cerri, S., Milanese, C., & Mastroberardino, P. G. (2019). Endocytic iron trafficking and mitochondria in Parkinson's disease. *The International Journal of Biochemistry & Cell Biology, 110*, 70–74. https://doi.org/10.1016/j.biocel.2019.02.009.

Chang, Y.-Z. (2019). *Brain iron metabolism and CNS diseases.*

Chen, Y., Qin, C., Huang, J., Tang, X., Liu, C., Huang, K., et al. (2020). The role of astrocytes in oxidative stress of central nervous system: A mixed blessing. *Cell Proliferation, 53*(3). https://doi.org/10.1111/cpr.12781.

DeGregorio-Rocasolano, N., Martí-Sistac, O., & Gasull, T. (2019). Deciphering the iron side of stroke: Neurodegeneration at the crossroads between iron dyshomeostasis, excitotoxicity, and ferroptosis. *Frontiers in Neuroscience, 13*, 85. https://doi.org/10.3389/fnins.2019.00085.

Dekens, D. W., De Deyn, P. P., Sap, F., Eisel, U. L. M., & Naudé, P. J. W. (2020). Iron chelators inhibit amyloid-β-induced production of lipocalin 2 in cultured astrocytes. *Neurochemistry International, 132*, 104607. https://doi.org/10.1016/j.neuint.2019.104607.

Diouf, I. (2020). Cerebrospinal fluid ceruloplasmin levels predict cognitive decline and brain atrophy in people with underlying β-amyloid pathology. *Neurobiology of Disease, 7*.

Dlouhy, A. C., Bailey, D. K., Steimle, B. L., Parker, H. V., & Kosman, D. J. (2019). Fluorescence resonance energy transfer links membrane ferroportin, hephaestin but not ferroportin, amyloid precursor protein complex with iron efflux. *Journal of Biological Chemistry, 294*(11), 4202–4214. https://doi.org/10.1074/jbc.RA118.005142.

Duca, G., Anghel, L., & Erhan, R. V. (2018). Structural aspects of lactoferrin and serum transferrin observed by ftir spectroscopy. *Chemistry Journal of Moldova, 13*(1), 111–116. https://doi.org/10.19261/cjm.2018.482.

Duck, K. A., Simpson, I. A., & Connor, J. R. (2017). Regulatory mechanisms for iron transport across the blood-brain barrier. *Biochemical and Biophysical Research Communications, 494*(1–2), 70–75. https://doi.org/10.1016/j.bbrc.2017.10.083.

Eckenroth, B. E., Steere, A. N., Chasteen, N. D., Everse, S. J., & Mason, A. B. (2011). How the binding of human transferrin primes the transferrin receptor potentiating iron release at endosomal pH. *Proceedings of the National Academy of Sciences, 108*(32), 13089–13094. https://doi.org/10.1073/pnas.1105786108.

Fedele, A. (2015). Sanfilippo syndrome: Causes, consequences, and treatments. *The Application of Clinical Genetics, 269*. https://doi.org/10.2147/TACG.S57672.

Fernandes, M. A., Hanck-Silva, G., Baveloni, F. G., Oshiro Junior, J. A., de Lima, F. T., Eloy, J. O., et al. (2020). A review of properties, delivery systems and analytical methods for the characterization of monomeric glycoprotein transferrin. *Critical Reviews in Analytical Chemistry*, 1–12. https://doi.org/10.1080/10408347.2020.1743639.

García-Montoya, I. A., Cendón, T. S., Arévalo-Gallegos, S., & Rascón-Cruz, Q. (2012). Lactoferrin a multiple bioactive protein: An overview. *Biochimica et Biophysica Acta (BBA) - General Subjects, 1820*(3), 226–236. https://doi.org/10.1016/j.bbagen.2011.06.018.

Guan, J., Wang, P., Lu, L., & Zhao, G. (2020). Association of plasma transferrin with cognitive decline in patients with mild cognitive impairment and Alzheimer's disease. *Frontiers in Aging Neuroscience, 12*, 38. https://doi.org/10.3389/fnagi.2020.00038.

Hametner, S., Wimmer, I., Haider, L., Pfeifenbring, S., Brück, W., & Lassmann, H. (2013). Iron and neurodegeneration in the multiple sclerosis brain: Iron in the MS Brain. *Annals of Neurology, 74*(6), 848–861. https://doi.org/10.1002/ana.23974.

Han, C., Liu, Y., Dai, R., Ismail, N., Su, W., & Li, B. (2020). Ferroptosis and its potential role in human diseases. *Frontiers in Pharmacology, 11*, 239. https://doi.org/10.3389/fphar.2020.00239.

Hare, D., Ayton, S., Bush, A., & Lei, P. (2013). A delicate balance: Iron metabolism and diseases of the brain. *Frontiers in Aging Neuroscience, 5*. https://doi.org/10.3389/fnagi.2013.00034.

Healy, S., McMahon, J. M., & FitzGerald, U. (2017). Modelling iron mismanagement in neurodegenerative disease in vitro: Paradigms, pitfalls, possibilities & practical considerations. *Progress in Neurobiology, 158*, 1–14. https://doi.org/10.1016/j.pneurobio.2017.08.004.

Heon-Roberts, R., Nguyen, A. L. A., & Pshezhetsky, A. V. (2020). Molecular bases of neurodegeneration and cognitive decline, the major burden of Sanfilippo disease. *Journal of Clinical Medicine, 9*(2), 344. https://doi.org/10.3390/jcm9020344.

Hirose, M. (2000). The structural mechanism for iron uptake and release by transferrins. *Bioscience Biotechnology and Biochemistry, 64*(7), 1328–1336. https://doi.org/10.1271/bbb.64.1328.

Hodgson, R., Kennedy, B. K., Masliah, E., Scearce-Levie, K., Tate, B., Venkateswaran, A., et al. (2020). Aging: Therapeutics for a healthy future. *Neuroscience & Biobehavioral Reviews, 108*, 453–458. https://doi.org/10.1016/j.neubiorev.2019.11.021.

Hodroge, Rackowski, L., Lefebvre, T., Karim, Z., Vitry, S., Gonnet, F., et al. (2016). Alteration of cerebral iron metabolism in Sanfilippo syndrome. *Molecular Genetics and Metabolism, 117*(2), S97. https://doi.org/10.1016/j.ymgme.2015.12.413.

Hohnholt, M. C., & Dringen, R. (2013). Uptake and metabolism of iron and iron oxide nanoparticles in brain astrocytes. *Biochemical Society Transactions, 41*(6), 1588–1592. https://doi.org/10.1042/BST20130114.

Ingrassia, R., Garavaglia, B., & Memo, M. (2019). DMT1 expression and iron levels at the crossroads between aging and neurodegeneration. *Frontiers in Neuroscience, 13*, 575. https://doi.org/10.3389/fnins.2019.00575.

Jeffrey, P. D., Bewley, M. C., MacGillivray, R. T. A., Mason, A. B., Woodworth, R. C., & Baker, E. N. (1998). Ligand-Induced conformational change in transferrins: Crystal structure of the open form of the N-terminal half-molecule of human transferrin †,‡. *Biochemistry, 37*(40), 13978–13986. https://doi.org/10.1021/bi9812064.

Jiang, R., Hua, C., Wan, Y., Jiang, B., Hu, H., Zheng, J., et al. (2015). Hephaestin and ceruloplasmin play distinct but interrelated roles in iron homeostasis in mouse brain. *Journal of Nutrition, 145*(5), 1003–1009. https://doi.org/10.3945/jn.114.207316.

Joppe, K., Roser, A.-E., Maass, F., & Lingor, P. (2019). The contribution of iron to protein aggregation disorders in the central nervous system. *Frontiers in Neuroscience, 13*(15), 155–165. https://doi.org/10.3389/fnins.2019.00015.

Kawabata, H. (2019). Transferrin and transferrin receptors update. *Free Radical Biology and Medicine, 133*, 46–54. https://doi.org/10.1016/j.freeradbiomed.2018.06.037.

Knutson, M. D. (2019). Non-transferrin-bound iron transporters. *Free Radical Biology and Medicine, 133*, 101–111. https://doi.org/10.1016/j.freeradbiomed.2018.10.413.

Kumari, A. (2018). Mucopolysaccharidoses. In *Sweet biochemistry* (pp. 75–84). Elsevier. https://doi.org/10.1016/B978-0-12-814453-4.00015-7.

Lambert, L. A. (2012). Molecular evolution of the transferrin family and associated receptors. *Biochimica et Biophysica Acta (BBA) - General Subjects, 1820*(3), 244–255. https://doi.org/10.1016/j.bbagen.2011.06.002.

Leung, W., Singh, I., McWilliams, S., Stockler, S., & Ipsiroglu, O. S. (2020). Iron deficiency and sleep – a scoping review. *Sleep Medicine Reviews, 51*, 101274. https://doi.org/10.1016/j.smrv.2020.101274.

Li, J., Cao, F., Yin, H., Huang, Z., Lin, Z., Mao, N., et al. (2020). Ferroptosis: Past, present and future. *Cell Death and Disease, 11*(2), 88. https://doi.org/10.1038/s41419-020-2298-2.

Licher, S., Darweesh, S. K. L., Wolters, F. J., Fani, L., Heshmatollah, A., Mutlu, U., et al. (2019). Lifetime risk of common neurological diseases in the elderly population. *Journal of Neurology, Neurosurgery and Psychiatry, 90*(2), 148–156. https://doi.org/10.1136/jnnp-2018-318650.

Li, H., & Qian, Z. M. (2002). Transferrin/transferrin receptor-mediated drug delivery. *Medicinal Research Reviews, 22*(3), 225–250. https://doi.org/10.1002/med.10008.

Liu, J.-L., Fan, Y.-G., Yang, Z.-S., Wang, Z.-Y., & Guo, C. (2018). Iron and Alzheimer's disease: From pathogenesis to therapeutic implications. *Frontiers in Neuroscience, 12*(632), 10–23. https://doi.org/10.3389/fnins.2018.00632.

MacGillivray, R. T. A., Bewley, M. C., Smith, C. A., He, Q.-Y., Mason, A. B., Woodworth, R. C., et al. (2000). Mutation of the iron ligand His 249 to glu in the N-lobe of human transferrin abolishes the dilysine "trigger" but does not significantly affect iron release † , ‡. *Biochemistry, 39*(6), 1211–1216. https://doi.org/10.1021/bi991522y.

MacGillivray, R. T. A., Moore, S. A., Chen, J., Anderson, B. F., Baker, H., Luo, Y., et al. (1998). Two high-resolution crystal structures of the recombinant N-lobe of human transferrin reveal a structural change implicated in iron release † , ‡. *Biochemistry, 37*(22), 7919–7928. https://doi.org/10.1021/bi980355j.

McCarthy, R. C., & Kosman, D. J. (2015). Iron transport across the blood–brain barrier: Development, neurovascular regulation and cerebral amyloid angiopathy. *Cellular and Molecular Life Sciences, 72*(4), 709–727. https://doi.org/10.1007/s00018-014-1771-4.

McWhirter, L., Ritchie, C., Stone, J., & Carson, A. (2020). Functional cognitive disorders: A systematic review. *The Lancet Psychiatry,* 7(2), 191–207. https://doi.org/10.1016/S2215-0366(19)30405-5.

Mills, E., Dong, X., Wang, F., & Xu, H. (2010). Mechanisms of brain iron transport: Insight into neurodegeneration and CNS disorders. *Future Medicinal Chemistry, 2*(1), 51–64. https://doi.org/10.4155/fmc.09.140.

Möller, H. E., Bossoni, L., Connor, J. R., Crichton, R. R., Does, M. D., Ward, R. J., et al. (2019). Iron, myelin, and the brain: Neuroimaging meets neurobiology. *Trends in Neurosciences, 42*(6), 384–401. https://doi.org/10.1016/j.tins.2019.03.009.

Muhoberac, B. B., & Vidal, R. (2019). Iron, ferritin, hereditary ferritinopathy, and neurodegeneration. *Frontiers in Neuroscience, 13,* 1195. https://doi.org/10.3389/fnins.2019.01195.

Nasrabady, S. E., Rizvi, B., Goldman, J. E., & Brickman, A. M. (2018). White matter changes in Alzheimer's disease: A focus on myelin and oligodendrocytes. *Acta Neuropathologica Communications, 6*(1), 22. https://doi.org/10.1186/s40478-018-0515-3.

Page, M. G. P. (2019). The role of iron and siderophores in infection, and the development of siderophore antibiotics. *Clinical Infectious Diseases, 69*(Suppl. ment_7), S529–S537. https://doi.org/10.1093/cid/ciz825.

Pardridge, W. M. (2017). Delivery of biologics across the blood–brain barrier with molecular trojan horse technology. *BioDrugs, 31*(6), 503–519. https://doi.org/10.1007/s40259-017-0248-z.

Pardridge, W. M. (2020). Blood-brain barrier and delivery of protein and gene therapeutics to brain. *Frontiers in Aging Neuroscience, 11,* 373. https://doi.org/10.3389/fnagi.2019.00373.

Piao, Y.-S., Lian, T.-H., Hu, Y., Zuo, L.-J., Guo, P., Yu, S.-Y., et al. (2017). Restless legs syndrome in Parkinson disease: Clinical characteristics, abnormal iron metabolism and altered neurotransmitters. *Scientific Reports,* 7(1), 10547. https://doi.org/10.1038/s41598-017-10593-7.

Picca, A., Mankowski, R. T., Kamenov, G., Anton, S. D., Manini, T. M., Buford, T. W., et al. (2019). Advanced age is associated with iron dyshomeostasis and mitochondrial DNA damage in human skeletal muscle. *Cells, 8*(12), 1525. https://doi.org/10.3390/cells8121525.

Pinheiro, R. G. R., Granja, A., Loureiro, J. A., Pereira, M. C., Pinheiro, M., Neves, A. R., et al. (2020). Quercetin lipid nanoparticles functionalized with transferrin for Alzheimer's disease. *European Journal of Pharmaceutical Sciences, 148,* 105314. https://doi.org/10.1016/j.ejps.2020.105314.

Prakash, A., Dhaliwal, G. K., Kumar, P., & Majeed, A. B. A. (2017). Brain biometals and Alzheimer's disease – boon or bane? *International Journal of Neuroscience, 127*(2), 99–108. https://doi.org/10.3109/00207454.2016.1174118.

Puy, V., Darwiche, W., Trudel, S., Gomila, C., Lony, C., Puy, L., et al. (2018). Predominant role of microglia in brain iron retention in Sanfilippo syndrome, a pediatric neurodegenerative disease. *Glia, 66*(8), 1709–1723. https://doi.org/10.1002/glia.23335.

Qian, Z., & Ke, Y. (2019). Brain iron transport. *Biological Reviews, 94*(5), 1672–1684. https://doi.org/10.1111/brv.12521.

Ropele, S., Enzinger, C., & Fazekas, F. (2017). Iron mapping in multiple sclerosis. *Neuroimaging Clinics of North America, 27*(2), 335–342. https://doi.org/10.1016/j.nic.2016.12.003.

Rouault, T. A. (2013). Iron metabolism in the CNS: Implications for neurodegenerative diseases. *Nature Reviews Neuroscience, 14*(8), 551–564. https://doi.org/10.1038/nrn3453.

Roy Sarkar, S., & Dutta, R. (2019). Neurological diseases associated with brain iron accumulation. *Acta Scientific Neurology, 3*(1), 53–61. https://doi.org/10.31080/ASNE.2020.03.neurological-diseases-associated-with-brain-iron-accumulation.

Sabbir, M. G. (2018). Loss of Ca2+/calmodulin dependent protein kinase kinase 2 leads to aberrant transferrin phosphorylation and trafficking: A potential biomarker for Alzheimer's disease. *Frontiers in Molecular Biosciences, 5*, 99. https://doi.org/10.3389/fmolb.2018.00099.

Sánchez-Castañeda, C., Squitieri, F., Di Paola, M., Dayan, M., Petrollini, M., & Sabatini, U. (2015). The role of iron in gray matter degeneration in Huntington's disease: A magnetic resonance imaging study: The role of iron in gray matter loss in HD. *Human Brain Mapping, 36*(1), 50–66. https://doi.org/10.1002/hbm.22612.

Scassellati, C., Bonvicini, C., Benussi, L., Ghidoni, R., & Squitti, R. (2020). Neurodevelopmental disorders: Metallomics studies for the identification of potential biomarkers associated to diagnosis and treatment. *Journal of Trace Elements in Medicine & Biology, 60*, 126499. https://doi.org/10.1016/j.jtemb.2020.126499.

Sfera, A., Bullock, K., Price, A., Inderias, L., & Osorio, C. (2018). Ferrosenescence: The iron age of neurodegeneration? *Mechanism of Ageing and Development, 174*, 63–75. https://doi.org/10.1016/j.mad.2017.11.012.

Shamsi, A., Mohammad, T., Khan, M. S., Shahwan, M., Husain, F. M., Rehman, M. T., et al. (2019). Unraveling binding mechanism of Alzheimer's drug rivastigmine tartrate with human transferrin: Molecular docking and multi-spectroscopic approach towards neurodegenerative diseases. *Biomolecules, 9*(9), 495. https://doi.org/10.3390/biom9090495.

Simpson, I. A., Ponnuru, P., Klinger, M. E., Myers, R. L., Devraj, K., Coe, C. L., et al. (2015). A novel model for brain iron uptake: Introducing the concept of regulation. *Journal of Cerebral Blood Flow and Metabolism, 35*(1), 48–57. https://doi.org/10.1038/jcbfm.2014.168.

Singh, Y. P., Pandey, A., Vishwakarma, S., & Modi, G. (2019). A review on iron chelators as potential therapeutic agents for the treatment of Alzheimer's and Parkinson's diseases. *Molecular Diversity, 23*(2), 509–526. https://doi.org/10.1007/s11030-018-9878-4.

Siotto, M., Filippi, M. M., Simonelli, I., Landi, D., Ghazaryan, A., Vollaro, S., et al. (2019). Oxidative stress related to iron metabolism in relapsing remitting multiple sclerosis patients with low disability. *Frontiers in Neuroscience, 13*(86), 174–182. https://doi.org/10.3389/fnins.2019.00086.

Sokolov, A. V., Voynova, I. V., Kostevich, V. A., Vlasenko, A. Y., Zakharova, E. T., & Vasilyev, V. B. (2017). Comparison of interaction between ceruloplasmin and lactoferrin/transferrin: To bind or not to bind. *Biochemistry, 82*(9), 1073–1078. https://doi.org/10.1134/S0006297917090115.

Stankiewicz, J. M., Neema, M., & Ceccarelli, A. (2014). Iron and multiple sclerosis. *Neurobiology of Aging, 35*, S51–S58. https://doi.org/10.1016/j.neurobiolaging.2014.03.039.

Sun, Y., Pham, A. N., Hare, D. J., & Waite, T. D. (2018). Kinetic modeling of pH-dependent oxidation of dopamine by iron and its relevance to Parkinson's disease. *Frontiers in Neuroscience, 12*(859), 68–85. https://doi.org/10.3389/fnins.2018.00859.

Taniguchi, M., Okayama, Y., Hashimoto, Y., Kitaura, M., Jimbo, D., Wakutani, Y., et al. (2008). Sugar chains of cerebrospinal fluid transferrin as a new biological marker of Alzheimer's disease. *Dementia and Geriatric Cognitive Disorders, 26*(2), 117–122. https://doi.org/10.1159/000147479.

Testi, C., Boffi, A., & Montemiglio, L. C. (2019). Structural analysis of the transferrin receptor multifaceted ligand(s) interface. *Biophysical Chemistry, 254*, 106242. https://doi.org/10.1016/j.bpc.2019.106242.

Todorich, B., Zhang, X., Slagle-Webb, B., Seaman, W. E., & Connor, J. R. (2008). Tim-2 is the receptor for H-ferritin on oligodendrocytes. *Journal of Neurochemistry, 107*(6), 1495–1505. https://doi.org/10.1111/j.1471-4159.2008.05678.x.

Uranga, R. M., & Salvador, G. A. (2018). Unraveling the burden of iron in neurodegeneration: Intersections with amyloid beta peptide pathology. *Oxidative Medicine and Cellular Longevity, 2018*, 1–12. https://doi.org/10.1155/2018/2850341.

Vellieux, G., & d'Ortho, M.-P. (2020). Le syndrome des jambes sans repos. *La Revue de Medecine Interne, 41*(4), 258–264. https://doi.org/10.1016/j.revmed.2020.01.001.

Wally, J., Halbrooks, P. J., Vonrhein, C., Rould, M. A., Everse, S. J., Mason, A. B., et al. (2006). The crystal structure of iron-free human serum transferrin provides insight into inter-lobe communication and receptor binding. *Journal of Biological Chemistry, 281*(34), 24934–24944. https://doi.org/10.1074/jbc.M604592200.

Wang, M., Lai, T. P., Wang, L., Zhang, H., Yang, N., Sadler, P. J., et al. (2015). "Anion clamp" allows flexible protein to impose coordination geometry on metal ions. *Chemical Communications, 51*(37), 7867–7870. https://doi.org/10.1039/C4CC09642H.

Wang, L., Yin, Y.-L., Liu, X.-Z., Shen, P., Zheng, Y.-G., Lan, X.-R., et al. (2020). Current understanding of metal ions in the pathogenesis of Alzheimer's disease. *Translational Neurodegeneration, 9*(1), 10. https://doi.org/10.1186/s40035-020-00189-z.

Wawer, A. A., Jennings, A., & Fairweather-Tait, S. J. (2018). Iron status in the elderly: A review of recent evidence. *Mechanism of Ageing and Development, 175*, 55–73. https://doi.org/10.1016/j.mad.2018.07.003.

White, K. N., Conesa, C., Sánchez, L., Amini, M., Farnaud, S., Lorvoralak, C., et al. (2012). The transfer of iron between ceruloplasmin and transferrins. *Biochimica et Biophysica Acta (BBA) - General Subjects, 1820*(3), 411–416. https://doi.org/10.1016/j.bbagen.2011.10.006.

Yanase, K., Uemura, N., Chiba, Y., Murakami, R., Fujihara, R., Matsumoto, K., et al. (2020). Immunoreactivities for hepcidin, ferroportin, and hephaestin in astrocytes and choroid plexus epithelium of human brains. *Neuropathology, 40*(1), 75–83. https://doi.org/10.1111/neup.12611.

Yang, N., Zhang, H., Wang, M., Hao, Q., & Sun, H. (2012). Iron and bismuth bound human serum transferrin reveals a partially-opened conformation in the N-lobe. *Scientific Reports, 2*(1), 999. https://doi.org/10.1038/srep00999.

Yu, S., Cao, C., Zuo, L., Chen, Z., Lian, T., Wang, F., et al. (2018). Clinical features and dysfunctions of iron metabolism in Parkinson disease patients with hyper echogenicity in substantia nigra: A cross-sectional study. *BMC Neurology, 18*(1), 9. https://doi.org/10.1186/s12883-018-1016-5.

Zanella, I., Filosto, M., & Biasiotto, G. (Eds.). (2020). *Iron and neurodegeneration*. Frontiers Media SA. https://doi.org/10.3389/978-2-88963-451-4.

Zuo, L.-J., Yu, S.-Y., Hu, Y., Wang, F., Piao, Y.-S., Lian, T.-H., et al. (2016). Serotonergic dysfunctions and abnormal iron metabolism: Relevant to mental fatigue of Parkinson disease. *Scientific Reports, 6*(1), 19. https://doi.org/10.1038/s41598-016-0018-z.

CHAPTER EIGHT

ABCG: a new fold of ABC exporters and a whole new bag of riddles!

Atanu Banerjee[a], Alexis Moreno[b], Jorgaq Pata[b], Pierre Falson[b,*] and Rajendra Prasad[a,c,*]

[a]Amity Institute of Biotechnology, Amity University Haryana, Gurgaon, Haryana, India
[b]Drug Resistance & Membrane Proteins Team, Molecular Microbiology and Structural Biochemistry Laboratory, CNRS-Lyon 1 University UMR5086, Institut de Biologie et Chimie des Protéines, Lyon, France
[c]Amity Institute of Integrative Sciences and Health, Amity University Haryana, Gurgaon, Haryana, India
*Corresponding authors: E-mail: pierre.falson@ibcp.fr, rprasad@ggn.amity.edu

Contents

1. ABC transporters and their relevance 164
   1.1 ABC transporter superfamily 164
   1.2 General organization 166
   1.3 Basis of transport mechanism 170
2. Structure of type I and type II exporters 171
   2.1 Type I structures 171
   2.2 Type II structures 172
      2.2.1 ABCG5/G8 172
      2.2.2 ABCG2 173
   2.3 Substrate-binding pockets 176
3. Yeast ABCG/PDR transporters: contribution toward understanding the enigma of non-catalytic nucleotide-binding sites 177
4. Concluding thoughts and future perspectives 182

Acknowledgments 183

References 183

Abstract

ATP-binding cassette (ABC) superfamily comprises membrane transporters that power the active transport of substrates across biological membranes. These proteins harness the energy of nucleotide binding and hydrolysis to fuel substrate translocation via an alternating-access mechanism. The primary structural blueprint is relatively conserved in all ABC transporters. A transport-competent ABC transporter is essentially made up of two nucleotide-binding domains (NBDs) and two transmembrane domains (TMDs). While the NBDs are conserved in their primary sequence and form at their

Advances in Protein Chemistry and Structural Biology, Volume 123
ISSN 1876-1623
https://doi.org/10.1016/bs.apcsb.2020.09.006

interface two nucleotide-binding sites (NBSs) for ATP binding and hydrolysis, the TMDs are variable among different families and form the translocation channel. Transporters catalyzing the efflux of substrates from the cells are called exporters. In humans, they range from A to G subfamilies, with the B, C and G subfamilies being involved in chemoresistance. The recently elucidated structures of ABCG5/G8 followed by those of ABCG2 highlighted a novel structural fold that triggered extensive research. Notably, suppressor genetics in the orthologous yeast Pleiotropic Drug Resistance (PDR) subfamily proteins have pointed to a crosstalk between TMDs and NBDs modulating substrate export. Considering the structural information provided by their neighbors from the G subfamily, these studies provide mechanistic keys and posit a functional role for the non-hydrolytic NBS found in several ABC exporters. The present chapter provides an overview of structural and functional aspects of ABCG proteins with a special emphasis on the yeast PDR systems.

Abbreviations

ABC	ATP binding cassette
ADP	Adenosine-5′-diphosphate
ATP	Adenosine-5′-triphosphate
Cdr1p	*Candida* drug resistance protein 1
CFTR	Cystic fibrosis transmembrane conductance regulator
CnH	Connecting helix
CpH	Coupling helix
cryo-EM	Cryo electron microscopy
ECD	Extra cellular domain
ECF	Energy-coupled factor
IFC	Inward-facing conformation
MRP1	Multidrug resistance-associated protein 1
NBD	Nucleotide binding domain
NBS	Nucleotide binding site
NTP	Nucleoside triphosphate
OFC	Outward-facing conformation
PDR	Pleiotropic drug resistance
P-gp	P-glycoprotein
P_i	Inorganic phosphate
SUR	Sulfonylurea receptor
TMD	Transmembrane domain
TMH	Transmembrane helix

1. ABC transporters and their relevance

1.1 ABC transporter superfamily

The ABC transporter superfamily is one of the most omnipresent family of membrane proteins found in all kingdoms of life. It includes importers (not detailed here) involved in the influx of nutrients, mostly found in prokaryotes and plants or in the particular energy-coupled factor (ECF) subfamily, and exporters that manage the efflux of many endobiotics (lipids, metabolites, pheromones, ...) and xenobiotics (antibiotics, antifungals, anticancer drugs, antigens) (Orelle, Mathieu, & Jault, 2019; Higgins, 2001; Rees, Johnson, & Lewinson, 2009). Some exceptions remain such as the Cystic Fibrosis Transmembrane Receptor, CFTR, which acts as a chloride channel (Rich et al., 1990; Vergani, Nairn, & Gadsby, 2003) and the sulfonylurea receptors SUR1 (ABCC8) and SUR2 (ABCC9) that are part of a potassium channel complex (Bryan et al., 2007). In bacteria, ABC transporters like $LptB_2FG$ can also act as a lipopolysaccharide channel, and others like MacA as mechano-transducers, in concert with additional periplasmic and outer membrane proteins (Ford & Beis, 2019).

These proteins share a common sequence signature, called the C motif (detailed below) located in their two NBDs (Fig. 1). Based on sequence

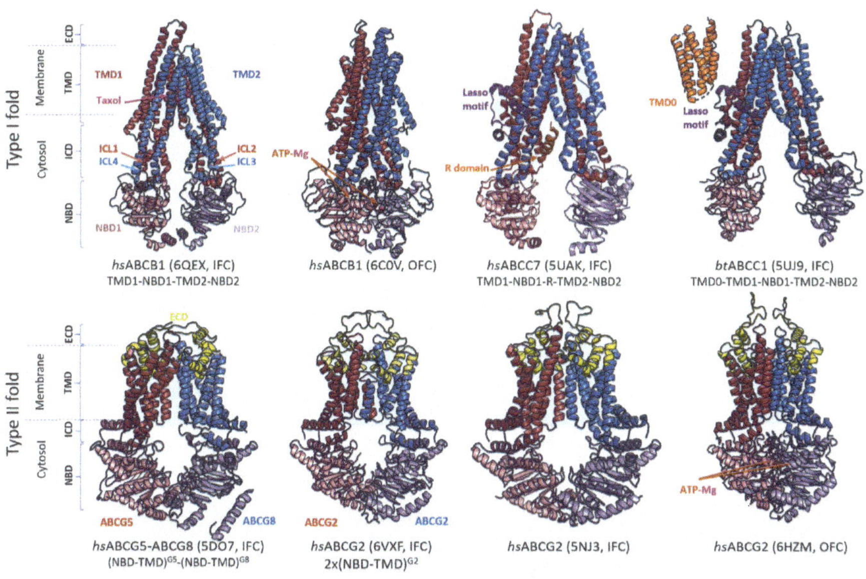

Fig. 1 ***Cartoon representations of type I and type II ABC exporters structures, in inward- and outward-facing conformations (IFC, OFC).*** Human ABCB1, ABCC7 and ABCCC1 on the top belong to the type I fold. They are made of a single polypeptide organized in a pseudo-dimer of two halves (*red* and *blue*) forming the different regions, nucleotide-binding domains (NBDs), intracellular domain (ICD) with intracellular loops (ICLs), transmembrane domain (TMD) and extracellular domain (ECD). ABCC7 and ABCC1 display an additional regulatory domain and a transmembrane domain, respectively. Human ABCG5/ABCG8 and ABCG2 belong to the type II fold. They are made of two polypeptides making each half of the hetero- or homo-dimer. PDB codes are indicated.

similarities, eukaryotic proteins are classified into 9 subfamilies from ABCA to ABCI, all comprising exporters except for the ABCE, ABCF and ABCI subfamilies which include soluble ABC proteins devoid of transmembrane domain (Jeong, Kim, Kang, & Lee, 2017; Navarro-Quiles, Mateo-Bonmatí, & Micol, 2018; Murina et al., 2019). This nomenclature was first adopted by the HUGO Gene Nomenclature Committee for human transporters which are divided into 7 subfamilies from ABCA to ABCG. The classification was further extended for non-mammalian proteins with the addition of the ABCH subfamily found in insects (Dean & Annilo, 2005) and some fishes (Luckenbach, Fischer, & Sturm, 2014; Popovic, Zaja, Loncar, & Smital, 2010), and of the ABCI subfamily which is specific to plants (Verrier et al., 2008). The Pleiotropic Drug Resistance (PDR) subfamily is orthologous to the human ABCG and specific to plants and fungi. It is the largest one found in yeast, with 10 PDR transporters among 30 ABC proteins in *S. cerevisiæ*, 9 among 26 in *Candida albicans*, and 7 among 25 in *Candida glabrata* (Kumari et al., 2018; Prasad, Khandelwal, & Banerjee, 2016).

1.2 General organization

ABC pumps are made of two regions, one protruding in the cytosol in which ATP binds and is hydrolyzed and the other in the membrane through which solutes are translocated from one compartment to another, both functions dependent on each other. Each region is made at least of two domains, leading to the general topology $(NBD\text{-}TMD)_2$ per transporter. These domains are arranged either in full transporters in a single polypeptide $TMD_1\text{-}NBD_1\text{-}TMD_2\text{-}NBD_2$ or $NBD_1\text{-}TMD_1\text{-}NBD_2\text{-}TMD_2$, or in half transporters in two identical or different polypeptides $(TMD\text{-}NBD)_2$, $(NBD\text{-}TMD)_2$, $TMD_1\text{-}NBD_1 + TMD_2\text{-}NBD_2$, $NBD_1\text{-}TMD_1 + NBD_2\text{-}TMD_2$. In prokaryotes each domain of a single transporter can also be found as a single polypeptide. The PDR subfamily, as the members of the human G subfamily, display the so-called "reverse topology" (NBD-TMD) compared to all others. Almost all of them are full plasma membrane-bound transporters.

The NBDs are well conserved in members of the ABC superfamily. They display highly conserved sequences required for ATP binding and hydrolysis, as detailed in Fig. 2 and Table 1. Characteristic of the NTPases, these amino-acid sequences were initially designated α and β folds, and further renamed Walker A and Walker B motifs, respectively (Saraste, Sibbald, & Wittinghofer, 1990; Walker, Saraste, Runswick, & Gay, 1982). The Walker A -or P-loop- GxxGxGKS/T, where x is any amino acid, binds the

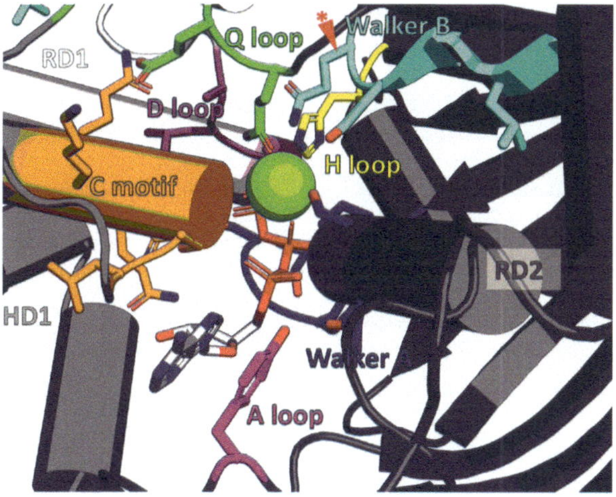

Fig. 2 ***Nucleotide binding and hydrolysis motifs found in ABC transporters.*** Cartoon representation of the ATP-bound human P-gp E1201Q mutant (PDB ID: 6COV) (Kim & Chen, 2018) illustrating the different consensus peptides involved in binding and hydrolysis of the nucleotide. A glutamine residue (*red star*) in the Walker B motif replaces the glutamate 1201 to prevent ATP hydrolysis and maintain the protein in a pre-hydrolytic state. ATP is shown as sticks and magnesium as a sphere. HD1/HD2: helical domain 1/2 (ABCα subdomain), RD1/RD2: RecA-like domain 1/2 (including ABCβ subdomain).

phosphate groups of the nucleotide, thanks to the contribution of amide groups from the backbone and the lateral chain of the lysine residue. The conserved serine/threonine residue participates in the coordination of the magnesium in complex with the nucleotide. The Walker B motif, hhhhDE where h is a hydrophobic residue, is also involved in the coordination of the metal. It is immediately followed by a catalytic glutamate residue that polarivzes a water molecule required for the hydrolysis of the nucleotide (Oldham & Chen, 2011; Szöllősi, Rose-Sperling, Hellmich, & Stockner, 2018). Other conserved motifs were further discovered. The conserved glutamine of the Q-loop consensus sequence h(h/Q)Q̲ binds the magnesium and senses the nucleotide γ-phosphate (Yang et al., 2011), the histidine residue of the so-called H-loop stabilizes the phosphate hydrolysis (Zaitseva, Jenewein, Jumpertz, Holland, & Schmitt, 2005), and the aspartate residue of the D-loop SALD̲ positions the attacking water molecule (Grossmann et al., 2014; Thomas & Tampé, 2020). Altogether, these motifs define the RecA-like motor core of ABC-type ATPases, AAA+ and helicases (Ye, Osborne, Groll, & Rapoport, 2004).

Table 1 Conserved sequence motifs of the nucleotide binding domains of ABC proteins.

Motif	ABC consensus sequence	PDR catalytic sequence	PDR non-catalytic sequence	Function
A loop	Aromatic residue	Aromatic residue	Aromatic residue	Aromatic residue interacting with the adenine ring of ATP
Walker A or P-loop	GxxGxGK (S/T)	GxxGxGK (S/T)	GxxGxG**C** (S/T)	ATP/GTP binding (α- and β-phosphate binding)
Q-loop	Q	Q	**E**	TMD-NBD communication, γ-phosphate sensor
ABC signature motif or C-loop	LSGGQ (K/R)Q	VSGGERKR	L**NVE**QRKR	ATP binding, NBD-NBD communication
Walker B	hhhhDE	hhhhDE	h**xxW**DN	ATP hydrolysis (coordination of catalytic water and Mg^{2+})
D-loop	hLD	hLD	hLD	NBD-NBD communication
H-loop or switch region	H(Q/R)	H(Q/R)	**Y**(Q/R)	ATP hydrolysis

Residues are identified by their 1-letter code, h = hydrophobic residue, x = any amino acid.
Based on Eggensperger and Tampé (2015).

A hallmark of ABC proteins is the presence of two additional subdomains, one α-helical, ABCα, and another made of antiparallel β-sheets, ABCβ (Cedric Orelle et al., 2010; Smith et al., 2002) (Fig. 3). ABCα contains the specific ABC consensus sequence (or Signature sequence or C-motif) LSGGQ (Hyde et al., 1990) that pins and orients the ATP (Moncalian et al., 2004; Rees et al., 2009). The glutamine residue of this sequence comes in close contact to the ribose moiety of the nucleotide. The ABCβ subdomain bears the conserved A-loop that consists of an aromatic residue that stabilizes the aromatic moiety of the nucleotide through π-π-stacking interactions.

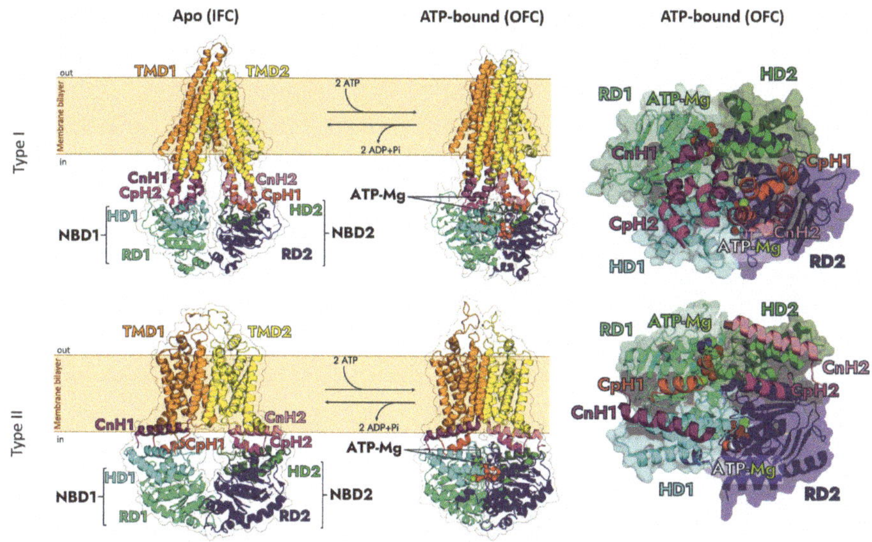

Fig. 3 ***Domain organization of type I and type II ABC exporters.*** **Left panels:** IFC and OFC of ABCB1 (top, PDB ID: 6QEX and 6C0V, respectively) (Alam et al., 2019; Kim & Chen, 2018) and of ABCG2 (bottom, PDB ID: 6VXF and 6HBU, respectively) (Manolaridis et al., 2018; Orlando & Liao, 2020) illustrate the differences in interdomains connections between the two types of exporters and the major conformational switch involved in substrate transport. Proteins are represented as cartoon with their corresponding surface outlined and colored by domains and structural features. ATP-Mg^{2+} are shown as spheres. **Right panels:** Close view of the ATP-bound dimerized NBDs, as seen from the membrane, with surrounding semi-transparent surface. The coupling helices of both types of exporters make contacts with both the RecA-like and helical subdomains. In type I exporters (top), the connecting helices interact with the RecA-like domain of its opposite NBD. In contrast, the connecting helices of type II exporters (bottom) interact with the helical domain of its corresponding NBD. CnH1/2: connecting helix 1/2, CpH1/2: coupling helix 1/2, HD1/HD2: helical domain 1/2 (ABCα subdomain), NBD1/2: nucleotide binding domain 1/2, RD1/RD2: RecA-like domain 1/2 (including ABCβ subdomain), TMD1/2: transmembrane domain 1/2.

RecA-like, ABCα, and ABCβ subdomains form the minimal nucleotide-binding unit that requires dimerization to be functional. When they merge during the transport cycle, these subdomains generate two head-to-tail nucleotide-binding sites (NBSs) at their interface. Each NBS is made of motifs of the RecA-like subdomain and A-loop of one NBD and the C motif of the second one. Both NBS bind and hydrolyze ATP but some heterodimers and full transporters display divergent motifs in one NBS preventing the nucleotide to be hydrolyzed at this site. These divergent sequences are called deviant or degenerated. They are displayed in Table 1. This is the case, for example of the human heterodimer ABCG5/G8 displayed in Fig. 1, or that of the full transporters Pleiotropic drug resistance protein 5 (Pdr5p) of *S. cerevisiæ* or *Candida* drug resistance protein 1 (Cdr1p) of *C. albicans*. Initially thought to be degenerated to save energy, mutagenesis studies of Pdr5p and Cdr1p now point to a specific role of this non-hydrolytic site that modulates both the ATPase and drug-transport activities (see Section 3) (Banerjee et al., 2020, 2018; Ernst et al., 2008; Prasad, Balzi, Banerjee, & Khandelwal, 2019; Saini, Gaur, & Prasad, 2006).

TMDs are generally made of 6 helices going through the membrane, however some ABC pumps can have an additional membrane region, such as the TMD0 of ABCC1 (Fig. 1). As for NBDs, TMDs generate a functional site at their interface, of which the shape depends on the type of pumps (see Section 2.3). Contrary to NBDs residues, TMDs show poor conservation. This is because their interaction with the surrounding lipids can be done with various aliphatic or aromatic residues and because residues forming the translocating pocket vary with the type of substrate transported. A deep mutagenesis study carried out on the 250 residues forming the membrane region of Cdr1p allowed to sort about 1/3 critical for antifungals, dyes and steroids transport (Baghel et al., 2017; Nim et al., 2016). Interestingly, these studies suggest a molecular basis of polyspecificity that characterizes this pump close to that of Mdr1p, the major drug/H^+ antiporter of *C. albicans*. Mdr1p also confers multidrug resistance to the yeast, and the same global mutagenesis strategy carried out revealed that the structural basis for its polyspecificity is an extended capacity brought by residues located at the periphery of a binding core to accommodate compounds differing in size and type (Redhu et al., 2018).

1.3 Basis of transport mechanism

Functionally, ABC transporters follow the much earlier conceptualized and general alternating access mechanism (Jardetzky, 1966). Following this concept, the pump undergoes two main conformational changes by which a substrate in the internal compartment (e.g. cytoplasm) binds to the membrane cavity and is then translocated to the external compartment (e.g. extracellular space). These two states correspond to the inward- and outward-facing conformations, IFC and OFC, illustrated in Figs. 1 and 3.

Roughly, the transport cycle carried out by ABC pumps starts in IFC with the binding of the substrate, also called allocrite to distinguish from ATP-Mg which is the substrate of the ATPase activity (Holland, 2011), to a cavity into the TMDs accessible from the inner compartment. The substrate binding triggers/stabilizes the fusion of the two NBDs that generates the NBSs on which bind the nucleotide-Mg complexes. This binding locks the pump in an occluded state by which the substrate cannot escape back to the inner compartment and triggers its translocation through an IFC – OFC transition. The OFC exposes the membrane cavity to the outer space that promotes release of the substrate. The substrate release triggers another conformational change that triggers ATP hydrolysis and resets the pump to its initial IFC.

The stoichiometry of hydrolyzed ATP versus transported molecules remains unclear. Several mechanisms have been proposed over time and most recent evidences suggest a binding and hydrolysis of the two ATP molecules in a sequential and asymmetric manner (Orelle et al., 2019). However, this cannot be a consensus mechanism since, as introduced above, several ABC transporters harbor a non-hydrolytic NBS. Photolabeling and structural experiments performed in the multidrug resistance-associated protein 1, MRP1/ABCC1 (Wang et al., 2020) and in the bacterial TM287/288 from *Thermotoga maritima* (Furuta, Sato, & Sakurai, 2016; Hohl et al., 2014; Hutter et al., 2019), together with electrophysiological and structural studies of the human and zebrafish CFTR (Basso, Vergani, Nairn, & Gadsby, 2003; Vergani, Lockless, Nairn, & Gadsby, 2005; Zhang, Yu, Yeh, & Hwang, 2018; Zhang, Yu, et al., 2018) have shown that the NBDs dimer has a looser interface at the non-hydrolytic NBS as opposed to the catalytic one where ATP is quickly hydrolyzed.

2. Structure of type I and type II exporters

Based on their architecture, ABC exporters are so far classified into two distinct folds named type I and type II as displayed in Figs. 1 and 3. Several structures in both types have been obtained in these states.

2.1 Type I structures

In 2006, Kaspar Locher and his team were the first to release an OFC structure of a type I exporter, Sav1866 from *Staphylococcus aureus*, resolved by X-ray crystallography in complex with ADP (Dawson & Locher, 2006). Geoffrey Chang and his team then released several X-ray apo, IFC and OFC structures of MsbA from *Escherichia coli*, *Vibrio cholerœ* and *Salmonella typhimurium*, in complex with AMP-PNP, a non-hydrolyzable analog of ATP (Ward, Reyes, Yu, Roth, & Chang, 2007). They then released the

X-ray IFC structures of the mouse P-gp, apo or in complex with 2 different cyclic peptides (Aller et al., 2009), acting as inhibitors of the transport of the dye Hœchst 33342 and the anticancer drug doxorubicin (Martinez et al., 2014). They completed this panel with other close IFC X-ray structures of the same protein (Ward et al., 2013). The first structure of a heterodimeric ABC exporter, with one non-catalytic NBS, was released by the group of Markus Seeger, TM287/288 (Hohl, Briand, Grütter, & Seeger, 2012). Jue Chen and her team released the X-ray IFC structure of the P-gp from *Cænorhabditis elegans* (Jin, Oldham, Zhang, & Chen, 2012) and more recently by cryo-electron microscopy (cryo-EM) those of the bovine MRP1/ABCC1 (Johnson & Chen, 2017) and human CFTR/ABCC7, apo and in complex with ATP-Mg (Liu, Zhang, Csanády, Gadsby, & Chen, 2017; Zhang, Yu, et al., 2018). This was recently followed by the cryo-EM structure of the human P-gp/ABCB1 apo or in complex with the inhibitor zosuquidar by Kaspar Locher and is team (Alam et al., 2018). Other X-ray and cryo-EM structures were also released, such as that of ATM1, either from *S. cerevisiæ*, apo and in complex with glutathione (Srinivasan, Pierik, & Lill, 2014) or from *Novosphingobium aromaticivorans* (Lee, Yang, Zhitnitsky, Lewinson, & Rees, 2014).

All these structures display the same overall topology. The membrane region is linked to the cytosolic nucleotide-binding region by a large intracellular domain (ICD) made of a pair of long hairpin-like loops (Figs. 1 and 3). In IFC, the drug-binding pocket can be as large as 6000 Å^3. Another remarkable feature exists in the inter-domain communication, in between each moiety/monomer, and between TMDs and NBDs. The first one is generated by a domain swapping by which the coupling helix (CpH) connecting membrane helices 4 and 5 of one moiety/monomer protrudes outside the membrane to come in contact with the NBD of the other moiety/monomer. The second one is the connecting helix (CnH), made of a similar contact between a second helical hairpin connecting membrane helices 2 and 3 and the NBD, but herein, within the same moiety/monomer (Fig. 3).

2.2 Type II structures

2.2.1 ABCG5/G8

In 2016, the teams of Helen Hobs and Daniel Rosenbaum released the crystal structure of the human sterol transporter, ABCG5/G8, the first of the G subfamily and homologs (Lee et al., 2016). ABCG5/G8 mediates the extraction of sterols in liver and intestines and mutations cause sitosterolemia, a disorder characterized by an accumulation of sterols in blood and tissues

(Berge et al., 2000). The protein, made of two different moieties NBD-TMD G5 and G8, revealed a fold deeply differing from that of type I. This new fold is characterized by an absence of domain swapping, NBDs in close-contact to the TMDs, the latter generating at their interface a large but narrow substrate-binding cavity. The pump also displays a large extracellular domain (ECD) between TM5 and TM6, in which, notably, two half-membrane helices penetrate and go out the outer leaflet of the membrane as a fold never observed before in the ABC exporters. In vicinity of these helices, the authors identified densities in the crystallographic map that possibly fit with cholesterol molecules. It was therefore proposed that the ECD may facilitate the exit of sterols. The crosstalk between NBDs and TMDs, a key feature of the transport cycle, is mediated by a bundle of coupling helix (CpH), connecting helix (CnH) and an E-helix (E corresponding to a conserved glutamate residue) through an array of H-bonds and salt bridges, altogether making a small ICD. A relay of polar residues extends from this bundle to the TMD interface, to confer flexibility and to lower the barrier energy for the different domain motions, a hypothesis supported by mutations in this relay that disrupt the function of the transporter (Lu et al., 2001).

2.2.2 ABCG2

ABCG2 was initially identified as the breast cancer resistance protein, BCRP, for its protector role across physiological barriers by expelling xenobiotics (Gutmann, Hruz, Zimmermann, Beglinger, & Drewe, 2005; Mao, 2008). ABCG2 was also localized in kidney proximal tubular cells in which it mediates the efflux of urate, for which the Q141K polymorphism increases the risk of gout (Qiu et al., 2014; Woodward et al., 2009). Together with at least ABCB1 and ABCC1, ABCG2 confers a multidrug resistance phenotype by expelling anticancer drugs (Tamaki, Ierano, Szakacs, Robey, & Bates, 2011).

The functional protein comprises two identical subunits displaying the typical NBD-TMD topology of that subfamily. ABCG2 exports a wide range of natural compounds, including porphyrins (Jonker et al., 2002; Krishnamurthy et al., 2004, 2007) for which the ECD of ABCG2 is required for export (Desuzinges-Mandon et al., 2010), folic acid (Chen et al., 2003) and urate (Woodward et al., 2009). ABCG2 also exports different classes of anti-tumor drugs (Doyle & Ross, 2003; Elkind et al., 2005; Robey et al., 2004) and fluorescent dyes (Özvegy, Váradi, & Sarkadi, 2002). For a complete review see Krishnamurthy & Schuetz (2006); Robey et al. (2018); Sarkadi, Ozvegy-Laczka, Német, and Váradi (2004).

Several cryo-EM structures of ABCG2 came up recently. The first one was an apo structure in complex with two antigen binding fragments (Fab) of the monoclonal antibody 5D3 (Taylor et al., 2017). The structure confirmed the general new type II fold discovered with ABCG5/G8. The TMD interface is formed by membrane helices 2 and 5 of each monomer and, due to a backbone shift of these helices, an inward-facing slit-like cavity is observed. A specific feature of ABCG2 in that conformation is a second small cavity below the ECD and above the main drug-binding cavity, separated from the latter by a plug formed by a di-leucine motif (L554-L555) and inaccessible in this state (Fig. 4). This small cavity is limited by the ECD, which acts as a lid. This small cavity may accommodate substrates with low affinity.

In a follow-up article, Kaspar Locher and his team showed that the main membrane cavity binds ligands, resolving the structure of ABCG2 in complex with inhibitors (Jackson et al., 2018). Two molecules of MZ29, a new inhibitor derived from well-known Ko143 (van Loevezijn, Allen, Schinkel, & Koomen, 2001), were localized, occupying the whole main cavity near the twofold symmetry axis of ABCG2 (Fig. 4). The binding is assured by

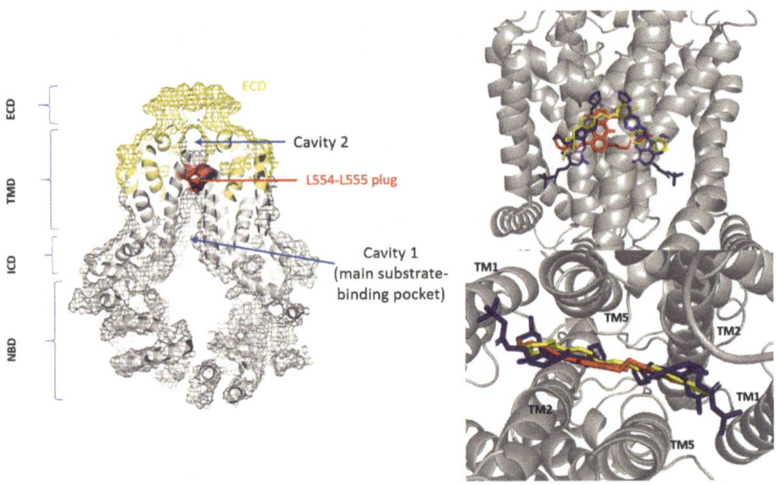

Fig. 4 ***Details of the substrate-binding pocket of ABCG2.*** **Left panel:** mesh representation of ABCG2 (5NJ3) (Taylor et al., 2017) highlighting cavities 1 and 2. So far, no substrate has been found binding to the upper one. **Right panels:** zoomed cartoon representation of ABCG2 structures in complex either with the inhibitor MZ29 (*blue*) (6FFC) (Jackson et al., 2018), or the substrate mitoxantrone (*red*) (6XVI) or imatinib (*yellow*) (6XVH) (Orlando & Liao, 2020), seen in the plan of the membrane (top) or seen from the NBDs (bottom).

hydrophobic interactions involving residues L555, F431, F432 and M549, and H-bonds involving residues T435 and N436. By contrast, another inhibitor, the MB136 which is a derivative of tariquidar (Roe et al., 1999), was localized in a single copy in the homodimer, in a region overlapping that of MZ29. Residues F432, F439 and V546 were found to interact with the compound. In this IFC conformation no NBS is generated, as for type I exporters. However, the cytoplasmic extremity of each NBD is in contact with the other via a NPDXF motif, as also observed with ABCG5/G8.

Further, the same group released the structure of ABCG2 bound to a substrate in OFC, or bound to two ATP-Mg in OFC, revealing for the first time such state of an ABCG subfamily transporter (Manolaridis et al., 2018). The substrate used for the study, the estrone-3-sulfate, was found to bind near to the twofold symmetry axis of ABCG2, in the same cavity where the two MZ29 were also localized. This demonstrated the polyspecific binding capacity of ABCG2. Mutations of certain residues implicated in binding such as N536A and F439A abolish the transport activity, while V546F impairs the transport and increases the ATPase activity. Of note, these residues were also identified to prevent the binding of inhibitor. Mutants Y432A and A540F of ABCG5/8, equivalent to F439A and V546F in ABCG2, abolish the transport of sterol, hinting at a common architecture of this binding pocket among G-subfamily transporters (Lee et al., 2016).

Other structural data came from a study earlier in 2020 (Orlando & Liao, 2020), reporting structures of ABCG2, firstly in apo state in a different conformation than that of initially released (Fig. 1), and also with bound drugs, imatinib, mitoxantrone and SN38 (Fig. 4). This new conformation (PDB: 6VFX) displays substantial differences with the previous structure (PDB 5NJ3). There is a rearrangement of the membrane helices, close to an OFC, with the large substrate-binding cavity fully collapsed. This is due to a shift of the membrane helix 5, in which residues are close to those of membrane helix 1 and 2 of the other monomer. At the interface between the two monomers, a cluster of sulfur-containing residues is observed, M541, M548, M549 and C544. Their side chains, along with that of F545 are oriented toward the opposing monomer, in contrast with previous structures. Regarding the substrate-bound structures, they look close to that of ABCG2 in complex with MZ29 or 5D3-bound ones, including residues interacting with substrates, as F432, F439 and F545. The authors brought evidences that this apo-closed structure having a collapsed substrate-binding cavity is relevant in cells. In this context, it seems that substrate or 5D3 binding causes a rearrangement of the TMDs to give the shape of the binding cavity initially observed. Another possibility would be that several conformations of TMDs are co-existing in cells, and substrate molecules bind to the proteins whose cavity is accessible.

2.3 Substrate-binding pockets

The structures of ABCG subfamily proteins revealed several differences compared to other ABC human transporters. It has been reported that ABCG2 confers to cancer cells a multidrug resistance phenotype, due to its ability to bind and extrude drugs of different classes. However, compared to other MDR transporters, like P-gp (ABCB1) or MRP1 (ABCC1), despite the common substrates, there are yet some specificities for each of the transporters.

With the 3D structures of these proteins having been resolved, we are now in a better position to understand the way that these proteins bind their substrates. The previously mentioned structural studies revealed how the ABCG2 binding pocket is optimal to accommodate flat polycyclic substrates, in a slit-like manner, whereas ABCB1 displays a globular pocket, where large substrates such as Taxol can be accommodated (Alam, Kowal, Broude, Roninson, & Locher, 2019; Aller et al., 2009) (Fig. 5). As for ABCC1, the structures have revealed a bipartite pocket, with one part accommodating hydrophobic substrates and the other part, positively charged, accommodating glutathione, since many substrates need to be conjugated to the latter to be transported (Johnson & Chen, 2017).

In the case of ABCG2, structures released so far show that substrates and inhibitors bind and overlap in the cavity situated in the crevice between monomers. However, more information is needed to fully understand the transport cycle and the transmission interface between TMD and NBD, and some interesting polymorphisms, like the mutations of R482, changing the substrate spectrum of the protein, even though this residue is not part of the binding pocket (Alqawi, Bates, & Georges, 2004; Ejendal, Diop, Schweiger, & Hrycyna, 2006), suggesting a possible allostery.

Furthermore, 3D structures, mutagenesis and molecular dynamics have shown the presence of a valve, or plug, through which, ABCG2 substrates must pass before getting transported outside of the cell (Khunweeraphong, Szöllősi, Stockner, & Kuchler, 2019). This plug is composed of residues G553 to T559, with two leucine residues, L554 and L555, standing on the top of the cavity, at the interface, and facing each other. This stretch of residues is highly conserved among ABCG and PDR members, and it

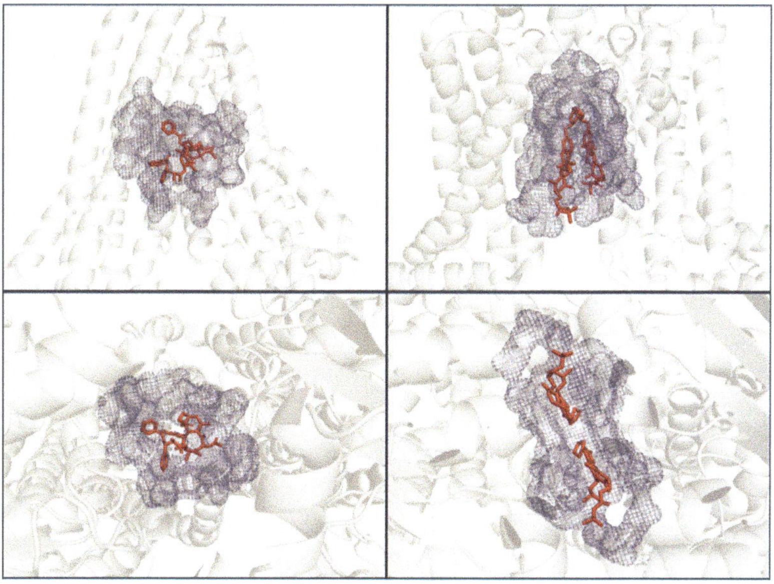

Fig. 5 ***Shapes of type I and type II substrate-binding pockets.*** **Left:** ABCB1 in complex either with taxol (*red*) (6QEX) (Alam et al., 2019) seen in the plane of membrane (top) or seen from NBDs (bottom), with the interacting residues represented as a mesh (*blue*), putting in evidence the globular shape of the type I substrate-binding pocket. **Right:** ABCG2 in complex with its inhibitor MZ29 (*red*) (6FFC) (Jackson et al., 2018) seen in the plane of membrane (top) or seen from NBDs (bottom), with the interacting residues represented as a mesh (*blue*), putting in evidence the slit-like shape of the type II substrate-binding pocket, accommodating flat hydrophobic compounds.

is required for the proper transport function of the protein. Substrate may need to pass through this valve before entering the before mentioned second cavity, which is limited by extracellular loops, the re-entry helix and the TM extremities. This region itself behaves as a lid, allowing the expulsion of the substrates and avoiding their re-entry in the upper cavity.

3. Yeast ABCG/PDR transporters: contribution toward understanding the enigma of non-catalytic nucleotide-binding sites

The yeast PDR transporters display many unique attributes at the level of sequence, structure as well as function. Beside the already mentioned reverse topology, these pumps are characterized by a large *N*-terminal cytosolic domain that includes the *N*-terminal NBD (Moreno, Banerjee, Prasad,

& Falson, 2019). Moreover, the most striking structural feature is the presence of a large ECD made of two long ECLs, ECL3 and partially homologous ECL6 (Lamping et al., 2010). Furthermore, PDR transporters are also asymmetric in nature, like many other clinically relevant counterparts such as CFTR/ABCC1, MRP1/ABCC7 and the Transporter associated with Antigen Processing TAP1/TAP2 proteins (Gao et al., 2000; Procko, Ferrin-O'Connell, Ng, & Gaudet, 2006; Sorum, Töröcsik, & Csanády, 2017). Even though two NBSs exist, only NBS1 participates in the ATP hydrolysis process. Interestingly, while this catalytic NBS comprises consensus sequences, NBS2, the non-catalytic NBS contains the so-called "deviant" sequences (Golin & Ambudkar, 2015), quite typical in the case of PDR transporters (Table 1).

It is important to mention that the "deviant site" sequences are conserved among the PDR transporters, implying a plausible functional importance of this non-catalytic NBS (Moreno et al., 2019). This claim is further strengthened by the observation that chimeras made of either two *N*-NBDs or two *C*-NBDs of *C. albicans* Cdr1p are both inactive, hinting to toward functional role of both NBDs in drug transport mechanism (Saini et al., 2006). Furthermore, a more recent attempt by Gupta et al. to restore symmetry in the NBS of asymmetric Pdr5p also witnessed similar fate, with varying order of loss of function observed in the mutants generated for each motif (Gupta, Kueppers, Hanekop, & Schmitt, 2014). Contrarily, the pieces of evidence from asymmetric proteins other than PDR ones are suggestive of negligible contribution from the non-catalytic NBS in the general catalysis mechanism and define the catalytic NBS as the primary facilitator of transport (Procko et al., 2006; Tsai, Jih, Shimizu, Li, & Hwang, 2010). Another remarkable feature of the PDR pumps is their constitutively active hydrolysis machinery, which is not stimulated by its substrates (Ernst, Kueppers, Stindt, Kuchler, & Schmitt, 2010; Prasad et al., 2015). This particular feature is in stark contrast to other ABC pumps like P-gp/ABCB1, which demonstrates manifold increase in ATP-hydrolysis on exposure to its substrates (Urbatsch & Senior, 1995). Together, the yeast PDR transporters contrast other ABC transporters on many fronts, and there is also a constant debate on the contribution of the non-catalytic NBS, wherein the PDR pumps support an active role unlike many other ABC pumps. In that line, the ease of genetic manipulation in yeasts has provided a strong impetus in solving the conundrum of active versus passive role for the non-catalytic NBS.

Following the suppressor genetics strategy, yeast cells expressing transport-defective single Cdr1p/Pdr5p mutants are exposed to toxic concentrations of xenobiotic substrates, forcing them to generate compensatory mutations alleviating the toxicity. Suppressor mutations are frequently recovered from a secondary site within the target protein. John Golin's group found early success with this strategy while studying Pdr5p transport. Initially the group identified the single mutation S558Y within the membrane helix 2 that rendered the pump-overexpressing yeast several-fold susceptible toward the favored Pdr5p's substrate cycloheximide, without however modifying the ATPase activity of Pdr5p (Sauna et al., 2008). Further, the authors submitted yeast cells expressing that mutant to the suppressor strategy and identified a secondary mutation N242K located close to the Q-loop of *N*-NBD, which is part of the non-catalytic NBS, which reversed the initial phenotype rendering the yeast expressing the Pdr5p S558Y-N242K double mutant, resistant to cycloheximide (Sauna et al., 2008) (Fig. 6). Following the same strategy, John Golin and his colleagues identified another second-site mutation K1016I from the N242K mutant, rescuing the yeast. K1016 is located close to the Signature sequence of the non-catalytic NBS (Downes et al., 2013). Interestingly, among all the suppressor mutations that alleviated the S558Y transport defect, a significant number of them were always located either close to or within the Q-loop of the non-catalytic NBS (Ananthaswamy et al., 2010). All these findings hinted toward a plausible role of the non-catalytic NBS in signal transmission or inter-domain crosstalk. Few studies that followed provided further strength to this claim. Furman et al. in their study with the D-loop mutants identified a D1042N mutation displaying uncoupling of ATPase activity and drug efflux (Furman et al., 2013). Notably the residue is also part of the non-catalytic NBS. Moreover, the study by Gupta et al. (discussed above) also described several deviant-to-consensus mutants demonstrating phenotypes like uncoupling of ATP hydrolysis and substrate transport (Gupta et al., 2014). The authors concluded that the substitutions in the non-catalytic NBS primarily led to defects in the crosstalk between the two NBSs and/or between the NBS(s) and TMD(s).

Investigation with the *C. albicans* Cdr1p counterpart by the group of Rajendra Prasad led to similar conclusions, pointing this time to the *C*-part of the non-hydrolytic NBS of the pump. Shah et al. identified three suppressors for one of the ICL1 mutant, E597A, that mapped to the non-catalytic NBS (Shah, Banerjee, et al., 2015; Shah, Rawal, et al., 2015). Out of the three, while two were located close to the D-loop, one mapped

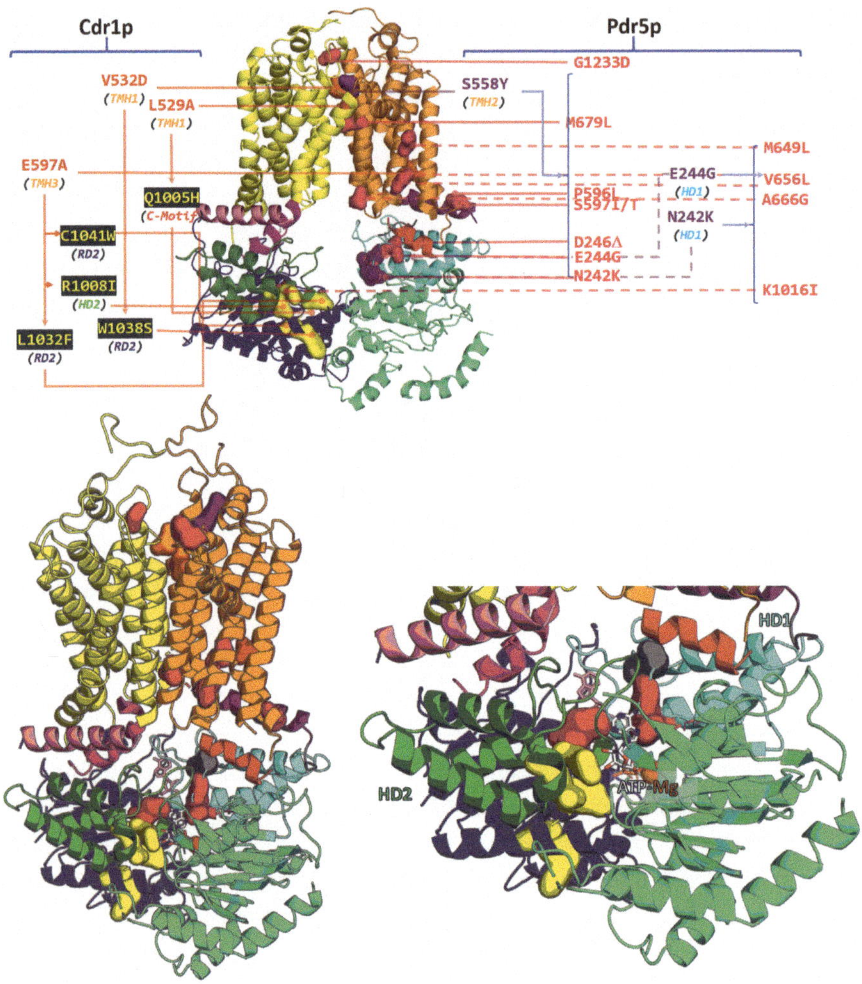

Fig. 6 ***Mapping of primary drug-sensitive and secondary rescuing mutants in yeast type II exporters Cdr1p and Pdr5p.*** **Upper panel:** 3D IFC models of Cdr1p and Pdr5p (Banerjee et al., 2020) with functional regions colored as described in Fig. 3. Primary drug-sensitizing mutants are S558Y (Sauna et al., 2008), N242K and E244G (Downes et al., 2013) in Pdr5p, and E597A (Shah, Rawal, et al., 2015; Shah, Rawal, et al., 2015), V532D (Banerjee et al., 2018) and L529A (Rawal et al., 2013) in Cdr1p. S558Y in Pdr5p is rescued by the secondary mutations N242K, E244G, D246Δ, S597I/T, M679L, P596L or G1233D; N242K is rescued by K1016I, V656L, P596L, L1367F, A670S, or I1063M; E244G is rescued by V656L (Ananthaswamy et al., 2010; Downes et al., 2013; Sauna et al., 2008). E597A, V532D and L529A in Cdr1p are rescued by C1041W, R1008I and L1032F for the first one and W1038S and Q1005H, respectively, for the last two (Banerjee et al., 2018, 2020). **Lower panel, left:** OFC model of ATP-Mg bound Cdr1p (Banerjee et al., 2020) colored as above with primary drug-sensitizing and suppressor mutants of Cdr1. Residues corresponding to those of Pdr5p are also displayed with the same color code. **Right:** zoom to the non-hydrolytic NBS.

close to the Signature sequence. Of note, E597A mutant also demonstrated uncoupling of ATPase activity and substrate transport, whereby the cells overexpressing the variant demonstrated reduced transport and active ATPase machinery. Two more recent studies by Rajendra Prasad and Pierre Falson groups further provided mechanistic insights into the active involvement of the non-catalytic NBS in the transport cycle of PDR pumps. The authors exploited suppressor genetics approach on two transport deficient TMD mutants, V532D and L529A. While V532D shows defect in both ATPase activity and substrate transport, L529A is ATPase competent but shows substrate binding and transport defects. When the V532D expressing *Saccharomyces cerevisiae* strain was exposed to ketoconazole, a suppressor strain survived, harboring the second-site mutation W1038S (Banerjee et al., 2018) (Fig. 6). The position of the residue is close to the D-loop of the non-catalytic NBS. The transport was restored in the suppressor strain V532D-W1038S for most of the substrates, and its ATPase activity was further reduced and showed hypersensitivity toward ketoconazole. Homology model built using ABCG5/G8 as the reference (Nim et al., 2016) showed loss of certain interactions at the level of dimer interface when the serine residue replaces tryptophan at the 1038 position. The authors attributed the resulting increased flexibility as a cause for the change in NBS-TMD crosstalk further leading to reclamation of the transport in the suppressor protein (Banerjee et al., 2018). A more elaborated study was conducted with the L529A mutant and its suppressor L529A-Q1005H, the secondary mutation also happening to be within the signature sequence of the non-catalytic NBS (Banerjee et al., 2020) (Fig. 6). Herein, it was observed that while both the mutant and suppressor proteins were ATPase competent, the single Q1005H mutant while reduced in the ATPase activity showed approximately twofold increased resistance toward majority of the tested substrates when compared to the wild-type protein. The observations made the authors hypothesize a rather direct control of the non-catalytic NBS in substrate-translocation via ATP binding at this site. The authors performed extensive molecular modeling with the help of available asymmetric transporter structures in ATP bound/unbound states as reference. In the ATP-bound structures of MRP1 and CFTR/ABCC7 (Johnson & Chen, 2018; Zhang, Liu, & Chen, 2017, 2018), the authors noted that the glutamine of the Signature sequence interacts with the ribose of the ATP through an H-bond. In contrast, when the position gets occupied by a histidine residue, as the case of H1350 in the non-catalytic NBS of CFTR/ABCC7, there is a noticeable displacement of the histidine and neighbors from the ribose by about 1–2 Å, indicating a rather loose ATP-bound state. Given that the primary defect of substrate binding in case of the L529A variant wasn't restored in the L529A-Q1005H suppressor protein, the authors chose to obtain a mechanistic overview of the possible structural changes that could result in the functional compensation using the ABCG5/G8 structure as a reference. The analysis revealed that Q1005 along with its proximal region indeed contribute to the closing/opening of the non-catalytic NBS, and as a result generates the transmission interface with the coupling and connecting helices, which were identified in the ABCG5/G8 structure (Lee et al., 2016). The loose ATP-bound state, as mentioned earlier in the case of Q1005H substitution, would thus impact the conformational changes at the transmembrane domains leading to transport restoration in the L529A-Q1005H protein (Banerjee et al., 2020).

Mapping all these primary drug-sensitive and secondary rescuing mutants found in the two PDR pumps allows visualization of a structural path from RD2 and HD2 in the *C*-terminal NBD to the *N*-terminal TMD through CpH1 and CnH1 (Fig. 6). As most of the secondary mutations are localized at the interface of the two NBDs regions forming the non-hydrolytic NBS, and taking into account the drastic change of side chain in each case, it is probable that they alter the affinity of the nucleotide at this site in a way that reduces the lifetime of the nucleotide-bound state for the NBS and/or modify its position with respect to the TMD for restoring the translocation process. The non-catalytic NBS, therefore seems essential to allow or to stabilize a functional organization of the different regions and such role seems to be mediated by the two helicoidal domains, HD1 and HD2 located on both sides of the non-hydrolyzed ATP (Fig. 6, right lower panel). Whether the same holds true for certain other asymmetric (and symmetrical?) pumps as well, is a question worth asking.

4. Concluding thoughts and future perspectives

ABC transporters with their immense physiological and clinical implications hold the limelight among all the diverse transporter protein superfamilies. The structural diversity in the ABC-superfamily as revealed in the past decade further adds to its importance as models for understanding transport mechanisms in the living system. Besides, the extensive diversity in the substrate repertoire remains a pertinent question for protein biochemists and the transporter research community. As it is true for all other proteins, the mechanistic keys always lie within the transporter structures. Despite the difficulty in unveiling membrane protein structures, the past decade has witnessed an overwhelming number of ABC transporter structures deposited in the databases, partly due to the significant advancement made on the cryo-EM front. While on the one hand, the increase in structural details has answered many questions related to the transport mechanism of ABC transporters, new questions have also emerged. The structures of ABCG proteins highlighted a novel fold within the ABC exporters and certain important linchpin sites, for instance, the connecting and coupling helices which are essential in the transport mechanism. Since several ABCG exporters function as drug transporters, in-depth details regarding their functioning are need of the hour. In parallel, the insights from the ABCG counterparts in yeast, i.e. the PDR subfamily have aided in understanding the inter- and intra-domain crosstalk operative in the ABCG exporters. Furthermore, the asymmetry in NBDs and its functional consequences has also been revealed by the extensive suppressor analysis employed with Cdr1p and Pdr5p transporters as highlighted in the chapter. Structures of the toxicity rescuing suppressor proteins should help understanding how the non-hydrolytic NBS modulates the transport of substrates at a molecular level. In addition, future studies may also exploit this combinatorial approach to address further questions about the structure and function of this very important subfamily of ABC exporters and possibly other subfamilies as well. The findings would not only advance the overall knowledge of transport mediated by ABC superfamily proteins but also make way for pharmacological interventions.

Acknowledgments

RP acknowledges funding from the Department of Biotechnology grant no. [BT/PR14117/BRB/10/1420/2015]. AB acknowledges funding support from the SERB grant no. SRG/2019/000514. PF acknowledges the *Centre National de la Recherche Scientifique* (CNRS), Lyon University and the French Research Agency (ANR-CLAMP2- 18-CE11-0002-01) for funding. AM and JP PhDs were funded by the Région Auvergne-Rhône-Alpes and the French Ministère de l'Education Nationale, de l'Enseignement Supérieur et de la Recherche, respectively.

References

Alam, A., Kowal, J., Broude, E., Roninson, I., & Locher, K. P. (2019). Structural insight into substrate and inhibitor discrimination by human P-glycoprotein. *Science, 363*(6428), 753–756. https://doi.org/10.1126/science.aav7102.

Alam, A., Küng, R., Kowal, J., McLeod, R. A., Tremp, N., Broude, E. V., et al. (2018). Structure of a zosuquidar and UIC2-bound human-mouse chimeric ABCB1. *Proceedings of the National Academy of Sciences, 115*(9), E1973–E1982. https://doi.org/10.1073/pnas.1717044115.

Aller, S. G., Yu, J., Ward, A., Weng, Y., Chittaboina, S., Zhuo, R., et al. (2009). Structure of P-glycoprotein reveals a molecular basis for poly-specific drug binding. *Science, 323*(5922), 1718–1722. https://doi.org/10.1126/science.1168750.

Alqawi, O., Bates, S., & Georges, E. (2004). Arginine482 to threonine mutation in the breast cancer resistance protein ABCG2 inhibits rhodamine 123 transport while increasing binding. *Biochemical Journal, 382*(Pt 2), 711–716. https://doi.org/10.1042/BJ20040355.

Ananthaswamy, N., Rutledge, R., Sauna, Z. E., Ambudkar, S. V., Dine, E., Nelson, E., et al. (2010). The signaling interface of the yeast multidrug transporter Pdr5 adopts a cis conformation, and there are functional overlap and equivalence of the deviant and canonical Q-loop residues. *Biochemistry, 49*(21), 4440–4449. https://doi.org/10.1021/bi100394j.

Baghel, P., Rawal, M. K., Khan, M. F., Sen, S., Siddiqui, M. H., Chaptal, V., et al. (2017). Multidrug ABC transporter Cdr1 of Candida albicans harbors specific and overlapping binding sites for human steroid hormones transport. *Biochimica et Biophysica Acta (BBA) – Biomembranes, 1859*(10), 1778–1789. https://doi.org/10.1016/j.bbamem.2017.05.011.

Banerjee, A., Moreno, A., Khan, M. F., Nair, R., Sharma, S., Sen, S., et al. (2020). Cdr1p highlights the role of the non-hydrolytic ATP-binding site in driving drug translocation in asymmetric ABC pumps. *Biochimica et Biophysica Acta (BBA) – Biomembranes, 1862*(2), 183131. https://doi.org/10.1016/j.bbamem.2019.183131.

Banerjee, A., Shah, A. H., Redhu, A. K., Moreno, A., Falson, P., & Prasad, R. (2018). W1038 near D-loop of NBD2 is a focal point for inter-domain communication in multidrug transporter Cdr1 of *Candida albicans*. *Biochimica et Biophysica Acta (BBA) – Biomembranes, 1860*(5), 965–972. https://doi.org/10.1016/j.bbamem.2018.01.022.

Basso, C., Vergani, P., Nairn, A. C., & Gadsby, D. C. (2003). Prolonged nonhydrolytic interaction of nucleotide with CFTR's NH_2 - terminal nucleotide binding domain and its role in channel gating. *The Journal of General Physiology, 122*(3), 333–348. https://doi.org/10.1085/jgp.200308798.

Berge, K. E., Tian, H., Graf, G. A., Yu, L., Grishin, N. V., Schultz, J., et al. (2000). Accumulation of dietary cholesterol in sitosterolemia caused by mutations in adjacent ABC transporters. *Science, 290*(5497), 1771–1775. https://doi.org/10.1126/science.290.5497.1771.

Bryan, J., Muñoz, A., Zhang, X., Düfer, M., Drews, G., Krippeit-Drews, P., et al. (2007). ABCC8 and ABCC9: ABC transporters that regulate K+ channels. *Pfluegers Archiv European Journal of Physiology, 453*(5), 703–718. https://doi.org/10.1007/s00424-006-0116-z.

Chen, Z.-S., Robey, R. W., Belinsky, M. G., Shchaveleva, I., Ren, X.-Q., Sugimoto, Y., et al. (2003). Transport of methotrexate, methotrexate polyglutamates, and 17beta-estradiol 17-(beta-D-glucuronide) by ABCG2: Effects of acquired mutations at R482 on methotrexate transport. *Cancer Research, 63*(14), 4048–4054.

Dawson, R. J. P., & Locher, K. P. (2006). Structure of a bacterial multidrug ABC transporter. *Nature, 443*(7108), 180–185. https://doi.org/10.1038/nature05155.

Dean, M., & Annilo, T. (2005). Evolution of the atp-binding cassette (ABC) transporter superfamily in vertebrates. *Annual Review of Genomics and Human Genetics, 6*(1), 123–142. https://doi.org/10.1146/annurev.genom.6.080604.162122.

Desuzinges-Mandon, E., Arnaud, O., Martinez, L., Huché, F., Di Pietro, A., & Falson, P. (2010). ABCG2 transports and transfers heme to albumin through its large extracellular loop. *Journal of Biological Chemistry, 285*(43), 33123–33133. https://doi.org/10.1074/jbc.M110.139170.

Downes, M. T., Mehla, J., Ananthaswamy, N., Wakschlag, A., Lamonde, M., Dine, E., et al. (2013). The transmission interface of the saccharomyces cerevisiae multidrug transporter Pdr5: Val-656 located in intracellular loop 2 plays a major role in Drug resistance. *Agents, Antimicrobial and Chemotherapy, 57*(2), 1025–1034. https://doi.org/10.1128/AAC.02133-12.

Doyle, L. A., & Ross, D. D. (2003). Multidrug resistance mediated by the breast cancer resistance protein BCRP (ABCG2). *Oncogene, 22*(47), 7340–7358. https://doi.org/10.1038/sj.onc.1206938.

Eggensperger, S., & Tampé, R. (2015). The transporter associated with antigen processing: A key player in adaptive immunity. *Biological Chemistry, 396*(9–10), 1059–1072. https://doi.org/10.1515/hsz-2014-0320.

Ejendal, K. F. K., Diop, N. K., Schweiger, L. C., & Hrycyna, C. A. (2006). The nature of amino acid 482 of human ABCG2 affects substrate transport and ATP hydrolysis but not substrate binding. *Protein Science: A Publication of the Protein Society, 15*(7), 1597–1607. https://doi.org/10.1110/ps.051998406.

Elkind, N. B., Szentpétery, Z., Apáti, Á., Özvegy-Laczka, C., Várady, G., Ujhelly, O., et al. (2005). Multidrug transporter ABCG2 prevents tumor cell death induced by the epidermal growth factor receptor inhibitor iressa (ZD1839, gefitinib). *Cancer Research, 65*(5), 1770–1777. https://doi.org/10.1158/0008-5472.CAN-04-3303.

Ernst, R., Kueppers, P., Klein, C. M., Schwarzmueller, T., Kuchler, K., & Schmitt, L. (2008). A mutation of the H-loop selectively affects rhodamine transport by the yeast multidrug ABC transporter Pdr5. *Proceedings of the National Academy of Sciences of the United States of America, 105*(13), 5069–5074. https://doi.org/10.1073/pnas.0800191105.

Ernst, R., Kueppers, P., Stindt, J., Kuchler, K., & Schmitt, L. (2010). Multidrug efflux pumps: Substrate selection in ATP-binding cassette multidrug efflux pumps-first come, first served? *FEBS Journal, 277*(3), 540–549. https://doi.org/10.1111/j.1742-4658.2009.07485.x.

Ford, R. C., & Beis, K. (2019). Learning the ABCs one at a time: Structure and mechanism of ABC transporters. *Biochemical Society Transactions, 47*(1), 23–36. https://doi.org/10.1042/BST20180147.

Furman, C., Mehla, J., Ananthaswamy, N., Arya, N., Kulesh, B., Kovach, I., et al. (2013). The deviant ATP-binding site of the multidrug efflux pump Pdr5 plays an active role in the transport cycle. *Journal of Biological Chemistry, 288*(42), 30420–30431. https://doi.org/10.1074/jbc.M113.494682.

Furuta, T., Sato, Y., & Sakurai, M. (2016). Structural dynamics of the heterodimeric ABC transporter TM287/288 induced by ATP and substrate binding. *Biochemistry, 55*(48), 6730–6738. https://doi.org/10.1021/acs.biochem.6b00947.

Gao, M., Cui, H.-R., Loe, D. W., Grant, C. E., Almquist, K. C., Cole, S. P. C., et al. (2000). Comparison of the functional characteristics of the nucleotide binding domains of multidrug resistance protein 1. *Journal of Biological Chemistry, 275*(17), 13098–13108. https://doi.org/10.1074/jbc.275.17.13098.

Golin, J., & Ambudkar, S. V. V. (2015). The multidrug transporter Pdr5 on the 25th anniversary of its discovery: An important model for the study of asymmetric ABC transporters. *Biochemical Journal, 467*(3), 353–363. https://doi.org/10.1042/BJ20150042.

Grossmann, N., Vakkasoglu, A. S., Hulpke, S., Abele, R., Gaudet, R., & Tampé, R. (2014). Mechanistic determinants of the directionality and energetics of active export by a heterodimeric ABC transporter. *Nature Communications, 5*(1), 1–10. https://doi.org/10.1038/ncomms6419.

Gupta, R. P., Kueppers, P., Hanekop, N., & Schmitt, L. (2014). Generating symmetry in the asymmetric ATP-binding cassette (ABC) transporter Pdr5 from *Saccharomyces cerevisiae.*

Journal of Biological Chemistry, 289(22), 15272–15279. https://doi.org/10.1074/jbc.M114.553065.
Gutmann, H., Hruz, P., Zimmermann, C., Beglinger, C., & Drewe, J. (2005). Distribution of breast cancer resistance protein (BCRP/ABCG2) mRNA expression along the human GI tract. *Biochemical Pharmacology, 70*(5), 695–699. https://doi.org/10.1016/j.bcp.2005.05.031.
Higgins, C. F. (2001). ABC transporters: Physiology, structure and mechanism – an overview. *Research in Microbiology, 152*(3), 205–210. https://doi.org/10.1016/S0923-2508(01)01193-7.
Hohl, M., Briand, C., Grütter, M. G., & Seeger, M. A. (2012). Crystal structure of a heterodimeric ABC transporter in its inward-facing conformation. *Nature Structural & Molecular Biology, 19*(4), 395–402. https://doi.org/10.1038/nsmb.2267.
Hohl, M., Hurlimann, L. M., Bohm, S., Schoppe, J., Grutter, M. G., Bordignon, E., et al. (2014). Structural basis for allosteric cross-talk between the asymmetric nucleotide binding sites of a heterodimeric ABC exporter. *Proceedings of the National Academy of Sciences, 111*(30), 11025–11030. https://doi.org/10.1073/pnas.1400485111.
Holland, I. B. (2011). ABC transporters, mechanisms and biology: An overview. *Essays in Biochemistry, 50*(1), 1–17. https://doi.org/10.1042/bse0500001.
Hutter, C. A. J., Timachi, M. H., Hürlimann, L. M., Zimmermann, I., Egloff, P., Göddeke, H., et al. (2019). The extracellular gate shapes the energy profile of an ABC exporter. *Nature Communications, 10.* https://doi.org/10.1038/s41467-019-09892-6.
Hyde, S. C., Emsley, P., Mimmack, M. M., Gileadi, U., Pearcet, S. R., Gallaghert, M. P., et al. (1990). Structural model of AlP-binding proteins associated with cystic fibrosis, multidrug resistance and bacterial transport. *Nature, 346*, 362–365.
Jackson, S. M., Manolaridis, I., Kowal, J., Zechner, M., Taylor, N. M. I., Bause, M., et al. (2018). Structural basis of small-molecule inhibition of human multidrug transporter ABCG2. *Nature Structural & Molecular Biology, 25*(4), 333–340. https://doi.org/10.1038/s41594-018-0049-1.
Jardetzky, O. (1966). Simple allosteric model for membrane pumps. *Nature, 211*(5052), 969–970. https://doi.org/10.1038/211969a0.
Jeong, C.-B., Kim, H.-S., Kang, H.-M., & Lee, J.-S. (2017). ATP-binding cassette (ABC) proteins in aquatic invertebrates: Evolutionary significance and application in marine ecotoxicology. *Aquatic Toxicology, 185*, 29–39. https://doi.org/10.1016/j.aquatox.2017.01.013.
Jin, M. S., Oldham, M. L., Zhang, Q., & Chen, J. (2012). Crystal structure of the multidrug transporter P-glycoprotein from *Caenorhabditis elegans*. *Nature, 490*(7421), 566–569. https://doi.org/10.1038/nature11448.
Johnson, Z. L., & Chen, J. (2017). Structural basis of substrate recognition by the multidrug resistance protein MRP1. *Cell, 168*(6), 1075–1085.e9. https://doi.org/10.1016/j.cell.2017.01.041.
Johnson, Z. L., & Chen, J. (2018). ATP binding enables substrate release from multidrug resistance protein 1. *Cell, 172*(1–2), 81–89. https://doi.org/10.1016/j.cell.2017.12.005.
Jonker, J. W., Buitelaar, M., Wagenaar, E., Van Der Valk, M. A., Scheffer, G. L., Scheper, R. J., et al. (2002). The breast cancer resistance protein protects against a major chlorophyll-derived dietary phototoxin and protoporphyria. *Proceedings of the National Academy of Sciences of the United States of America, 99*(24), 15649–15654. https://doi.org/10.1073/pnas.202607599.
Khunweeraphong, N., Szöllősi, D., Stockner, T., & Kuchler, K. (2019). The ABCG2 multidrug transporter is a pump gated by a valve and an extracellular lid. *Nature Communications, 10.* https://doi.org/10.1038/s41467-019-13302-2.

Kim, Y., & Chen, J. (2018). Molecular structure of human P-glycoprotein in the ATP-bound, outward-facing conformation. *Science, 359*(6378), 915–919. https://doi.org/10.1126/science.aar7389.

Krishnamurthy, P., Ross, D. D., Nakanishi, T., Bailey-Dell, K., Zhou, S., Mercer, K. E., et al. (2004). The stem cell marker Bcrp/ABCG2 enhances hypoxic cell survival through interactions with heme. *Journal of Biological Chemistry, 279*(23), 24218–24225. https://doi.org/10.1074/jbc.M313599200.

Krishnamurthy, P., & Schuetz, J. D. (2006). Role of ABCG2/BCRP in biology and medicine. *Annual Review of Pharmacology and Toxicology, 46*, 381–410. https://doi.org/10.1146/annurev.pharmtox.46.120604.141238.

Krishnamurthy, P., Xie, T., & Schuetz, J. D. (2007). The role of transporters in cellular heme and porphyrin homeostasis. *Pharmacology & Therapeutics, 114*(3), 345–358. https://doi.org/10.1016/j.pharmthera.2007.02.001.

Kumari, S., Kumar, M., Khandelwal, N. K., Kumari, P., Varma, M., Vishwakarma, P., et al. (2018). ABC transportome inventory of human pathogenic yeast Candida glabrata: Phylogenetic and expression analysis. *PLoS One, 13*(8), 1–21. https://doi.org/10.1371/journal.pone.0202993.

Lamping, E., Baret, P. V., Holmes, A. R., Monk, B. C., Goffeau, A., & Cannon, R. D. (2010). Fungal PDR transporters: Phylogeny, topology, motifs and function. *Fungal Genetics and Biology, 47*(2), 1–36. https://doi.org/10.1016/j.fgb.2009.10.007.

Lee, J.-Y., Kinch, L. N., Borek, D. M., Wang, J., Wang, J., Urbatsch, I. L., et al. (2016). Crystal structure of the human sterol transporter ABCG5/ABCG8. *Nature, 533*(7604), 561–564. https://doi.org/10.1038/nature17666.

Lee, J. Y., Yang, J. G., Zhitnitsky, D., Lewinson, O., & Rees, D. C. (2014). Structural basis for heavy metal detoxification by an Atm1-type ABC exporter. *Science, 343*(6175), 1133–1136. https://doi.org/10.1126/science.1246489.

Liu, F., Zhang, Z., Csanády, L., Gadsby, D. C., & Chen, J. (2017). Molecular structure of the human CFTR ion channel. *Cell, 169*(1), 85–95. https://doi.org/10.1016/j.cell.2017.02.024.

van Loevezijn, A., Allen, J. D., Schinkel, A. H., & Koomen, G. J. (2001). Inhibition of BCRP-mediated drug efflux by fumitremorgin-type indolyl diketopiperazines. *Bioorganic & Medicinal Chemistry Letters, 11*(1), 29–32. https://doi.org/10.1016/s0960-894x(00)00588-6.

Luckenbach, T., Fischer, S., & Sturm, A. (2014). Current advances on ABC drug transporters in fish. *Comparative Biochemistry and Physiology - Part C: Toxicology & Pharmacology, 165*, 28–52. https://doi.org/10.1016/j.cbpc.2014.05.002.

Lu, K., Lee, M.-H., Hazard, S., Brooks-Wilson, A., Hidaka, H., Kojima, H., et al. (2001). Two genes that map to the STSL locus cause sitosterolemia: Genomic structure and spectrum of mutations involving sterolin-1 and sterolin-2, encoded by ABCG5 and ABCG8, respectively. *The American Journal of Human Genetics, 69*(2), 278–290.

Manolaridis, I., Jackson, S. M., Taylor, N. M. I., Kowal, J., Stahlberg, H., & Locher, K. P. (2018). Cryo-EM structures of a human ABCG2 mutant trapped in ATP-bound and substrate-bound states. *Nature, 563*(7731), 426–430. https://doi.org/10.1038/s41586-018-0680-3.

Mao, Q. (2008). BCRP/ABCG2 in the placenta: Expression, function and regulation. *Pharmaceutical Research, 25*(6), 1244–1255. https://doi.org/10.1007/s11095-008-9537-z.

Martinez, L., Arnaud, O., Henin, E., Tao, H., Chaptal, V., Doshi, R., et al. (2014). Understanding polyspecificity within the substrate-binding cavity of the human multidrug resistance P-glycoprotein. *FEBS Journal, 281*(3), 673–682. https://doi.org/10.1111/febs.12613.

Moncalian, G., Lengsfeld, B., Bhaskara, V., Hopfner, K.-P., Karcher, A., Alden, E., et al. (2004). The rad50 signature motif: Essential to ATP binding and biological function.

Journal of Molecular Biology, 335(4), 937–951. https://doi.org/10.1016/j.jmb.2003.11.026.

Moreno, A., Banerjee, A., Prasad, R., & Falson, P. (2019). PDR-like ABC systems in pathogenic fungi. *Research in Microbiology, 170*(8), 417–425. https://doi.org/10.1016/j.resmic.2019.09.002.

Murina, V., Kasari, M., Takada, H., Hinnu, M., Saha, C. K., Grimshaw, J. W., et al. (2019). ABCF ATPases involved in protein synthesis, ribosome assembly and antibiotic resistance: structural and functional diversification across the Tree of Life. *Journal of Molecular Biology, 431*(18), 3568–3590. https://doi.org/10.1016/j.jmb.2018.12.013.

Navarro-Quiles, C., Mateo-Bonmatí, E., & Micol, J. L. (2018). ABCE proteins: >From molecules to development. *Frontiers in Plant Science, 9*. https://doi.org/10.3389/fpls.2018.01125.

Nim, S., Lobato, L. G., Moreno, A., Chaptal, V., Rawal, M. K., Falson, P., et al. (2016). Atomic modelling and systematic mutagenesis identify residues in multiple drug binding sites that are essential for drug resistance in the major Candida transporter Cdr1. *Biochimica et Biophysica Acta Biomembranes, 1858*(11), 2858–2870. https://doi.org/10.1016/j.bbamem.2016.08.011.

Oldham, M. L., & Chen, J. (2011). Snapshots of the maltose transporter during ATP hydrolysis. *Proceedings of the National Academy of Sciences of the United States of America, 108*(37), 15152–15156. https://doi.org/10.1073/pnas.1108858108.

Orelle, C., Alvarez, F. J. D., Oldham, M. L., Orelle, A., Wiley, T. E., Chen, J., et al. (2010). Dynamics of alpha-helical subdomain rotation in the intact maltose ATP-binding cassette transporter. *Proceedings of the National Academy of Sciences of the United States of America, 107*(47), 20293–20298. https://doi.org/10.1073/pnas.1006544107.

Orelle, C., Mathieu, K., & Jault, J.-M. (2019). Multidrug ABC transporters in bacteria. *Research in Microbiology, 170*(8), 381–391. https://doi.org/10.1016/j.resmic.2019.06.001.

Orlando, B. J., & Liao, M. (2020). ABCG2 transports anticancer drugs via a closed-to-open switch. *Nature Communications, 11*(1), 2264. https://doi.org/10.1038/s41467-020-16155-2.

Özvegy, C., Váradi, A., & Sarkadi, B. (2002). Characterization of drug transport, ATP hydrolysis, and nucleotide trapping by the human ABCG2 multidrug transporter modulation of substrate specificity by a point mutation. *Journal of Biological Chemistry, 277*(50), 47980–47990. https://doi.org/10.1074/jbc.M207857200.

Popovic, M., Zaja, R., Loncar, J., & Smital, T. (2010). A novel ABC transporter: The first insight into zebrafish (*Danio rerio*) ABCH1. *Marine Environmental Research, 69*, S11–S13. https://doi.org/10.1016/j.marenvres.2009.10.016.

Prasad, R., Balzi, E., Banerjee, A., & Khandelwal, N. K. (2019). All about CDR transporters: Past, present, and future. *Yeast, 36*(4), 223–233. https://doi.org/10.1002/yea.3356.

Prasad, R., Banerjee, A., Khandelwa, N. K., Dhamgaye, S., Khandelwal, N. K., & Dhamgaye, S. (2015). The ABCs of Candida albicans multidrug transporter Cdr1. *Eukaryotic Cell, 14*(12), 1154–1164. https://doi.org/10.1128/EC.00137-15.

Prasad, R., Khandelwal, N. K., & Banerjee, A. (2016). Yeast ABC transporters in lipid trafficking. *Genetics, Fungal and Biology, 93*, 25–34. https://doi.org/10.1016/j.fgb.2016.05.008.

Procko, E., Ferrin-O'Connell, I., Ng, S.-L., & Gaudet, R. (2006). Distinct structural and functional properties of the ATPase sites in an asymmetric ABC transporter. *Molecular Cell, 24*(1), 51–62. https://doi.org/10.1016/j.molcel.2006.07.034.

Qiu, Y., Liu, H., Qing, Y., Yang, M., Tan, X., Zhao, M., et al. (2014). The ABCG2 gene Q141K polymorphism contributes to an increased risk of gout: A meta-analysis of 2185 cases. *Modern Rheumatology, 24*(5), 829–834. https://doi.org/10.3109/14397595.2013.875639.

Rawal, M. K., Khan, M. F., Kapoor, K., Goyal, N., Sen, S., Saxena, A. K., et al. (2013). Insight into pleiotropic drug resistance ATP-binding cassette pump drug transport through mutagenesis of Cdr1p transmembrane domains. *Journal of Biological Chemistry, 288*(34), 24480–24493. https://doi.org/10.1074/jbc.M113.488353.

Redhu, A. K., Banerjee, A., Shah, A. H., Moreno, A., Rawal, M. K., Nair, R., et al. (2018). Molecular basis of substrate polyspecificity of the Candida albicans Mdr1p multidrug/H+ antiporter. *Journal of Molecular Biology, 430*(5), 682–694. https://doi.org/10.1016/j.jmb.2018.01.005.

Rees, D. C., Johnson, E., & Lewinson, O. (2009). ABC transporters: The power to change. *Nature Reviews Molecular Cell Biology, 10*(3), 218–227. https://doi.org/10.1038/nrm2646.

Rich, D. P., Anderson, M. P., Gregory, R. J., Cheng, S. H., Paul, S., Jefferson, D. M., et al. (1990). Expression of cystic fibrosis transmembrane conductance regulator corrects defective chloride channel regulation in cystic fibrosis airway epithelial cells. *Nature, 347*(6291), 358–363. https://doi.org/10.1038/347358a0.

Robey, R. W., Pluchino, K. M., Hall, M. D., Fojo, A. T., Bates, S. E., & Gottesman, M. M. (2018). Revisiting the role of ABC transporters in multidrug-resistant cancer. *Nature Reviews Cancer, 18*(7), 452–464. https://doi.org/10.1038/s41568-018-0005-8.

Robey, R. W., Steadman, K., Polgar, O., Morisaki, K., Blayney, M., Mistry, P., et al. (2004). Pheophorbide a is a specific probe for ABCG2 function and inhibition. *Cancer Research, 64*(4), 1242–1246. https://doi.org/10.1158/0008-5472.can-03-3298.

Roe, M., Folkes, A., Ashworth, P., Brumwell, J., Chima, L., Hunjan, S., et al. (1999). Reversal of P-glycoprotein mediated multidrug resistance by novel anthranilamide derivatives. *Bioorganic & Medicinal Chemistry Letters, 9*(4), 595–600. https://doi.org/10.1016/s0960-894x(99)00030-x.

Saini, P., Gaur, N. A., & Prasad, R. (2006). Chimeras of the ABC drug transporter Cdr1p reveal functional indispensability of transmembrane domains and nucleotide-binding domains, but transmembrane segment 12 is replaceable with the corresponding homologous region of the non-drug transporter Cdr3p. *Microbiology, 152*(5), 1559–1573. https://doi.org/10.1099/mic.0.28471-0.

Saraste, M., Sibbald, P. R., & Wittinghofer, A. (1990). The P-loop—a common motif in ATP- and GTP-binding proteins. *Trends in Biochemical Sciences, 15*(11), 430–434. https://doi.org/10.1016/0968-0004(90)90281-F.

Sarkadi, B., Ozvegy-Laczka, C., Német, K., & Váradi, A. (2004). ABCG2—a transporter for all seasons. *FEBS Letters, 567*(1), 116–120. https://doi.org/10.1016/j.febslet.2004.03.123.

Sauna, Z. E., Bohn, S. S., Rutledge, R., Dougherty, M. P., Cronin, S., May, L., et al. (2008). Mutations define cross-talk between the N-terminal nucleotide-binding domain and transmembrane helix-2 of the yeast multidrug transporter Pdr5 possible conservation of a signaling interface for coupling atp hydrolysis to drug transport. *Journal of Biological Chemistry, 283*(50), 35010–35022. https://doi.org/10.1074/jbc.M806446200.

Shah, A. H., Banerjee, A., Rawal, M. K., Saxena, A. K., Mondal, A. K., & Prasad, R. (2015). ABC transporter cdr1p harbors charged residues in the intracellular loop and nucleotide-binding domain critical for protein trafficking and drug resistance. *FEMS Yeast Research, 15*(5), 1–10. https://doi.org/10.1093/femsyr/fov036.

Shah, A. H., Rawal, M. K., Dhamgaye, S., Komath, S. S., Saxena, A. K., & Prasad, R. (2015). Mutational analysis of intracellular loops identify cross talk with nucleotide binding domains of yeast ABC transporter Cdr1p. *Scientific Reports, 5*(May), 1–17. https://doi.org/10.1038/srep11211.

Smith, P. C., Karpowich, N., Millen, L., Moody, J. E., Rosen, J., Thomas, P. J., et al. (2002). ATP binding to the motor domain from an ABC transporter drives formation of a nucleotide sandwich dimer. *Molecular Cell, 10*(1), 139–149.

Sorum, B., Töröcsik, B., & Csanády, L. (2017). Asymmetry of movements in CFTR's two ATP sites during pore opening serves their distinct functions. *ELife, 6*. https://doi.org/10.7554/eLife.29013.

Srinivasan, V., Pierik, A. J., & Lill, R. (2014). Crystal structures of nucleotide-free and glutathione-bound mitochondrial ABC transporter Atm1. *Science, 343*(6175), 1137–1140. https://doi.org/10.1126/science.1246729.

Szöllősi, D., Rose-Sperling, D., Hellmich, U. A., & Stockner, T. (2018). Comparison of mechanistic transport cycle models of ABC exporters. *Biochimica et Biophysica Acta – Biomembranes, 1860*(4), 818–832. https://doi.org/10.1016/j.bbamem.2017.10.028.

Tamaki, A., Ierano, C., Szakacs, G., Robey, R. W., & Bates, S. E. (2011). The controversial role of ABC transporters in clinical oncology. *Essays in Biochemistry, 50*(1), 209–232. https://doi.org/10.1042/bse0500209.

Taylor, N. M. I., Manolaridis, I., Jackson, S. M., Kowal, J., Stahlberg, H., & Locher, K. P. (2017). Structure of the human multidrug transporter ABCG2. *Nature, 546*(7659), 504–509. https://doi.org/10.1038/nature22345.

Thomas, C., & Tampé, R. (2020). Structural and mechanistic principles of ABC transporters. *Annual Review of Biochemistry, 89*(1), 605–636. https://doi.org/10.1146/annurev-biochem-011520-105201.

Tsai, M.-F., Jih, K.-Y., Shimizu, H., Li, M., & Hwang, T.-C. (2010). Optimization of the degenerated interfacial ATP binding site improves the function of disease-related mutant cystic fibrosis transmembrane conductance regulator (CFTR) channels. *Journal of Biological Chemistry, 285*(48), 37663–37671. https://doi.org/10.1074/jbc.M110.172817.

Urbatsch, I. L., & Senior, A. E. (1995). Effects of lipids on ATPase activity of purified Chinese hamster P-glycoprotein. *Archives of Biochemistry and Biophysics, 316*(1), 135–140. https://doi.org/10.1006/abbi.1995.1020.

Vergani, P., Lockless, S. W., Nairn, A. C., & Gadsby, D. C. (2005). CFTR channel opening by ATP-driven tight dimerization of its nucleotide-binding domains. *Nature, 433*(7028), 876–880. https://doi.org/10.1038/nature03313.

Vergani, P., Nairn, A. C., & Gadsby, D. C. (2003). On the mechanism of MgATP-dependent gating of CFTR Cl– channels. *The Journal of General Physiology, 121*(1), 17–36. https://doi.org/10.1085/jgp.20028673.

Verrier, P. J., Bird, D., Burla, B., Dassa, E., Forestier, C., Geisler, M., et al. (2008). Plant ABC proteins – a unified nomenclature and updated inventory. *Trends in Plant Science, 13*(4), 151–159. https://doi.org/10.1016/j.tplants.2008.02.001.

Walker, J. E., Saraste, M., Runswick, M. J., & Gay, N. J. (1982). Distantly related sequences in the alpha- and beta-subunits of ATP synthase, myosin, kinases and other ATP-requiring enzymes and a common nucleotide binding fold. *The EMBO Journal, 1*(8), 945–951.

Wang, L., Johnson, Z. L., Wasserman, M. R., Levring, J., Chen, J., & Liu, S. (2020). Characterization of the kinetic cycle of an ABC transporter by single-molecule and cryo-EM analyses. *ELife, 9*, e56451. https://doi.org/10.7554/eLife.56451.

Ward, A., Reyes, C. L., Yu, J., Roth, C. B., & Chang, G. (2007). Flexibility in the ABC transporter MsbA: Alternating access with a twist. *Proceedings of the National Academy of Sciences of the United States of America, 104*(48), 19005–19010. https://doi.org/10.1073/pnas.0709388104.

Ward, A. B., Szewczyk, P., Grimard, V., Lee, C.-W., Martinez, L., Doshi, R., et al. (2013). Structures of P-glycoprotein reveal its conformational flexibility and an epitope on the nucleotide-binding domain. *Proceedings of the National Academy of Sciences of the United States of America, 110*(33), 13386–13391. https://doi.org/10.1073/pnas.1309275110.

Woodward, O. M., Köttgen, A., Coresh, J., Boerwinkle, E., Guggino, W. B., & Köttgen, M. (2009). Identification of a urate transporter, ABCG2, with a common functional polymorphism causing gout. *Proceedings of the National Academy of Sciences of the United States of America, 106*(25), 10338–10342. https://doi.org/10.1073/pnas.0901249106.

Yang, R., Hou, Y., Campbell, C. A., Palaniyandi, K., Zhao, Q., Bordner, A. J., et al. (2011). Glutamine residues in Q-loops of multidrug resistance protein MRP1 contribute to ATP binding via interaction with metal cofactor. *Biochimica et Biophysica Acta (BBA) – Biomembranes, 1808*(7), 1790–1796. https://doi.org/10.1016/j.bbamem.2011.02.004.

Ye, J., Osborne, A. R., Groll, M., & Rapoport, T. A. (2004). RecA-like motor ATPases—lessons from structures. *Biochimica et Biophysica Acta Bioenergetics, 1659*(1), 1–18. https://doi.org/10.1016/j.bbabio.2004.06.003.

Zaitseva, J., Jenewein, S., Jumpertz, T., Holland, I. B., & Schmitt, L. (2005). H662 is the linchpin of ATP hydrolysis in the nucleotide-binding domain of the ABC transporter HlyB. *The EMBO Journal, 24*(11), 1901–1910. https://doi.org/10.1038/sj.emboj.7600657.

Zhang, Z., Liu, F., & Chen, J. (2017). Conformational changes of CFTR upon phosphorylation and ATP binding. *Cell, 170*(3), 483–491. https://doi.org/10.1016/j.cell.2017.06.041.

Zhang, Z., Liu, F., & Chen, J. (2018). Molecular structure of the ATP-bound, phosphorylated human CFTR. *Proceedings of the National Academy of Sciences of the United States of America, 115*(50), 12757–12762. https://doi.org/10.1073/pnas.1815287115.

Zhang, J., Yu, Y.-C., Yeh, J.-T., & Hwang, T.-C. (2018). Functional characterization reveals that zebrafish CFTR prefers to occupy closed channel conformations. *PLoS One, 13*(12). https://doi.org/10.1371/journal.pone.0209862.

CHAPTER NINE

Serum albumin: clinical significance of drug binding and development as drug delivery vehicle

Saad Tayyab[a,*] and Shevin Rizal Feroz[b]
[a]Biochemistry Programme, Institute of Biological Sciences, Faculty of Science, University of Malaya, Kuala Lumpur, Malaysia
[b]Department of Biological Sciences and Biotechnology, Faculty of Science and Technology, Universiti Kebangsaan Malaysia, Bangi, Selangor, Malaysia
*Corresponding author: E-mail: saadtayyab2004@yahoo.com

Contents

1. Clinical implications of drug–serum albumin binding 194
   1.1 Drug–drug interactions and displacement 194
      1.1.1 Criteria of clinically significant drug–drug interactions 195
      1.1.2 Examples of clinically significant drug–drug interactions 197
   1.2 Diseased states 199
2. HSA-based drug delivery systems 201
   2.1 Cellular receptors of HSA 202
   2.2 Noncovalent drug–albumin association 204
   2.3 Covalent drug–albumin conjugates 206
   2.4 Albumin fusions 208
   2.5 HSA nanosystems 211
Acknowledgments 213
References 213

Abstract

Human serum albumin, the primary transport and reservoir protein in the human circulatory system, interacts with numerous endogenous and exogenous ligands of varying structural characteristics. The mode of binding of drugs to albumin is central to understanding their pharmacokinetic profiles and has a major influence on their *in vivo* efficacy. Altered drug binding to albumin due to drug-drug interactions or abnormal physiology may result in marked changes in the active drug concentration, thus affecting its pharmacokinetic and pharmacodynamic properties. The propensity of drug-drug interaction to be clinically significant as well as possible exploitation of such

Advances in Protein Chemistry and Structural Biology, Volume 123
ISSN 1876-1623
https://doi.org/10.1016/bs.apcsb.2020.08.003

interactions for therapeutic purposes is reviewed. Being the major organs of albumin metabolism, any impairment in the liver and kidney functions frequently alter the level of serum albumin, which affects the pharmacokinetic profiles of drugs and may have serious clinical implications. The natural function of serum albumin as a drug carrier is facilitated by its interaction with various cellular receptors. These receptors not only promote the uptake of drugs into cells but are also responsible for the extraordinarily long circulatory half-life of albumin. This property in combination with the presence of multiple ligand binding pockets have led to the emergence of serum albumin as an attractive vehicle for novel drug delivery systems. Here, we provide an overview of various albumin-based drug delivery strategies, classified according to their methods of drug attachment, and highlight their experimental and clinical successes.

1. Clinical implications of drug–serum albumin binding

Generally, normal drug binding to human serum albumin (HSA) is altered under two scenarios. The first is the result of drug–drug interactions, due to the presence of other ligands (typically other drugs) that are known to bind to HSA and leads to the displacement of the first drug from the protein. The second is the consequence of abnormal physiology in diseased states, where the binding capacity of HSA can vary significantly from the normal range. Under certain conditions, both events can alter the concentrations of the free/bound fractions of a drug in the plasma and thus, its pharmacodynamic and pharmacokinetic profiles.

1.1 Drug–drug interactions and displacement

Considering the fact that HSA is able to bind a wide variety of drugs in a rather promiscuous manner, it is not surprising that co-administration of drugs may sometimes lead to drug–drug interactions and the displacement of one of the drugs from the protein. This could arise either from direct binding of ligands to the same binding site on HSA (competitive displacement), or from conformational changes at the binding locus of the displaced drug when the displacing drug interacts with HSA at another site (allosteric displacement) (Yamasaki, Chuang, Maruyama, & Otagiri, 2013).

Depending on the property of the drug in question, its displacement from HSA can have a major influence on its pharmacological properties. Firstly, the concentration of its free form will be elevated in the plasma. Consequently, this exposes a greater amount of the drug to the action of detoxification enzymes that metabolize the drug into more soluble species for easier clearance. As a result, its *in vivo* half-life will be shorter. At the same time, assuming the drug that is being displaced is in its active form, this will result in a transient enhancement of pharmacological effect (both desired and adverse), as a greater amount of the drug is now available for interaction with its target receptor (Rimac, Debeljak, Bojic, & Miller, 2017). Although drug—drug interactions are common, the resulting displacement effect however, does not always translate into clinically significant physiological changes (Hochman, Tang, & Prueksaritanont, 2015). For the displacement to be significant, certain conditions must be satisfied, as elaborated below.

1.1.1 Criteria of clinically significant drug—drug interactions

The most important criterion for a drug—drug interaction to be classified as clinically significant is that the displaced drug should have a high protein binding capacity (greater than 90%) and a narrow therapeutic index (toxic, when present at concentrations slightly higher than the therapeutic range) (Benet & Hoener, 2002). Thus, a small reduction in the bound fraction of the drug will cause the concentration of its unbound fraction to be raised significantly, leading to its increased pharmacological activity or toxicity. Since only this fraction is pharmacologically active, any alteration to its degree of binding to HSA will necessarily affect its active concentration, even when the total drug concentration remains constant. Unfortunately, this can lead to cases of under/overdosing of patients, if the drug's therapeutic index is not properly considered (Bohnert & Gan, 2013). For instance, in subjects with higher than normal protein binding (reduced free drug concentration), the administered dose might be thought as sufficient (because the total drug concentration is unchanged), while in reality the active concentration is actually inadequate to exert the desired pharmacological effect. On the other hand, in cases where the protein binding is lower than normal (increased free drug concentration), higher clearance rate will generally restore the active concentration back to normal levels after a transient period of high pharmacological activity. However, the total drug concentration will now be lower than expected, which might lead to the patient to be administered a higher dose (even when the active concentration is normal), resulting in toxicity. In addition, in order for the displacement action to occur, the displacer drug itself should be highly protein-bound and be present at a high concentration to occupy its binding site(s) on HSA (Rimac et al., 2017).

Table 1 Classification of common drugs according to hepatic extraction ratio.

Low ER	Medium ER	High ER
Alfentanil	Midazolam	Alprenolol
Carbamazepine	Nifedipine	Diclofenac
Diazepam	Desipramine	Glyceryl trinitrate
Erythromycin		Imipramine
Isoniazid		Ketamine
Lorazepam		Lignocaine
Methadone		Meperidine
Phenobarbital		Metoprolol
Phenytoin		Morphine
Procainamide		Nitroglycerin
Rocuronium		Propofol
Tolbutamide		Propoxyphene
Theophylline		Propranolol
Warfarin		Tocainide
		Verapamil

Moreover, the extraction ratio (ER) of the displaced drug should also be considered (Table 1). The ER is defined as the rate of plasma removal of a drug as it passes through an organ of elimination (primarily the liver and kidneys) divided by the rate at which it enters the organ. In other words, the ER quantifies the efficiency of drug elimination from circulation by an organ. A drug with low ER (close to 0) mostly remains in plasma after a single pass through the organ, while one with a high ER (close to 1) will be mostly excreted. The influence of protein binding and displacement on a drug's clearance varies according to its ER (Roberts, Pea, & Lipman, 2013; Smith, Di, & Kerns, 2010). Generally, for drugs with low ER, only the free fraction is susceptible for elimination. Hence, with the increase in free drug concentration, the rate of elimination will also increase. Consequently, the concentration of the unbound fraction will remain relatively unchanged, and adjustments to the administered dosage will usually not be necessary (Fig. 1A). In contrast, a high ER drug will be cleared from the plasma when it enters organs of elimination irrespective of its binding to proteins, hence the protein-bound fraction is equally susceptible to clearance. Therefore, the clearance rate will be more or less constant, even if the concentration of the free fraction is increased due to drug displacement. Subsequently, this will result in elevated levels of the free drug and thus, increased pharmacological activity (Fig. 1B). However, this may also increase the risk of adverse effects associated with the drug.

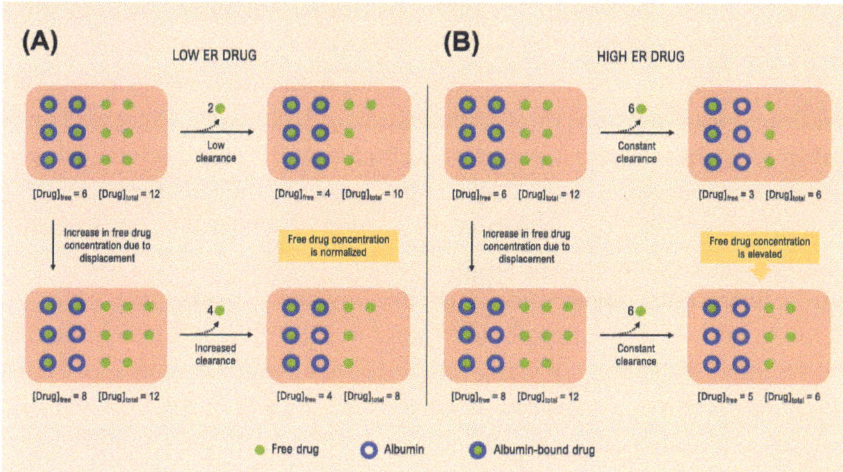

Fig. 1 Behavior of low and high ER drugs upon increase in the free drug concentration (reduced protein binding). (A) The clearance rate of a low ER drug increases in proportion to its free concentration, which normalizes the final free drug concentration. (B) The clearance rate of a high ER drug is relatively unaffected by changes to its free concentration, resulting in elevated free drug concentration.

1.1.2 Examples of clinically significant drug–drug interactions

Since HSA also binds various endogenous ligands in addition to drugs, it can be expected that interaction of compounds such as heme (or its derivatives) or fatty acids with HSA can influence the binding of drugs to this protein, and vice versa. For example, in conditions characterized by increased heme–HSA interaction, the binding of site I drugs to HSA has been shown to be weaker than normal. Conversely, presence of drugs that bind to site I of HSA has been shown to elevate plasma levels of ferric heme (Fe^{III}), which shares the same binding site (Ascenzi & Fasano, 2010). Here, bilirubin, a byproduct of heme catabolism requires special attention, which is displaced by various drugs including phenylbutazone, warfarin, furosemide, valproate, and sulfonamides (Kragh-Hansen, Chuang, & Otagiri, 2002). Hence, administration of such drugs should be avoided in individuals with elevated levels of plasma bilirubin such as newborns which are at risk of kernicterus. As such, suggestions have been made that therapeutics intended for use in newborns should be assessed for their ability to displace bilirubin (Amin, 2016). Fatty acids too can act as displacer molecules (both competitively and allosterically) as they are abundantly present in diet and bind to multiple sites on HSA. Indeed, their binding to HSA has been linked to decreased binding of drugs such as warfarin, diazepam, ibuprofen, furosemide, thiamylal and flurbiprofen (Kragh-Hansen et al., 2002).

Being highly bound to HSA (~99%) along with a narrow therapeutic index, drug–drug interactions involving warfarin seem to be easily explained by displacement events. However, it is also a low ER drug, which complicates the understanding of the clinical significance of these drug–drug interactions. For instance, the anticoagulant activity of warfarin (a site I ligand), has been reported to increase under the co-administration of phenylbutazone or bucolome (Hatakeyama, Niwano, Niwano, Kosukegawa, & Izumi, 2010). As both of these drugs bind to site I of HSA, it was initially assumed that the increased activity of warfarin was due to an increase in its active concentration upon being displaced. If we consider the fact that it has a low ER, then this explanation alone is not sufficient as its active concentration is expected to be fairly constant. Indeed, it is now established that enhancement of its activity is also contributed by inhibition of its metabolism, resulting in reduced plasma clearance (Yamasaki et al., 2013). Nevertheless, there are interactions involving warfarin that are explained only through displacement, e.g. warfarin–chloral hydrate and warfarin–clofibrate interactions (Fanali et al., 2012). There is also a report describing that patients who were on long-term warfarin medication, suffered from increased bleeding incidences, when administered with valproic acid (Yoon, Giraldo, & Wijdicks, 2011).

Nonetheless, drug–drug interactions can also be manipulated to produce desirable pharmacological effects. Diclofenac, an anti-inflammatory drug that is commonly used to treat patients with rheumatoid arthritis, binds specifically to site II of HSA and is classified as a high ER drug. It has been demonstrated that its pain-relieving effect is enhanced in patients receiving oral dosage of nabumetone, which is metabolized into the active 6-methoxy-2-naphthylacetic acid (6-MNA), also a site II ligand (Setoguchi et al., 2013). Additionally, it has also been observed that CYP2C9, the enzyme responsible for the metabolism of diclofenac is not inhibited by nabumetone/6-MNA (Zi et al., 2010). Thus, increased diclofenac activity caused by elevated active concentration in this case can be attributed to displacement by 6-MNA alone.

Clinical application of 6-MNA as a displacer is also useful in radiology. Iofetamine labeled with iodine-123, has long been used for cerebral imaging. However, its tendency to bind to HSA and α_1-acid glycoprotein limits its free plasma concentration. Concomitant administration of 6-MNA however, has led to the displacement of iofetamine from HSA and increased its free plasma concentration (Kawai et al., 2009). As a result, it can accumulate

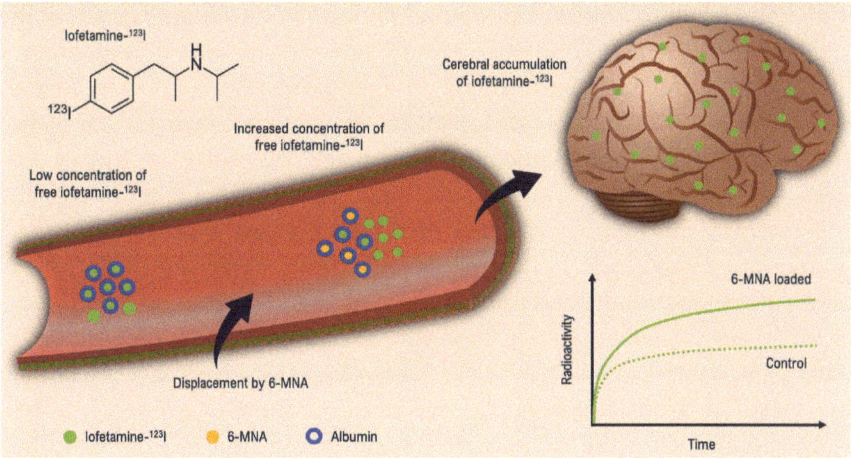

Fig. 2 Drug–drug interaction between iofetamine-^{123}I and 6-MNA. Displacement of iofetamine-^{123}I from site II of HSA by 6-MNA leads to an increase in the free concentration of the radioactive drug. More iofetamine-^{123}I is thus accumulated in the brain at a faster rate, producing better imaging results. The time–activity curves show that the cerebral radioactivity of iofetamine-^{123}I is higher in 6-MNA-loaded subjects compared to those administered an equivalent dose of only iofetamine-^{123}I.

faster and at higher concentrations at the target tissue (Fig. 2). Such phenomenon is beneficial as it translates into shorter imaging time and higher signal-to-noise ratio. Furthermore, patients are exposed to lower doses of radiation and the radioactive agent can be cleared faster (Rimac et al., 2017).

1.2 Diseased states

A few diseases, especially those that affect hepatic and renal functions, have been shown to alter the normal binding of drugs to HSA. This occurs mainly due to changes in the plasma concentration and structure of HSA, as well as accumulation of endogenous substances that can potentially interfere with drug–HSA interactions.

Among the many functions of the liver, synthesis of plasma proteins including HSA and metabolism of xenobiotic compounds for elimination have an exceptionally large influence on the pharmacokinetic profile of a drug. For example, plasma albumin concentration can be reduced in individuals with impaired hepatic function. This decreases the availability of sites for drug binding, leading to increased likelihood of unwanted displacement events and higher than normal free drug concentrations. In a previous study, free tiagabine fraction was found to be 43% higher in patients suffering from liver disease than that present in normal subjects upon administration of the same dose of the drug (Lau et al., 1997). Aside from hypoalbuminemia, a drug's distribution can also be altered in liver patients due to reduced clearance of metabolites. For instance, patients with hepatitis/cirrhosis displayed elevated levels of unbound phenytoin compared to healthy individuals (Lau et al., 1997). This observation was attributed to the liver's reduced capacity to metabolize and remove bilirubin from the plasma, which inhibited the binding of phenytoin to HSA. Interestingly, in an earlier study, analysis of sera of patients with cirrhotic liver indicated reduced albumin binding capacity at site II (Kober, Jenner, Sjöholm, Borgå, & Odar-Cederlöf, 1978). Since then, this parameter has been adopted as a measure of severity of liver diseases and to evaluate the effectiveness of treatments (Klammt et al., 2008).

Our kidneys are also responsible for the elimination of excess and unwanted substances from the body. As such, conditions that negatively affect renal function are also synonymous with altered drug–albumin binding. Patients with nephrotic syndrome for example, frequently display decreased drug binding to HSA due to hypoalbuminemia. In many cases, a massive drop in HSA levels (below 20 mg/mL) has been reported (Harris & Ismail, 1994); in contrast, the normal HSA concentration in adults is ~40 mg/mL. Compared with healthy individuals, patients with this syndrome usually have very high active drug concentrations; notably, levels of free clofibrate and phenytoin in plasma have been measured to be 200% and 90% higher, respectively, in such patients (Yamasaki et al., 2013). Although decreased albumin binding of drugs is also common in patients with chronic renal failure and uremia, this is not contributed by reduced albumin levels alone. Instead, build-up of uremic toxins such as hippuric acid, indole acetate, indoxyl sulfate and 3-carboxy-4-methyl-5-propyl-2-furanpropanoic acid (CMPF) have also been suggested to inhibit the binding of drugs to HSA via displacement actions in these patients (Watanabe, Miyamoto, Otagiri, & Maruyama, 2011). This is reasonable as hippuric acid and indole-containing molecules mainly bind to site II of HSA, while CMPF selectively binds to site I (Sakai, Takadate, & Otagiri, 1995). Another interesting example of how renal conditions can affect drug–albumin binding can be seen in patients undergoing hemodialysis. Several studies have reported that hemodialysis patients have higher levels of oxidized albumin, which serves as a biomarker of oxidative stress (Nakatani et al., 2018). This is clinically significant as this form of HSA has an altered conformation with lower drug binding capacity, especially toward site II ligands (Mera et al., 2005).

Structural changes in HSA is also common in hyperglycemic conditions, such as in diabetes. Levels of glycated HSA can increase up to 30% in diabetic individuals, compared to ~5% in healthy adults (Rondeau & Bourdon, 2011). Available crystallographic data have established that site I can accommodate two molecules of glucose, with one of them covalently bonded to Lys-195 at the opening of the binding site and partially blocking it (Wang et al., 2013). Since glycation of HSA has been observed at both sites I and II, this can potentially result in reduced drug binding (Anguizola et al., 2013). Nevertheless, the degree to which glycation of HSA influences the pharmacokinetics of drugs is still unclear.

2. HSA-based drug delivery systems

For a drug to exert its adequate pharmacological effect, it must be available at the target site at an appropriate concentration and frequency without inducing adverse effects. Since most drugs are small molecules with non-specific distribution, this presents several challenges in terms of their pharmacokinetics (Sleep, Cameron, & Evans, 2013). For instance, they are often subject to rapid renal elimination, especially if the molecular weight is well below the filtration threshold in the kidneys. Additionally, such drugs also have short plasma half-lives due to enzymatic modifications producing more hydrophilic, and thus easier to clear, metabolites.

As a result, many drug delivery strategies have been developed to overcome these challenges, each with their advantages and shortcomings. For example, a common method to increase the half-life of peptide drugs is by conjugation with polyethylene glycol (PEGylation). Although several PEGylated drugs have been successfully tested and marketed, issues such as accumulation of high levels of PEG in organs and inhibition of activity remain (Bidwell, 2012). Similarly, while liposomes have high biocompatibility and can self-assemble, their application is restricted by limitations in encapsulation capacity, solubility in biological fluids and short half-lives (Damiati et al., 2019). In view of the above, several drug delivery systems incorporating HSA have been developed to take advantage of its desirable drug transport properties such as the presence of multiple drug binding sites, a very long plasma half-life (almost 20 days) and ability to interact with cellular receptors (Larsen, Kuhlmann, Hvam, & Howard, 2016).

2.1 Cellular receptors of HSA

One of the appeals of HSA-based drug delivery strategies is the specific interactions between HSA and cellular receptors (summary in Table 2) that not only facilitate uptake of drugs into cells via transcytosis, but also help in the recycling of HSA.

One of the albumin receptors found on the surface of vascular endothelial cells is albondin, also known as the Gp60 receptor. Binding of HSA to albondin increases the membrane permeability of the cell to accept the HSA in circulation. This interaction involves the activation of a membrane protein called caveolin, which leads to the formation of caveolae vesicles that contain the albumin molecule (Tiruppathi, Song, Bergenfeldt, Sass, & Malik, 1997). The action of albondin can account for almost half of the total albumin transported across the vascular endothelium. Presence of anti-albondin antibodies essentially blocks the binding of HSA to the cell surface (Schnitzer & Oh, 1994). Hence, albondin is regarded as one of the major receptors, responsible for the cellular transcytosis of HSA as well as preventing it from degradation in the lysosome.

Osteonectin (other names include SPARC and BM-40) is a receptor found in a wide variety of cell types and is highly expressed in malignant cells (Podhajcer et al., 2008). Interestingly, anti-osteonectin antibodies also

Table 2 Major cellular receptors of albumin and their biological functions with respect to albumin.

Receptor	Function	Tissue	References
Albondin (Gp60)	Transcytosis of native albumin via caveolae	Vascular endothelium	Hoogenboezem and Duvall (2018)
Osteonectin (SPARC)	Transcytosis of native albumin	Endothelium, skeletal muscle, vascular smooth muscle cells, fibroblasts, tumor cells	Podhajcer et al. (2008)
FcRn	Recycling of native albumin	Endothelium, kidneys, lungs, gut, blood brain barrier, antigen presenting cells	Pyzik et al. (2019); Sleep (2014)
Gp18/Gp30	Degradation of modified albumin	Endothelium, macrophages, fibroblasts	Merlot et al. (2014)
Cubilin/ Megalin	Renal reabsorption of albumin	Kidney tubule cells, gut, placenta	Larsen et al. (2016)

recognize albondin, implying that both receptors have similar structural features for binding to albumin (Schnitzer & Oh, 1992). Accumulation of albumin-bound drugs in cancerous tissues has partly been associated with the interaction of osteonectin with albumin. In particular, a clinical trial involving Abraxane®, a paclitaxel/HSA nanoparticle, has suggested that treatment response is correlated with the level of osteonectin expression (Desai et al., 2008).

The key to the long half-life of HSA is its interaction with the neonatal Fc receptor (FcRn), a glycoprotein which is present in various tissues including the vascular, brain and renal endothelia as well as the intestinal and lung epithelia (Sleep, 2014). The affinity of FcRn toward HSA is low under physiological conditions. However, within the acidic environment of the endosome, the binding affinity between the two proteins is high, thus saving the albumin from lysosomal degradation (Andersen et al., 2012). The salvaged albumin is then recycled to the extracellular space where the environment is less suited for tight binding, thereby prolonging its half-life.

In contrast to albondin and FcRn, the two most important receptors involved in the degradation of albumin are the Gp18 and Gp30 surface glycoproteins. They are highly expressed in liver endothelial cells as well as in peritoneal macrophages and display increased affinity to modified HSA over native HSA (Sleep, 2014). The modifications in HSA can occur naturally, such as oxidation of albumin during episodes of inflammation and diabetes-induced glycation. Hence, the Gp18 and Gp30 receptors are important in the elimination of old, damaged, and altered albumins by directing them to the lysosome (Merlot, Kalinowski, & Richardson, 2014). However, these receptors also show high affinity toward chemically-modified albumin. For instance, the binding of Gp18 and Gp30 to albumin treated with maleic anhydride and gold-albumin complexes is about 1000-fold greater than that to unmodified albumin (Schnitzer, Sung, Horvat, & Bravo, 1992).

The cubilin/megalin complex is a multi-ligand receptor complex, present in the kidney proximal tubules and is involved in the reabsorption of filtered proteins. The glycoprotein cubilin is only found on the cell membrane, and thus requires megalin (which extends into the cytoplasm) as a co-receptor to facilitate the endocytosis of ligands, which include albumin (Larsen et al., 2016). Therefore, the cubilin/megalin receptor complex is important for the reabsorption and rescue of albumin from renal elimination. Several animal studies have shown that deficiency in cubilin and/or megalin led to albuminuria (Amsellem et al., 2010; Weyer et al., 2011). Similarly, in humans, individuals with mutation in the cubilin gene show elevated protein levels in urine (Gräsbeck, 2006).

2.2 Noncovalent drug–albumin association

Arguably the simplest drug delivery strategy that exploits the natural transport function of HSA is to generate drug derivatives with enhanced affinity to the protein via noncovalent associations. Although this approach may appear simplistic, it is proven effective in improving the bioavailability of various drugs without compromising their safety profile.

A particularly successful method is acylation of drugs, where a modified form of the drug is created that interacts favorably with HSA at its fatty acid binding sites. Insulin detemir (Levemir®), an insulin analog modified by the addition of myristic acid at Lys B29, is marketed by Novo Nordisk for the treatment of diabetes. Upon subcutaneous injection, it binds rapidly and strongly to HSA in circulation at subdomain IIIA, extending its half-life from ~5 min to 5–7 h (Home & Kurtzhals, 2006). Another long-acting insulin analog by Novo Nordisk is insulin degludec (Tresiba®), which contains a palmitic diacid that is attached to insulin via a glutamyl linker (Fig. 3). With its long acyl group, the half-life of insulin degludec extends to almost a full day, allowing it to be used for a once-daily dosing (Tambascia & Eliaschewitz, 2015). Precisely the same modification has also been incorporated into glucagon-like peptide-1 (GLP-1) at Lys26, leading to the development of liraglutide (Victoza®). Compared to native GLP-1, liraglutide is more resistant against degradation by dipeptidyl peptidase-4 (DPP-4), which extends its half-life from ~2 min to 11–15 h (Mehta, Marso, & Neeland, 2017). An improvement on liraglutide is the ultra-long-acting semaglutide (Ozempic®), whose Lys26 is instead attached to a stearic diacid. With a half-life of 160 h, a once-weekly injection of semaglutide is sufficient for most type 2 diabetic patients (Dhillon, 2018).

Aside from fatty acids, 4-(*p*-iodophenyl) butanoic acid derivatives have also been demonstrated as potential albumin binding agents that can be coupled to drugs to extend their half-lives. One particular derivative, 6-(4-(4-iodophenyl)butanamido)hexanoate, more popularly known as AlbuTag, has a relatively long half-life of 24 h in plasma. Covalent attachment of AlbuTag to the C-terminal Cys of a therapeutic single-chain variable fragment (scFv) antibody has shown success in prolonging its half-life from ~20 min to 40 h and resulting in higher uptake by tumor cells (Trüssel et al., 2009). Similarly, conjugation of AlbuTag to contrast agents such as gadopentetate

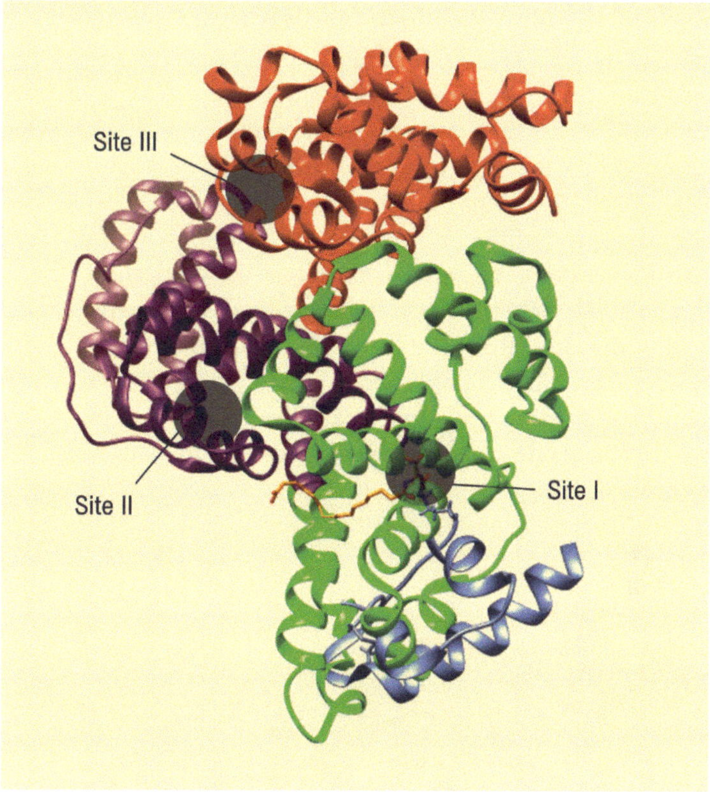

Fig. 3 Binding of insulin degludec (insulin is in blue and the attached palmitic diacid is in orange) to HSA is facilitated by hydrophobic interactions between palmitic diacid and the fatty acid binding sites of HSA. Domains I, II, and III of HSA are colored in red, green, and purple, respectively. Sites I, II, and III of HSA are located in subdomains IIA, IIIA, and IB, respectively. PDB IDs: HSA, 1E7H; insulin degludec, 4AJX.

dimeglumine that are used in biomedical imaging techniques, has improved half-life in the range of 50–100 fold (Zorzi, Linciano, & Angelini, 2019).

Peptides and proteins that can form specific interactions with albumin have also been utilized to improve the pharmacokinetic profiles of drugs. Ozoralizumab, developed by Ablynx, is a trivalent, bispecific Nanobody® that targets and inhibits TNF-α (Krah et al., 2016). However, due to its small size (38 kDa), it is susceptible for renal elimination. Hence, it has been designed to associate strongly with HSA via its albumin binding domain for extension of its *in vivo* half-life. It is currently at phase III of clinical trial for the treatment of rheumatoid arthritis and has shown high efficacy and tolerability in patients.

2.3 Covalent drug–albumin conjugates

Owing to its high efficiency and minimal adverse effects, conjugating prodrugs with albumin has emerged as a common strategy for drug delivery. The amino acid most often targeted for conjugation is Cys34, which possesses a free thiol group. This approach usually requires linkers such as maleimides to attach the drug to albumin. For example, aldoxorubicin, a maleimide derivative of the anticancer drug doxorubicin, links itself covalently with Cys 34 of HSA upon intravenous administration (Fig. 4). The acid-sensitive linker is then cleaved in the acidic environment at tumor sites, thus releasing the drug from the carrier (Kratz et al., 2002). Compared to unmodified doxorubicin, aldoxorubicin treatment has shown reduced side effects and produced better survival and response rates in clinical trials (Cranmer, 2019). Aside from Cys34, successful conjugation has also been demonstrated with Lys residues of HSA. Pre-clinical studies of the

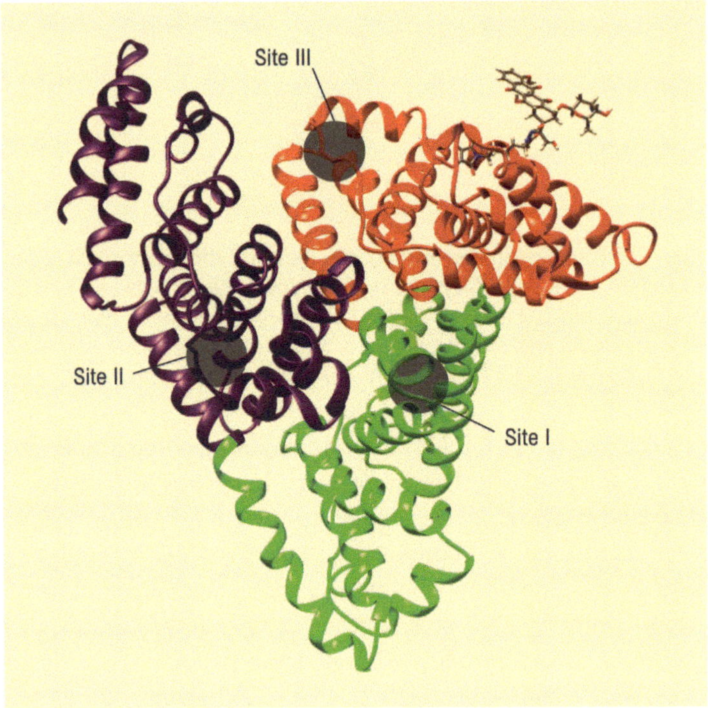

Fig. 4 Covalent conjugation of aldoxorubicin at Cys34 of HSA. Domains I, II, and III of HSA are colored in red, green, and purple, respectively. Sites I, II, and III of HSA are located in subdomains IIA, IIIA, and IB, respectively. HSA PDB ID:1BM0.

chemotherapeutic agent, methotrexate (MTX) conjugated with Lys residues of HSA (MTX–HSA) have shown improved pharmacological properties over free MTX in terms of uptake into cancer cells and circulatory half-life (Stehle et al., 1999). Enhanced therapeutic effects of MTX–HSA on three xenograft models have been attributed to its increased retention and permeability in solid tumors upon conjugation (Burger, Hartung, Stehle, Sinn, & Fiebig, 2001). In both phases I and II of clinical trials, the drug has presented favorable toxicological profile and response rate in patients suffering from urothelial carcinoma (Bolling et al., 2006).

Chemical conjugation with HSA has also been applied to nucleic acid therapeutics with applications in gene silencing. A common problem with the use of small interfering RNA (siRNA) is its poor pharmacokinetic property, which is usually mitigated by complex delivery strategies involving synthetic carriers (Sarett et al., 2017). A simpler alternative, as demonstrated by Lau et al. (2012), relies on conjugating modified siRNA with the free Cys34 of HSA. In this approach, the 3′ end of the sense strand is chemically modified with succinimidyl 4-[N-maleimidomethyl] cyclohexane-1-carboxylic acid (SMCC) to form an activated siRNA that is capable of forming a disulfide bridge with Cys34 of circulating HSA *in situ* upon administration. In BALB/c mice injected with the activated siRNA, higher concentration of the siRNA has been found in the myocardium compared to those treated with the unmodified siRNA; in addition, the mRNA level of the targeted gene (*IGF-IR*) is reduced by 35%.

Similar techniques have also been used to attach proteinaceous drugs to albumin. For instance, an analog of the YY peptide, which is believed to attenuate appetite by binding to the Y2 receptor (Y2R) in the hypothalamus, has been modified using several maleimide linkers including SMCC to facilitate its conjugation to HSA at Cys 34 (Ehrlich et al., 2013). The resulting peptide conjugates have maintained their selectivity toward Y2R in cell-based assays and exhibited significantly longer circulation time, leading to reduced food intake compared to unmodified and PEGylated peptides in animal studies. This method of conjugation to Cys34 to prolong the circulatory half-life has also been applied to other protein therapeutics such as the Kringle 5 peptide (Leger et al., 2004), insulin (Thibaudeau et al., 2005), designed ankyrin repeat proteins (Simon, Frey, Zangemeister-Wittke, & Plückthun, 2013) and the antiretroviral C34 peptide (Stoddart et al., 2008).

2.4 Albumin fusions

Another ingenious way to attach peptide and protein drugs to albumin is to exploit the gene expression machinery of yeast and mammalian cells. Advances in protein engineering have enabled the cloning of the gene for albumin and that of the target protein as a single open reading frame, and the resulting expressed product is a fusion of albumin and the target protein though peptide bonding (Fig. 5). The fusion is typically engineered at either the C- or N-terminal ends of albumin, but double fusions at both terminals have also been reported, creating bispecific fusions carrying two distinct drugs (McDonagh et al., 2012).

A major advantage of fusing therapeutic peptides and proteins to HSA is their greatly increased conformational stability and plasma half-life, which can range from 5-fold to 5000-fold (Sleep, 2014). Additionally, the drug manufacturing process can be streamlined, where different albumin fusions can be generated and isolated at the same time using albumin as the purification tag. A potential issue that may arise from fusing different proteins in a single polypeptide chain is the misfolding of the fused proteins due to the presence of additional amino acids residues, which can affect the

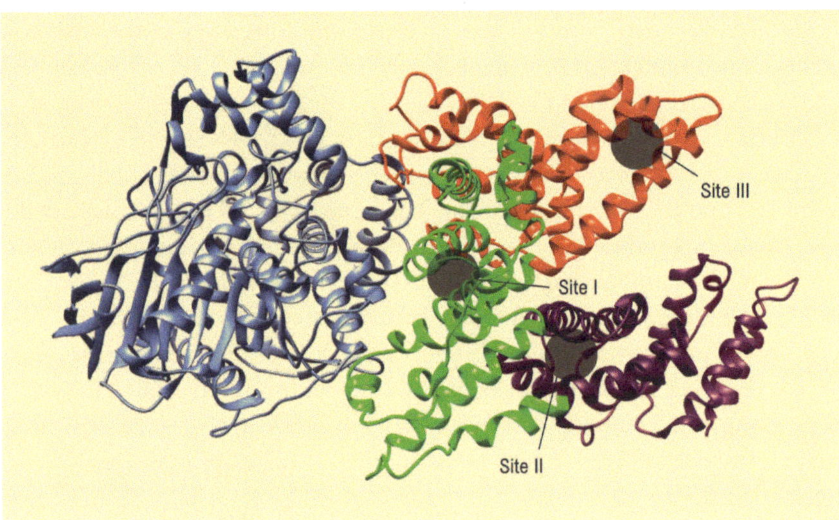

Fig. 5 The recombinant protein TV-1380 is generated by the genetic fusion of the C-terminal of truncated butyrylcholinesterase (blue) to the N-terminal of HSA. Domains I, II, and III of HSA are colored in red, green, and purple, respectively. Sites I, II, and III of HSA are located in subdomains IIA, IIIA, and IB, respectively. PDB IDs: HSA, 3JQZ; butyrylcholinesterase, 1P0I.

pharmacologic impact of the therapeutic protein. Nevertheless, this can be resolved by inserting a spacer sequence between the proteins, such as in the case of CSL689, the coagulation factor VIIa is fused to the N-terminal of HSA via a 31-amino acid sequence, comprising Ser and Gly residues (Schulte, 2008). Another concern is the possible emergence of antibodies against the fusion, which may negatively affect the drug's pharmacokinetic properties. Further, these antibodies may react with other proteins with similar structure as the fused therapeutic (Sleep, 2014). In this aspect, albumin fusions have fared better than many other forms of modified proteins. For instance, less than 1% of patients administered with the fusion drug albinterferon α-2b have shown presence of antibodies against the treatment, which were markedly lesser than the 20% of patients, who were treated with PEGylated interferon α (Rustgi, 2009).

In view of their beneficial traits, pharmaceutical companies have been actively developing HSA fusions as potential treatments for various diseases and conditions. Selected examples of these fusions, which are at various stages of clinical trials, with some successfully approved for commercial use, are listed in Table 3. Albiglutide, a fusion of GLP-1 with HSA, was the first albumin fusion product given approval by the FDA to be marketed for the treatment of type II diabetes. Compared to native GLP-1, the long-acting albiglutide, which contains modified GLP-1 that is resistant to DPP-4, has a considerably longer half-life of 4–7 days (Strohl, 2015). An innovative strategy in the development of albumin fusions is to attach two different proteins to either ends of HSA, as exemplified by MM-111, a bispecific antibody–HSA fusion containing anti-HER2 and anti-HER3 scFv antibodies (McDonagh et al., 2012). Designed primarily for the treatment of HER2-positive cancers such as breast, esophageal, and gastric cancers, this fusion increases the half-life of the antibodies to almost 90 h compared to less than 1 h, when the antibodies are given in their free forms (Richards et al., 2014). In addition to modifying the therapeutic protein in a fusion to improve half-life, efforts have also been taken to create fusions incorporating mutant HSA to achieve the same goal. Interestingly, a K573P HSA mutant developed by Novozymes showed 12-fold higher affinity toward FcRn, translating to a prolonged half-life over native HSA (Andersen et al., 2014). If such mutants can successfully be fused to therapeutic proteins, this may lead to further and substantial half-life extension of drugs.

Table 3 Examples of HSA fusions at various stages of clinical development.

Fusion name	Company	Structure	Indication	Status	References
Albiglutide (Tanzeum®/ Eperzan®)	GSK	Glucagon-like peptide-1 dimer fused to N-terminal of HSA	Type II diabetes	Approved for commercialization	Fala (2015)
Albinterferon (Zalbin™/ Jouleferon™)	Human Genome Sciences, Novartis	Interferon α-2b fused to C-terminal of HSA	Hepatitis C	Withdrawn after phase IIB	Vezali, Aghemo, and Colombo (2011)
Albutrepenonacog alfa (Idelvion®)	CSL Behring	Coagulation factor IX fused to N-terminal of HSA	Hemophilia B	Approved for commercialization	Lyseng-Williamson (2017)
Balugrastim (Egranli™)	Teva	Granulocyte colony-stimulating factor fused to C-terminal of HSA	Neutropenia	Withdrawn after phase III (non-safety related)	EMA (2014)
CSL689	CSL Behring	Coagulation factor VIIa	Hemophilia A and B	Phase II/III	NLM, NCT02484638
MM-111	Merrimack	Bispecific antibody fusion of anti-HER2–HSA–anti-HER3	Esophageal, gastric and breast cancer	Phase II completed	NLM, NCT01774851
TV-1106	Teva	Human growth hormone fused to C-terminal of HSA	Growth hormone deficiency	Phase III	NLM, NCT02410343
TV-1380	Teva	Butyrylcholinesterase fused to N-terminal of HSA	Cocaine addiction and overdose	Phase II completed	NLM, NCT01887366

2.5 HSA nanosystems

The approval of Abraxane® (*nab*-paclitaxel/ABI-007) for oncological treatment 15 years ago marked a milestone for HSA-based nanotherapy. Prior to its development, the poorly soluble paclitaxel was usually administered as a formulation containing Cremophor EL as the solvent. However, the use of Cremophor EL has been recognized as the cause of the clinical toxicities associated with paclitaxel therapies (Sleep, 2014). In contrast, the safety profile of Abraxane® is well regarded, thanks to the low immunogenicity and non-toxic nature of HSA. Originally marketed as treatment for metastatic breast cancer, it is now approved/undergoing clinical trials for the treatment of other types of cancers including non-small cell lung cancer, pancreatic cancer, ovarian cancer, bladder cancer and melanoma (Kratz, 2014). Structurally, the Abraxane® nanoparticle is ~130 nm in diameter, with the hydrophobic paclitaxel in its core, surrounded by a layer of HSA (Desai et al., 2008). Its efficacy against solid tumors is partly attributable to its size. The convoluted network of blood vessels resulting from angiogenesis and the thick extracellular matrix of tumor tissues present a challenge to effective drug delivery. Due to their small size, nanoparticles that are <10 nm have excellent penetrability, especially into deep tumors (Yang et al., 2019). However, they are also prone to rapid elimination, which hinders their accumulation in tumor tissues. On the other hand, while nanoparticles that are larger than 200 nm generally improve the pharmacokinetic properties of the drug cargo, they tend to have low diffusion rates into the tumor space. Hence, nanoparticle diameters in the range 10–200 nm are ideal for their entry into and accumulation in tumors due to the enhanced permeability and retention effect associated with solid tumors (Tan & Ho, 2018).

The commercial success of Abraxane® has led to its manufacturer, Celgene, to expand its portfolio of cancer targeting HSA nanoparticles. ABI-008 (*nab*-docetaxel) and ABI-009 (*nab*-rapamycin) are currently the most advanced nanosystems in terms of clinical status, with both undergoing phase 2 clinical trials for the treatment of breast, prostate and bladder cancers (Van de Sande, Cosyns, Willaert, & Ceelen, 2020). ABI-011, which carries thiocolchicine dimers as its cargo, has completed phase 1 trial, involving patients with advanced solid tumors or lymphomas. Interestingly, Celgene has also developed a new experimental HSA nanoparticle containing docetaxel (ABI-013), which has a diameter of 70 nm. Although it has only been tested preclinically, it has shown higher efficacy against xenograft models of human breast, colorectal, and lung cancers when compared with conventional solvent-based formulation of docetaxel (D'Cruz, Piacente, Trieu, Tao, & Desai, 2010).

In order to improve therapeutic targeting of the tumor microenvironment, several groups have experimented with incorporating modified HSA in the form of nanoparticles that have greater selectivity toward cancer cells. Since cancer cells usually overexpress certain receptors, the ligands for such receptors can thus be manipulated as targeting agents (Hoogenboezem & Duvall, 2018). As an example, glycyrrhetinic acid (GA) receptors are highly expressed in hepatoma cells. To specifically target these cells, a GA-modified HSA nanoparticle (170 nm diameter) has been developed with doxorubicin as the chemotherapeutic payload (Qi et al., 2015). Compared with the non-targeted nanoparticles (doxorubicin/HSA), the doxorubicin/GA-HSA nanoparticles have displayed increased tumor uptake, higher accumulation and greater cytotoxic effect against hepatoma cells. The same concept has also been applied for targeted delivery to non-tumor cells. To deliver methylprednisolone to oligodendrocytes for the treatment of spinal cord injury (SCI), the drug has been packaged in the form of HSA nanoparticles (190 nm diameter) and decorated with the NEP_{1-40} peptide (Lin et al., 2019). This peptide not only recognizes the Nogo protein on the surface of oligodendrocytes to facilitate the delivery of the nanoparticles, but also has intrinsic neuroregenerative activity. The synergistic therapeutic effects of methylprednisolone and the NEP_{1-40} peptide have been presented in a rat model of SCI, where the rats have exhibited improved locomotor scores, when treated with the nanoparticles.

A relatively recent innovation in the design of HSA nanosystems is the development of theranostic HSA nanoparticles, which have combined therapeutic and diagnostic functions (Lee et al., 2020; Hu et al., 2020). For instance, the frequent occurrence of hypoxia in solid tumors has been exploited for specific drug delivery by the development of a size-switchable nanosystem that responds to oxygen level (Yang et al., 2019). This theranostic nanosystem is composed of two distinct HSA conjugates: an oxaliplatin–HSA conjugate (HO) for chemotherapy, and a chlorin e6–HSA conjugate (HC) for photodynamic therapy and fluorescence imaging. These conjugates are cross-linked via an azobenzene linker that is hypoxia-responsive, forming HO/HC nanoparticles, which are 100–150 nm in diameter and stable under normal oxygen partial pressure. However, the azobenzene group is cleaved under a hypoxic environment, thus releasing the individual HO and HC nanoparticles that easily penetrate the tumor space due to their small size (<10 nm). Similarly, the near-infrared absorbing dye croconine has been used to induce the self-assembly of HSA, forming theranostic croconine/HSA nanoparticles, which can accumulate in tumors (Zou, Chang, & Yan, 2020). The pH-sensitive croconine dye allows for real-time photoacoustic pH imaging of the tumor. In addition, in the acidic microenvironment of tumors, the photothermal activity of the dye is activated, leading to ablation of tumors.

Acknowledgments

The authors would like to thank Khairul Azreena Bakar for her help in preparing the artwork. Facilities provided by the Institute of Biological Sciences, Faculty of Science, University of Malaya and the Department of Biological Sciences and Biotechnology, Faculty of Science and Technology, Universiti Kebangsaan Malaysia are gratefully acknowledged.

References

Amin, S. B. (2016). Bilirubin binding capacity in the preterm neonate. *Clinics in Perinatology, 43*(2), 241–257.

Amsellem, S., Gburek, J., Hamard, G., Nielsen, R., Willnow, T. E., Devuyst, O., et al. (2010). Cubilin is essential for albumin reabsorption in the renal proximal tubule. *Journal of the American Society of Nephrology, 21*(11), 1859–1867.

Andersen, J. T., Dalhus, B., Cameron, J., Daba, M. B., Plumridge, A., Evans, L., et al. (2012). Structure-based mutagenesis reveals the albumin-binding site of the neonatal Fc receptor. *Nature Communications, 3*, 610.

Andersen, J. T., Dalhus, B., Viuff, D., Ravn, B. T., Gunnarsen, K. S., Plumridge, A., et al. (2014). Extending serum half-life of albumin by engineering FcRn binding. *Journal of Biological Chemistry, 289*, 13492–13502.

Anguizola, J., Matsuda, R., Barnaby, O., Hoy, K. S., Wa, C., DeBolt, E., et al. (2013). Glycation of human serum albumin. *Clinica Chimica Acta, 425*, 64–76.

Ascenzi, P., & Fasano, M. (2010). Allostery in a monomeric protein: The case of human serum albumin. *Biophysical Chemistry, 148*(1–3), 16–22.

Benet, L. Z., & Hoener, B.-A. (2002). Changes in plasma protein binding have little clinical relevance. *Clinical Pharmacology and Therapeutics, 71*(3), 115–121.

Bidwell, G. L. (2012). Peptides for cancer therapy: A drug-development opportunity and a drug-delivery challenge. *Therapeutic Delivery, 3*(5), 609–621.

Bohnert, T., & Gan, L.-S. (2013). Plasma protein binding: From discovery to development. *Journal of Pharmaceutical Sciences, 102*(9), 2953–2994.

Bolling, C., Graefe, T., Lubbing, C., Jankevicius, F., Uktveris, S., Cesas, A., et al. (2006). Phase II study of MTX-HSA in combination with cisplatin as first line treatment in patients with advanced or metastatic transitional cell carcinoma. *Investigational New Drugs, 24*(6), 521–527.

Burger, A. M., Hartung, G., Stehle, G., Sinn, H., & Fiebig, H. H. (2001). Pre-clinical evaluation of a methotrexate–albumin conjugate (MTX-HSA) in human tumor xenografts *in vivo*. *International Journal of Cancer, 92*(5), 718–724.

Cranmer, L. D. (2019). Spotlight on aldoxorubicin (INNO-206) and its potential in the treatment of soft tissue sarcomas: Evidence to date. *OncoTargets and Therapy, 12*, 2047–2062.

D'Cruz, O., Piacente, M., Trieu, V., Tao, C., & Desai, N. (2010). ABI-013: A novel nanoparticle albumin-bound (nab) docetaxel analog with superior antitumor activity over docetaxel. *Cancer Research, 70*(8), 4420.

Damiati, S., Scheberl, A., Zayni, S., Damiati, S. A., Schuster, B., & Kompella, U. B. (2019). Albumin-bound nanodiscs as delivery vehicle candidates: Development and characterization. *Biophysical Chemistry, 251*, 106178.

Desai, N., Trieu, V., Hwang, L. Y., Wu, R., Soon-Shiong, P., & Gradishar, W. J. (2008). Improved effectiveness of nanoparticle albumin-bound (nab) paclitaxel versus polysorbate-based docetaxel in multiple xenografts as a function of HER2 and SPARC status. *Anti-Cancer Drugs, 19*(9), 899–909.

Dhillon, S. (2018). Semaglutide: First global approval. *Drugs, 78*(2), 275–284.

Ehrlich, G. K., Michel, H., Truitt, T., Riboulet, W., Pop-Damkov, P., Goelzer, P., et al. (2013). Preparation and characterization of albumin conjugates of a truncated peptide YY analogue for half-life extension. *Bioconjugate Chemistry, 24*(12), 2015–2024.

European Medicines Agency (EMA). (2014). *Assessment report: Egranli* (pp. 1–90). https://www.ema.europa.eu/en/documents/withdrawal-report/withdrawal-assessment-report-egranli_en.pdf.

Fala, L. (2015). Tanzeum (albiglutide): A once-weekly GLP-1 receptor agonist subcutaneous injection approved for the treatment of patients with type 2 diabetes. *American Health and Drug Benefits, 8*, 126–130.

Fanali, G., di Masi, A., Trezza, V., Marino, M., Fasano, M., & Ascenzi, P. (2012). Human serum albumin: >From bench to bedside. *Molecular Aspects of Medicine, 33*(3), 209–290.

Gräsbeck, R. (2006). Imerslund-Gräsbeck syndrome (selective vitamin B12 malabsorption with proteinuria). *Orphanet Journal of Rare Diseases, 1*, 17.

Harris, R. C., & Ismail, N. (1994). Extrarenal complications of the nephrotic syndrome. *American Journal of Kidney Diseases, 23*(4), 477–497.

Hatakeyama, Y., Niwano, S., Niwano, H., Kosukegawa, T., & Izumi, T. (2010). Risks and benefits of combined use of bucolome and warfarin in anticoagulation therapy. *International Heart Journal, 51*(6), 399–403.

Hochman, J., Tang, C., & Prueksaritanont, T. (2015). Drug–drug interactions related to altered absorption and plasma protein binding: Theoretical and regulatory considerations, and an industry perspective. *Journal of Pharmaceutical Sciences, 104*(3), 916–929.

Home, P., & Kurtzhals, P. (2006). Insulin detemir: From concept to clinical experience. *Expert Opinion on Pharmacotherapy, 7*(3), 325–343.

Hoogenboezem, E. N., & Duvall, C. L. (2018). Harnessing albumin as a carrier for cancer therapies. *Advanced Drug Delivery Reviews, 130*, 73–89.

Hu, X. L., Kwon, N., Yan, K. C., Sedgwick, A. C., Chen, G. R., He, X. P., et al. (2020). Bio-conjugated advanced materials for targeted disease theranostics. *Advanced Functional Materials, 30*(13), 1907906.

Kawai, K., Nishii, R., Shikano, N., Makino, N., Kuga, N., Yoshimoto, M., et al. (2009). Serum protein binding displacement: Theoretical analysis using a hypothetical radiopharmaceutical and experimental analysis with ^{123}I-*N*-isopropyl-*p*-iodoamphetamine. *Nuclear Medicine and Biology, 36*(1), 99–106.

Klammt, S., Mitzner, S. R., Stange, J., Loock, J., Heemann, U., Emmrich, J., et al. (2008). Improvement of impaired albumin binding capacity in acute-on-chronic liver failure by albumin dialysis. *Liver Transplantation, 14*(9), 1333–1339.

Kober, A., Jenner, A., Sjöholm, I., Borgå, O., & Odar-Cederlöf, I. (1978). Differentiated effects of liver cirrhosis on the albumin binding sites for diazepam, salicylic acid and warfarin. *Biochemical Pharmacology, 27*(23), 2729–2735.

Kragh-Hansen, U., Chuang, V. T. G., & Otagiri, M. (2002). Practical aspects of the ligand-binding and enzymatic properties of human serum albumin. *Biological and Pharmaceutical Bulletin, 25*(6), 695–704.

Krah, S., Schröter, C., Zielonka, S., Empting, M., Valldorf, B., & Kolmar, H. (2016). Single-domain antibodies for biomedical applications. *Immunopharmacology and Immunotoxicology, 38*(1), 21–28.
Kratz, F. (2014). A clinical update of using albumin as a drug vehicle – a commentary. *Journal of Controlled Release, 190*, 331–336.
Kratz, F., Warnecke, A., Scheuermann, K., Stockmar, C., Schwab, J., Lazar, P., et al. (2002). Probing the cysteine-34 position of endogenous serum albumin with thiol-binding doxorubicin derivatives. Improved efficacy of an acid-sensitive doxorubicin derivative with specific albumin-binding properties compared to that of the parent compound. *Journal of Medicinal Chemistry, 45*(25), 5523–5533.
Larsen, M. T., Kuhlmann, M., Hvam, M. L., & Howard, K. A. (2016). Albumin-based drug delivery: Harnessing nature to cure disease. *Molecular and Cellular Therapies, 4*, 3.
Lau, S., Graham, B., Cao, N., Boyd, B. J., Pouton, C. W., & White, P. J. (2012). Enhanced extravasation, stability and *in vivo* cardiac gene silencing via *in situ* siRNA-albumin conjugation. *Molecular Pharmaceutics, 9*(1), 71–80.
Lau, A. H., Gustavson, L. E., Sperelakis, R., Lam, N. P., El-Shourbagy, T., Qian, J. X., et al. (1997). Pharmacokinetics and safety of tiagabine in subjects with various degrees of hepatic function. *Epilepsia, 38*(4), 445–451.
Lee, S., Pham, T. C., Bae, C., Choi, Y., Kim, Y. K., & Yoon, J. (2020). Nano theranostics platforms that utilize proteins. *Coordination Chemistry Reviews, 412*, 213258.
Leger, R., Benquet, C., Huang, X., Quraishi, O., van Wyk, P., & Bridon, D. (2004). Kringle 5 peptide-albumin conjugates with anti-migratory activity. *Bioorganic and Medicinal Chemistry Letters, 14*(4), 841–845.
Lin, Y., Li, C., Li, J., Deng, R., Huang, J., Zhang, Q., et al. (2019). NEP_{1-40}-modified human serum albumin nanoparticles enhance the therapeutic effect of methylprednisolone against spinal cord injury. *Journal of Nanobiotechnology, 17*, 12.
Lyseng-Williamson, K. A. (2017). Coagulation factor IX (recombinant), albumin fusion protein (albutrepenonacog alfa; Idelvion®): A review of its use in haemophilia B. *Drugs, 77*(1), 97–106.
McDonagh, C. F., Huhalov, A., Harms, B. D., Adams, S., Paragas, V., Oyama, S., et al. (2012). Antitumor activity of a novel bispecific antibody that targets the ErbB2/ErbB3 oncogenic unit and inhibits heregulin-induced activation of ErbB3. *Molecular Cancer Therapeutics, 11*(3), 582–593.
Mehta, A., Marso, S. P., & Neeland, I. J. (2017). Liraglutide for weight management: A critical review of the evidence. *Obesity Science and Practice, 3*(1), 3–14.
Mera, K., Anraku, M., Kitamura, K., Nakajou, K., Maruyama, T., & Otagiri, M. (2005). The structure and function of oxidized albumin in hemodialysis patients: Its role in elevated oxidative stress via neutrophil burst. *Biochemical and Biophysical Research Communications, 334*(4), 1322–1328.
Merlot, A. M., Kalinowski, D. S., & Richardson, D. R. (2014). Unraveling the mysteries of serum albumin—more than just a serum protein. *Frontiers in Physiology, 5*, 299.
Nakatani, S., Yasukawa, K., Ishimura, E., Nakatani, A., Toi, N., Uedono, H., et al. (2018). Non-mercaptalbumin, oxidized form of serum albumin, significantly associated with renal function and anemia in chronic kidney disease patients. *Scientific Reports, 8*, 16796.
National Library of Medicine (NLM) NCT01774851: A study of MM-111 and paclitaxel with trastuzumab in patients HER2 positive carcinomas of the distal esophagus, gastro-esophageal (GE) junction and stomach. https://clinicaltrials.gov/ct2/show/NCT01774851. [Accessed 7 August 2020].
National Library of Medicine (NLM) NCT01887366: Efficacy and safety of TV-1380 as treatment for facilitation of abstinence in cocaine-dependent subjects. https://clinicaltrials.gov/ct2/show/NCT01887366. [Accessed 7 August 2020].

National Library of Medicine (NLM) NCT02410343: Study of TV-1106 in growth hormone-deficient adults. https://clinicaltrials.gov/ct2/show/NCT02410343. [Accessed 7 August 2020].
National Library of Medicine (NLM) NCT02484638: Study of recombinant factor VIIa fusion protein (rVIIa-FP, CSL689) for on-demand treatment of bleeding episodes in patients with hemophilia A or B with inhibitors. https://clinicaltrials.gov/ct2/show/NCT02484638. [Accessed 7 August 2020].
Podhajcer, O. L., Benedetti, L. G., Girotti, M. R., Prada, F., Salvatierra, E., & Llera, A. S. (2008). The role of the matricellular protein SPARC in the dynamic interaction between the tumor and the host. *Cancer and Metastasis Reviews, 27*, 523.
Pyzik, M., Sand, K. M., Hubbard, J. J., Andersen, J. T., Sandlie, I., & Blumberg, R. S. (2019). The neonatal Fc receptor (FcRn): A misnomer? *Frontiers in Immunology, 10*, 1540.
Qi, W. W., Yu, H. Y., Guo, H., Lou, J., Wang, Z. M., Liu, P., et al. (2015). Doxorubicin-loaded glycyrrhetinic acid modified recombinant human serum albumin nanoparticles for targeting liver tumor chemotherapy. *Molecular Pharmaceutics, 12*(3), 675–683.
Richards, D. A., Braiteh, F. S., Garcia, A. A., Denlinger, C. S., Conkling, P. R., Edenfield, W. J., et al. (2014). A phase 1 study of MM-111, a bispecific HER2/HER3 antibody fusion protein, combined with multiple treatment regimens in patients with advanced HER-positive solid tumors. *Journal of Clinical Oncology, 32*, 651.
Rimac, H., Debeljak, Z., Bojic, M., & Miller, L. (2017). Displacement of drugs from human serum albumin: From molecular interactions to clinical significance. *Current Medicinal Chemistry, 24*(18), 1930–1947.
Roberts, J. A., Pea, F., & Lipman, J. (2013). The clinical relevance of plasma protein binding changes. *Clinical Pharmacokinetics, 52*(1), 1–8.
Rondeau, P., & Bourdon, E. (2011). The glycation of albumin: Structural and functional impacts. *Biochimie, 93*(4), 645–658.
Rustgi, V. K. (2009). Albinterferon alfa-2b, a novel fusion protein of human albumin and human interferon alfa-2b, for chronic hepatitis C. *Current Medical Research and Opinion, 25*(4), 991–1002.
Sakai, T., Takadate, A., & Otagiri, M. (1995). Characterization of binding site of uremic toxins on human serum albumin. *Biological and Pharmaceutical Bulletin, 18*(12), 1755–1761.
Sarett, S. M., Werfel, T. A., Lee, L., Jackson, M. A., Kilchrist, K. V., Brantley-Sieders, D., et al. (2017). Lipophilic siRNA targets albumin in situ and promotes bioavailability, tumor penetration, and carrier-free gene silencing. *Proceedings of the National Academy of Sciences of the United States of America, 114*(32), E6490–E6497.
Schnitzer, J. E., & Oh, P. (1992). Antibodies to SPARC inhibit albumin binding to SPARC, gp60, and microvascular endothelium. *American Journal of Physiology, 263*(6), H1872–H1879.
Schnitzer, J. E., & Oh, P. (1994). Albondin-mediated capillary permeability to albumin. Differential role of receptors in endothelial transcytosis and endocytosis of native and modified albumins. *Journal of Biological Chemistry, 269*(8), 6072–6082.
Schnitzer, J. E., Sung, A., Horvat, R., & Bravo, J. (1992). Preferential interaction of albumin-binding proteins, gp30 and gp18, with conformationally modified albumins. Presence in many cells and tissues with a possible role in catabolism. *Journal of Biological Chemistry, 267*(34), 24544–24553.
Schulte, S. (2008). Use of albumin fusion technology to prolong the half-life of recombinant factor VIIa. *Thrombosis Research, 122*, S14–S19.
Setoguchi, N., Takamura, N., Fujita, K., Ogata, K., Tokunaga, J., Nishio, T., et al. (2013). A diclofenac suppository–nabumetone combination therapy for arthritic pain relief and a monitoring method for the diclofenac binding capacity of HSA site II in rheumatoid arthritis. *Biopharmaceutics and Drug Disposition, 34*(2), 125–136.

Simon, M., Frey, R., Zangemeister-Wittke, U., & Plückthun, A. (2013). Orthogonal assembly of a designed ankyrin repeat protein-cytotoxin conjugate with a clickable serum albumin module for half-life extension. *Bioconjugate Chemistry, 24*(11), 1955–1966.
Sleep, D. (2014). Albumin and its application in drug delivery. *Expert Opinion on Drug Delivery, 12*(5), 793–812.
Sleep, D., Cameron, J., & Evans, L. R. (2013). Albumin as a versatile platform for drug half-life extension. *Biochimica et Biophysica Acta, 1830*(12), 5526–5534.
Smith, D. A., Di, L., & Kerns, E. H. (2010). The effect of plasma protein binding on *in vivo* efficacy: Misconceptions in drug discovery. *Nature Reviews Drug Discovery, 9*(12), 929–939.
Stehle, G., Wunder, A., Schrenk, H. H., Hartung, G., Heene, D. L., & Sinn, H. (1999). Methotrexate–albumin conjugate causes tumor growth delay in Dunning R3327 HI prostate cancer-bearing rats. *Anti-Cancer Drugs, 10*(4), 405–411.
Stoddart, C. A., Nault, G., Galkina, S. A., Thibaudeau, K., Bakis, P., Bousquet-Gagnon, N., et al. (2008). Albumin-conjugated C34 peptide HIV-1 fusion inhibitor: Equipotent to C34 and T-20 *in vitro* with sustained activity in SCID-hu Thy/Liv mice. *Journal of Biological Chemistry, 283*(49), 34045–34052.
Strohl, W. R. (2015). Fusion proteins for half-life extension of biologics as a strategy to make biobetters. *BioDrugs, 29*(4), 215–239.
Tambascia, M. A., & Eliaschewitz, F. G. (2015). Degludec: The new ultra-long insulin analogue. *Diabetology and Metabolic Syndrome, 7*, 57.
Tan, Y. L., & Ho, H. K. (2018). Navigating albumin-based nanoparticles through various drug delivery routes. *Drug Discovery Today, 23*(5), 1108–1114.
Thibaudeau, K., Léger, R., Huang, X., Robitaille, M., Quraishi, O., Soucy, C., et al. (2005). Synthesis and evaluation of insulin-human serum albumin conjugates. *Bioconjugate Chemistry, 16*(4), 1000–1008.
Tiruppathi, C., Song, W., Bergenfeldt, M., Sass, P., & Malik, A. B. (1997). Gp60 activation mediates albumin transcytosis in endothelial cells by tyrosine kinase-dependent pathway. *Journal of Biological Chemistry, 272*(41), 25968–25975.
Trüssel, S., Dumelin, C., Frey, K., Villa, A., Buller, F., & Neri, D. (2009). New strategy for the extension of the serum half-life of antibody fragments. *Bioconjugate Chemistry, 20*(12), 2286–2292.
Van de Sande, L., Cosyns, S., Willaert, W., & Ceelen, W. (2020). Albumin-based cancer therapeutics for intraperitoneal drug delivery: A review. *Drug Delivery, 27*(1), 40–53.
Vezali, E., Aghemo, A., & Colombo, M. (2011). Interferon in the treatment of chronic hepatitis C: A drug caught between past and future. *Expert Opinion on Biological Therapy, 11*(3), 301–313.
Wang, Y., Yu, H., Shi, X., Luo, Z., Lin, D., & Huang, M. (2013). Structural mechanism of ring-opening reaction of glucose by human serum albumin. *Journal of Biological Chemistry, 288*(22), 15980–15987.
Watanabe, H., Miyamoto, Y., Otagiri, M., & Maruyama, T. (2011). Update on the pharmacokinetics and redox properties of protein-bound uremic toxins. *Journal of Pharmaceutical Sciences, 100*(9), 3682–3695.
Weyer, K., Storm, T., Shan, J., Vainio, S., Kozyraki, R., Verroust, P. J., et al. (2011). Mouse model of proximal tubule endocytic dysfunction. *Nephrology Dialysis Transplantation, 26*(11), 3446–3451.
Yamasaki, K., Chuang, V. T. G., Maruyama, T., & Otagiri, M. (2013). Albumin–drug interaction and its clinical implication. *Biochimica et Biophysica Acta, 1830*(12), 5435–5443.
Yang, G., Phua, S. Z. F., Lim, W. Q., Zhang, R., Feng, L., Liu, G., et al. (2019). A hypoxia-responsive albumin-based nanosystem for deep tumor penetration and excellent therapeutic efficacy. *Advanced Materials, 31*, 1901513.

Yoon, H. W., Giraldo, E. A., & Wijdicks, E. F. (2011). Valproic acid and warfarin: An underrecognized drug interaction. *Neurocritical Care, 15*(1), 182–185.

Zi, J., Liu, D., Ma, P., Huang, H., Zhu, J., Wei, D., et al. (2010). Effects of CYP2C9*3 and CYP2C9*13 on diclofenac metabolism and inhibition-based drug-drug interactions. *Drug Metabolism and Pharmacokinetics, 25*(4), 343–350.

Zorzi, A., Linciano, S., & Angelini, A. (2019). Non-covalent albumin-binding ligands for extending the circulating half-life of small biotherapeutics. *Medicinal Chemistry Communications, 10*, 1068–1081.

Zou, Q., Chang, R., & Yan, X. (2020). Self-assembling proteins for design of anticancer nanodrugs. *Chemistry—An Asian Journal, 15*(9), 1405–1419.

CHAPTER TEN

PTR2/POT/NPF transporters: what makes them tick?

Bala K. Prabhala[a], Moazur Rahman[b,c], Hussam H. Nour-eldin[d], Flemming Steen Jørgensen[e] and Osman Mirza[e,*]
[a]Institute of Physics, Chemistry and Pharmacy, University of Southern Denmark, Odense, Denmark
[b]School of Biological Sciences, University of the Punjab, Lahore, Punjab, Pakistan
[c]Drug Discovery and Structural Biology Group, Health Biotechnology Division, National Institute for Biotechnology and Genetic Engineering (NIBGE), Faisalabad, Punjab, Pakistan
[d]DynaMo Center, Copenhagen Plant Science Centre, Department of Plant and Environmental Sciences, University of Copenhagen, Frederiksberg, Denmark
[e]Department of Drug Design and Pharmacology, Faculty of Health and Medical Sciences, University of Copenhagen, Copenhagen, Denmark
*Corresponding author: E-mail: om@sund.ku.dk

Contents

1. Introduction 220
2. A plethora of natural substrates 221
   2.1 Ions and water 221
   2.2 Single amino acids and amino acid drug conjugates 222
   2.3 Peptides and peptide drug conjugates 223
   2.4 Organic anions, and a cation 227
   2.5 Uncharged substrates 227
   2.6 Similarity and diversity among the substrates 229
3. 3D-structure and transport mechanism 230
   3.1 A mechanism for peptide transport 231
   3.2 Do the non-peptide substrates fit in the transport mechanism? 233
References 234

Abstract

PTR2/POT/NPF are a family of primarily proton coupled transporters that belong to the major facilitator super family and are found across most kingdoms of life. They are involved in uptake of nutrients, hormones, ions and several orally administered drug molecules. A wealth of structural and functional data is available for this family; the similarity between the protein structural features have been discussed and investigated in detail on several occasions, however there are no reports on the unification of substrate information. In order to fill this gap, we have collected information about substrates across the entire PTR2/POT/NPF family in order to provide key insights into what makes a molecule a substrate and whether there are common features among confirmed

Advances in Protein Chemistry and Structural Biology, Volume 123
ISSN 1876-1623
https://doi.org/10.1016/bs.apcsb.2020.10.002

substrates. This review will be of particular interest for researchers in the field trying to probe the mechanisms responsible for the different selectivity of these transporters at a molecular resolution, and to design novel substrates.

1. Introduction

Import of nutrients from the extracellular space in the cell is an absolute necessity for sustenance of all forms of life. Some nutrients are able to cross the cellular membrane by passive diffusion while other have to be facilitated by specific transport proteins; the processes of facilitation may be energized, and in such cases either by primary or secondary energy sources. The proton-coupled peptide oligopeptide transporters (POTs) also termed PTR2s and a subgroup hereof called Nitrate transporter 1/Peptide transporter Family (NPF), collectively are a large family of transporters that utilize the electrochemical proton gradient for substrate uptake (Jørgensen et al., 2017; Leran et al., 2014). From here on we will refer to them as POTs or NPFs, the latter when referring to plant members.

POT members from human in particular have been in focus due to their role in drug uptake from the small intestine and the potential they hold in order to increase drug bioavailability (Ganapathy, Huang, Wang, Ganapathy, & Leibach, 1998; Sugawara et al., 2000). NPFs on the other hand have received great attention due to their role in distribution and deposition of a variety of compounds. POTs from livestock and fish are also being pursued as a means to create a more effective feed composition. PTR2 from yeast and bacterial POTs were among the first POTs to be characterized (Verri, Terova, Dabrowski, & Saroglia, 2011). However, in the recent years bacterial POTs in particular, have been pursued intensively as model systems to understand the more fragile human POTs (Beale et al., 2015; Boggavarapu, Jeckelmann, Harder, Ucurum, & Fotiadis, 2015; Doki et al., 2013; Guettou et al., 2014; Minhas & Newstead, 2019; Newstead et al., 2011; Ural-Blimke et al., 2019; Zhao et al., 2014).

Currently the PFAM database (family PFAM00854) (El-Gebali et al., 2019) lists more than 18,000 unique PTR2/POT/NPFs sequences. At times this family has been considered not to be, but is most frequently, included as a member of the major facilitator superfamily. Plant NPFs are divided into 8 clades (Leran et al., 2014) NPF1 to NPF8, based on phylogeny, and much work has been done to understand their respective functional differences with much still to be done. POTs from other kingdoms of life usually fall into two main groups: the PEPT-like POTs and the PHTs (will be discussed later). The PEPT1-like POTs dominate, by far, and are found in bacteria, fungi, protists and plants; the PHT appear to be specific to mammals (no reports of PHTs in lower vertebrates available so far). The phylogenetic differences among the PEPT-like POTs are interesting however have not yet been studied in detail.

2. A plethora of natural substrates

Considering the PTR2/POT/NPFs as whole, the substrate range is found to be exceptionally broad, that is, within the family, promiscuity is observed but several substrate specificities appear to have evolved as well. In this section, we present natural substrates and xenobiotics where applicable. Some of the structures of substrates discussed are presented in Fig. 1 (amino acids and conjugates) and 2 (peptides and conjugates).

2.1 Ions and water

Uptake of nitrate was among the first activities discovered for NPFs by NPF 6.3 from *Arabidopsis thaliana* (Tsay, Schroeder, Feldmann, & Crawford, 1993) and has since then been observed for 15 other NPFs (Corratgé-Faillie & Lacombe, 2017). NPF 6.3 has been shown to transport nitrate in an apparent dual affinity fashion (Liu & Tsay, 2003; Wang, Liu, & Crawford, 1998), however there is some ambiguity about this, as other groups show a single and low affinity process (Tsay et al., 1993). To our knowledge, none of the other NPFs have exhibited dual affinity. A single NPF, NPF3.1, has shown uptake of nitrite as well as nitrate (Pike et al., 2014). Apart from nitrate/nitrite it has been observed that NPF2.4 and 2.5 is able to facilitate passive efflux of chloride ions (Li, Byrt, et al., 2016; Li, Qiu, et al., 2016; Taochy et al., 2015).

An interesting observation was made by Chen et al. (2010) while studying fluid uptake in the intestine in mice; they found that a substantial amount of water uptake can be attributed to PEPT1, and that water uptake only occurs concomitant to peptide uptake. Thus, although not a specific substrate as such, it is an important observation that water molecules can be transported along with the primary substrate. The number of water molecules depends presumably on the space left by the primary substrate in the interdomain cavity.

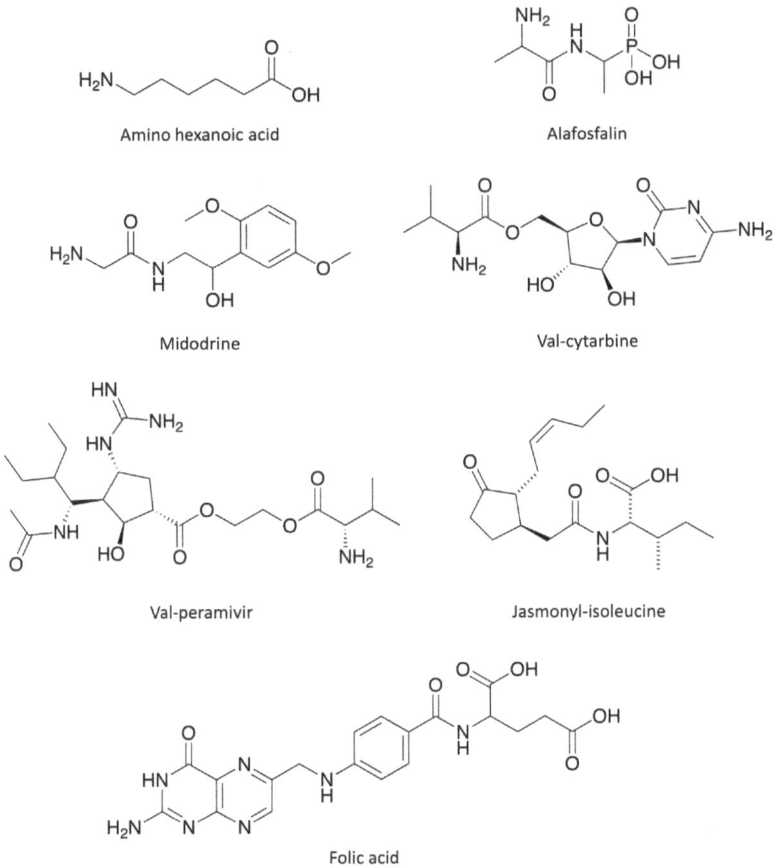

Fig. 1 Structures of amino acids and conjugates.

2.2 Single amino acids and amino acid drug conjugates

Generally, single amino acids are only rarely seen to be substrates of the PTR2/POT/NPFs. There is however a small subgroup of mammalian transporters called peptide histidine transporters 1 and 2 (PHT1 and PHT2) within the family that has been shown to be able to transport histidine coupled to the inward transport of protons. These transporters are very similar to other members of the family; however, they are able to transport histidine and smaller peptides but apparently not other amino acids (Bhardwaj, Herrera-Ruiz, Eltoukhy, Saad, & Knipp, 2006; Wang et al., 2019; Yamashita et al., 1997). Compared to other members of the family, the PHTs have not been studied in much detail. But their role within lysosomal maturation among other have been suggested (Bissa, Beedle, & Govindarajan, 2016).

Apart from the above-mentioned transporters, NPF8.3 from *A. thaliana* showed ability to transport histidine, but it was later also reported to transport di and tripeptides (Dietrich et al., 2004; Frommer, Hummel, & Rentsch, 1994). It should also be mentioned that aminohexanoic acid (also known as 6-aminocaproic acid, Fig. 1), although not a regular amino acid, was shown to be transported by YbgH from *Escherichia coli* (Casagrande et al., 2009; Zhao et al., 2014). Similarly, studies showed that PEPT1 is able to transport some omega-amino-fatty acids and that none of the functional groups could be removed without loss of transport (Doring et al., 1998).

Alafosfalin, an antibacterial compound that is sometimes considered a dipeptide analog, is aminomethylphosphonic acid with an alanine attached through an amide bond (Fig. 1). Several studies have reported transport of alafosfalin by mammalian and *E. coli* POTs (Harder et al., 2008; Neumann, Bruch, Gebauer, & Brandsch, 2004). Amino acid derivatization of drugs, i.e. forming prodrugs, as a means to enhance bioavailability through human PEPT1 has resulted in several synthetic substrates. Most if not all of these are formed by creation of an ester linkage to facilitate hydrolysis to free the drug compound after transport, some examples are valacyclovir and zidovudine (Friedrichsen et al., 2002; Han et al., 1998), peramivir-l-Ile/Val (Sun et al., 2018), L-DOPA-Phe (Tamai et al., 1998), Val-cytarabine (Sun et al., 2009) and midodrine (Tsuda et al., 2006) (Fig. 1). All of these compounds contain a free amino-terminus respectively.

The naturally occurring plant hormone jasmonoyl-isoleucine (Fig. 1) is important for plant defense and growth purposes among various others. Several NPFs, including NPF8.3, transport jasmonoyl-isoleucine (Chiba et al., 2015; Saito et al., 2015). Different from the prodrugs mentioned above, jasmonic acid is attached to isoleucine through an amide bond leaving a free carboxy-terminus (Fig. 1). One of the important anti-cancer drug and protease inhibitor bestatin containing a leucine coupled to hydroxy amino phenyl butanoic acid was observed to be a substrate of mammalian PEPT1 and YdgR from *E. coli* (Prabhala et al., 2017; Sala-Rabanal, Loo, Hirayama, Turk, & Wright, 2006). Similarly, we recently found that folic acid (Fig. 1), which contains a glutamyl residues with a free carboxy-terminus, is transported by the bacterial POT YdgR (unpublished results).

2.3 Peptides and peptide drug conjugates

Most if not all POT members characterized to date along with five NPF members are able to transport peptides (Corratgé-Faillie & Lacombe, 2017). A prototypical POT member transports peptide length of two and/or three amino acids, comprising virtually all combinations of amino acids, with some non-exclusive preferences for charges and hydrophobicity (Biegel et al., 2006; Jensen, Simonsen, et al., 2012; Vig et al., 2006). Some of the exceptions here are peptides consisting of only positively charged residues (Arg-Arg-Arg, Lys-Lys-Lys, Arg-Lys) and hydrophobic peptides such as Trp-Trp-Trp, Trp-Trp and Trp-Tyr which were not recognized as PEPT1 substrates or very poorly transported (Biegel et al., 2006; Vig et al., 2006).

Similar preferences for charged amino acid in peptides at the either of the termini were also observed in several bacterial POTs for e.g. presence of Lys 318 at the active site of $PepT_{SO}$ preferred negatively charged residues at the N terminus of the tested di and tripeptides (Prabhala et al., 2015) as observed from a pyranine based pH sensitive assay (Prabhala et al., 2014). Interestingly, this observation was confirmed when the crystal structure of a POT from *yersinia* (YePEPT) was published. In YePEPT, the presence of Lys314 in the proximity of active site imposed a preference of a negatively charged residues at the N terminal (Asp-Ala and Glu-Ala) of peptides while the positive charge at the either termini were not tolerated (Boggavarapu et al., 2015). Similarly, in YjdL (*E. coli*) presence of Asp392 in 5 Å vicinity yielded a preference for Ala-Lys over Lys-Ala. This preference was eliminated when Asp392 was mutated to a serine and additionally this also resulted in preference for a tripeptide (Jensen, Ismat, et al., 2012) and clearly establishes that active site of POTs can be tuned to the selective uptake of peptides.

PEPT1 has been investigated widely for improving the bioavailability of drugs by synthesis of di/tripeptide-drug conjugates (prodrugs). An antiretroviral drug such as lopinavir was coupled to a dipeptide His-Leu to form a conjugate His-Leu-lopinavir. Conjugation of lopinavir circumvented the efflux of lopinavir by P-gp and MRP-2 (Mandal, Pal, & Mitra, 2016) (Fig. 2). Similarly, acyclovir was conjugated to dipeptides Glu-Sar, D-Glu-Ala, Val-Val and Gly-Val via ester linkages (Anand, Patel, & Mitra, 2003; Thomsen et al., 2003). The dipeptides-acyclovir conjugates appeared to have affinity for PEPT1 and were observed to be better promoeities than valine itself. Dipeptide (Val-Val,Val-Phe, Gly-Sar, Val-Gly and Tyr-Val) functionalized polymeric nanoparticles (made of polylactic acid, poly ethylene glycol and glycolic acid) encapsulating acyclovir and docetaxel were observed to transported by PEPT1 providing promising alternatives to the existing mode of delivery of poorly absorbed drugs (Du et al., 2018; Gourdon et al., 2017). The neurotransmitter precursor L-DOPA

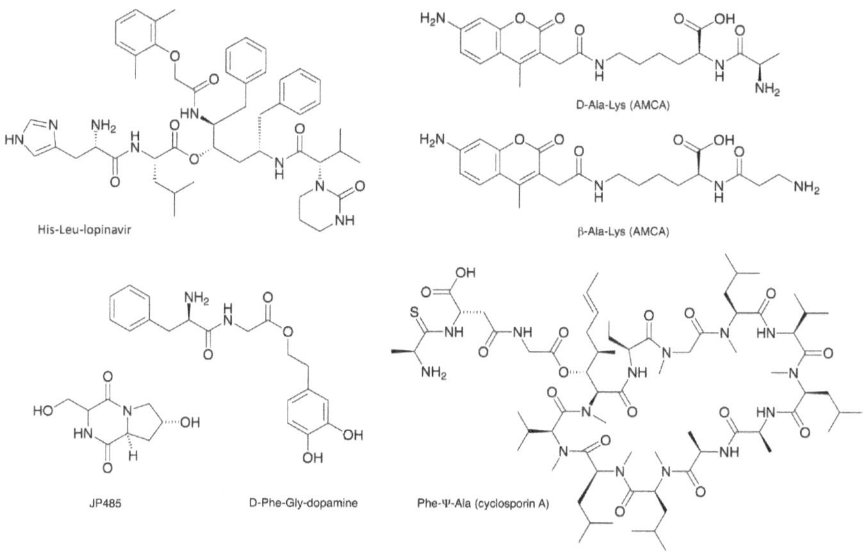

Fig. 2 Structures of peptides and conjugates.

(not a PEPT substrate) was transported through PEPT1 when conjugated with D-Phe-Gly (Fig. 2) (Wang et al., 2010). The transport and uptake of antimetabolite drug floxuridine was shown to be enhanced upon conjugation with dipeptides (Phe-Gly, Phe-Tyr and Ile-Gly) (Tsume, Hilfinger, & Amidon, 2008). Generally, cyclic peptides are low affinity and poor substrates for PepT1 such as JP845 (Miao et al., 2011). Pseudo-tripeptidization of JP485 improved the transport and affinity toward PEPT1 (Jiang et al., 2019). Peptides are generally rapidly catabolized in the cells and in order to detect the entire conjugate several peptides that are resistant to intracellular hydrolysis were either synthesized such as ß-Ala-Lys and Phe-Ψ-Ala. The thiopeptide Phe-Ψ-Ala has been extensively tested by coupling it to different drugs with poor bioavailability (propofol, zanamivir etc.) and gastrointestinal side effects (ibuprofen) and PEPT1 mediated transport was observed (Foley et al., 2018). Using this peptide as a pro-moiety cyclosporin A was transported in Pept1 dependent manner. This was the first instance where such a large peptides of MW > 1000 Da was reported to be transported by PEPT1 (Fig. 2) (Foley, Bermudez, Bailey, & Meredith, 2016).

The reporter substrate of POTs ß-Ala-Lys (AMCA), was an interesting development as it enabled to distinguish between different PepT isoforms. ß-Ala-Lys (AMCA) was known to be a substrate of PEPT2 but while D-Ala-Lys (AMCA) was observed to be transported by PEPT1 (Kottra,

Spanier, Verri, & Daniel, 2013). Although all the *E. coli* POTs investigated for their functional activity were able to transport ß-Ala-Lys (AMCA) (Ernst et al., 2009; Jensen, Ismat, et al., 2012; Prabhala et al., 2014; Sharma, Aduri, Iqbal, Prabhala, & Mirza, 2016) with YbgH being an exception (Casagrande et al., 2009).

ß-Ala-His (a cationic peptide, also known as carnosine) and its methylated analog (also known as anserine) are endogenously produced bioactive dipeptides that are resistant to cytoplasmic degradation. Both carnosine and anserine were observed to be substrates of PEPT1 (maximal transport at pH = 6) (Geissler, Zwarg, Knutter, Markwardt, & Brandsch, 2010; Takeo & Ikumi, 2013; Wenzel, Thwaites, & Daniel, 1995).

In addition to natural peptides and peptidomimetics drugs, some of the bacterial cell wall derived peptides such as muramyl dipeptide, Tri-DAP and fMLP were observed to be substrates of PEPT1 (Fig. 3). These peptides are speculated to induce chemokine production and activate the neutrophils to reach the site of inflammation (Dalmasso et al., 2010; Merlin et al., 1998; Nosworthy & Brunton, 2016).

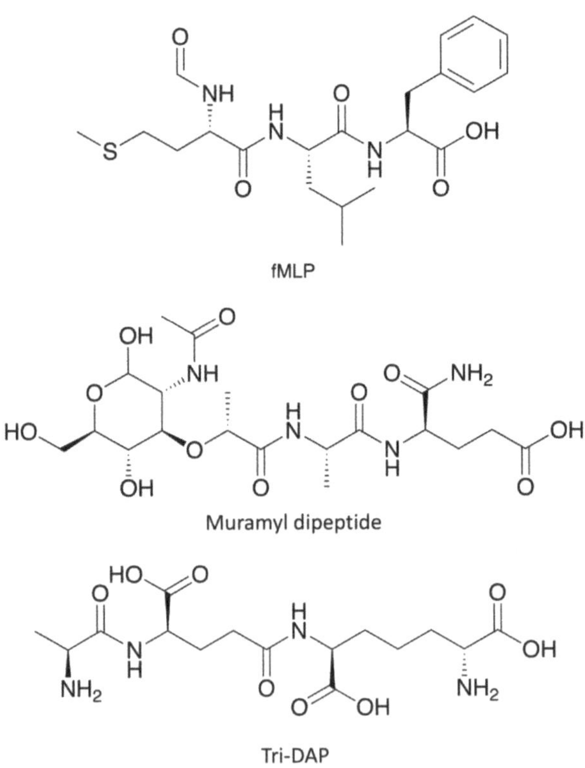

Fig. 3 Structures of bacterial cell wall derived peptides.

2.4 Organic anions, and a cation

The definition of an- and cations used here refers to molecules that are not derivates of amino acids, i.e. attached to an amino acid through the amide or carboxyl-group. Having stated this, to our knowledge no organic cations have been reported to be substrates of the PTR2/POT/NPFs. On the other hand, several organic anions are found to be substrates; this class of substrates is dominated by plants NPFs, many of whom are also known to be low-affinity nitrate transporters. Apart from dipeptides and tripeptides, NPFs have evolved for the transport of glucosinolates, that are derived from an amino acid, typically tryptophan or methionine, contain a sulfate and a thio-glucosyl moiety (bioactive secondary metabolites with antibiotic and anticancer properties) (Jørgensen, Xu, Crocoll, Ernst, et al., 2017; Nour-Eldin et al., 2012). Plant hormones like auxin and abscisic acid are also transported by a number of NPF members. Abscisic acid an important compound involved in dormancy, germination, cell division and biotic/abiotic stress (Kanno et al., 2012) and indole acetic acid (IAA/auxin, a plant growth hormone) is transported by 13 NPF members (Krouk et al., 2010) (Fig. 4). The tricyclic compounds such as giberellins, a plant hormone involved in root and shoot elongation was also transported by NPFs (Chiba et al., 2015; Wulff et al., 2019). Additionally, PepT1 was previously known to transport ß-lactam antibiotics like cefixime, ampicillin and ceftibuten (Takeo & Ikumi, 2013), ACE inhibitors like enalapril (Knutter et al., 2008; Zhu, Chen, Steel, Hediger, & Smith, 2000).

To our knowledge, strictosidine (structure not shown), an NPF substrate, is the only positively charged substrate found so far (Payne et al., 2017).

2.5 Uncharged substrates

Mammalian PEPT1 has been studied extensively due to its role in drug absorption and potential as a means to increase bioavailability of orally administered drugs. Generally high affinity peptide substrates should possess both free N and C termini, but it has also been observed that peptide bond is sufficient for a compound to be a substrate albeit with low affinity (Meredith et al., 2000). Recently, we showed that chloramphenicol, which is uncharged but contains a peptide bond (Fig. 4) is a substrate of YdgR from *E. coli* (Prabhala et al., 2018). In addition, the NPF MeCGTR1 from cassava

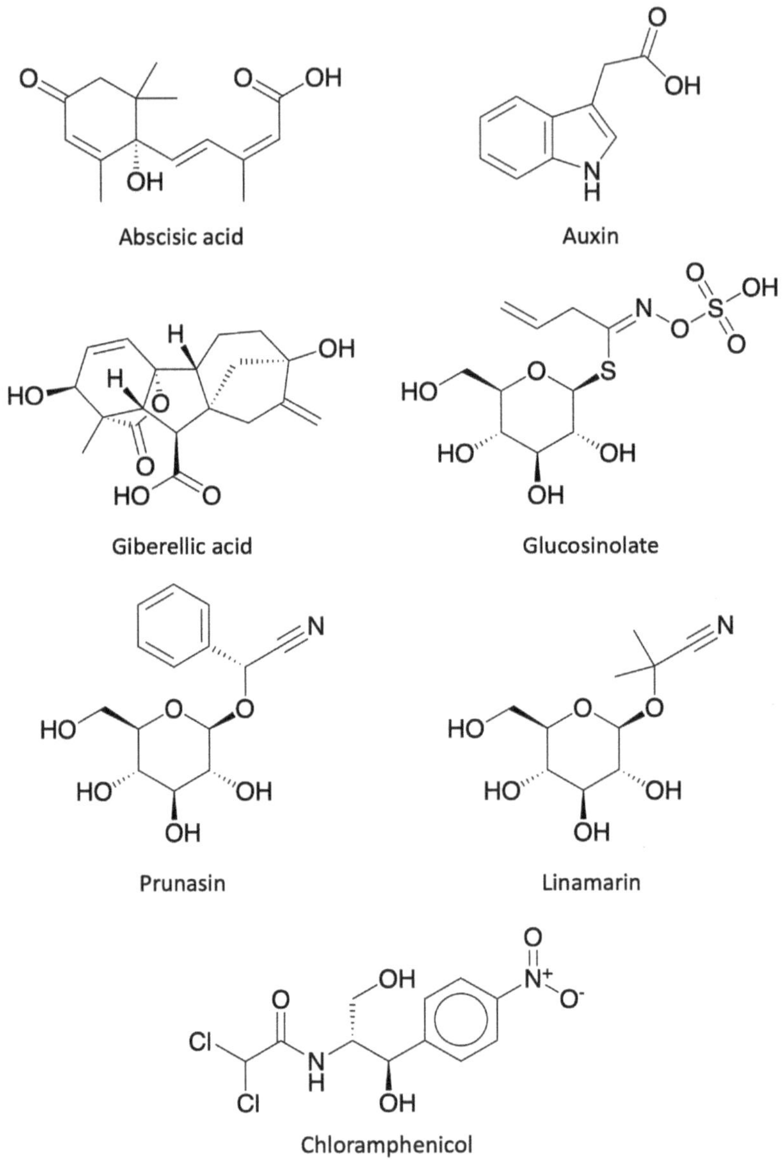

Fig. 4 Structures of substrates of NPF transporters.

transported the non-charged phenylalanine- and valine-derived cyanogenic glucosides prunasin and linamarin, respectively (Jørgensen, Xu, Crocoll, Ernst, et al., 2017).

2.6 Similarity and diversity among the substrates

In order to obtain a more quantitative view of the substrate diversity, we have performed a principal component (PC) analysis for 19 substrates based on the classical Lipinski Rule-of-Five descriptors, molecular weight (MW $\leq$ 500), lipophilicity (logP $\leq$ 5) and number of hydrogen-bond acceptors and donors (#HBA $\leq$ 10, #HBD $\leq$ 5) (Lipinski, Lombardo, Dominy, & Feeney, 2001). PC1 is primarily positively associated with the molecular weight whereas PC2 is negatively associated with the lipophilicity. Examples on peptide substrates are shown in Fig. 5.

Computational Details: The peptide substrates were extracted from the literature and comprised 19 compounds (shown in Figs. 1–4). Lipinski Rule-of-Five descriptors were calculated via a KNIME (version 3.5.3) workflow accessing the Schrödinger software (version 2019-1, Schrödinger, LLC, New York, 2019). The principal component (PC) analysis was performed by KNIME using the PCA and PCA Compute nodes (Berthold et al., 2011).

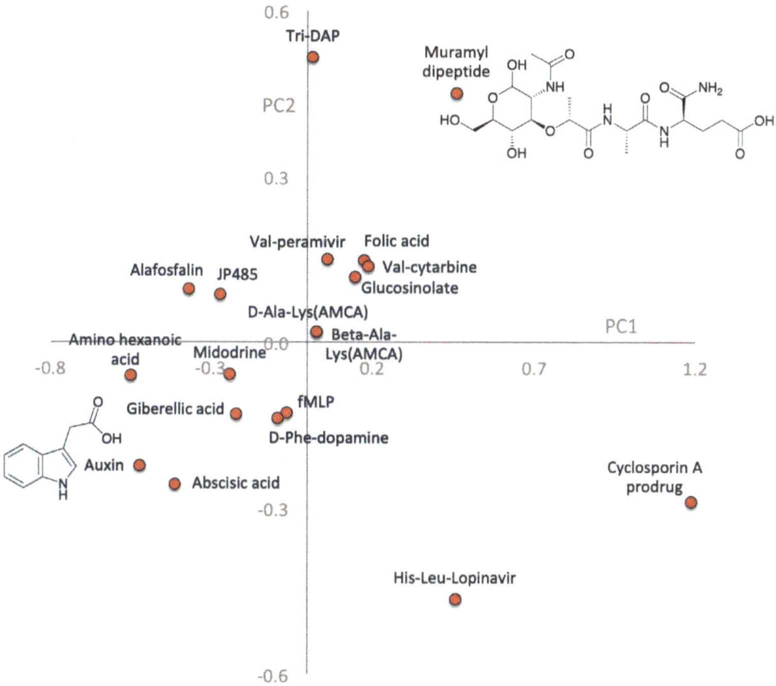

Fig. 5 PCA plot for the POT substrates discussed in the paper.

The majority of the peptides show a clear relationship between size (molecular weight) and polarity (lipophilicity) ranging from the smallest and most lipophilic compound auxin to the largest and less lipophilic compound muramyl dipeptide. The only two exceptions are the His-Leu-lopinavir and the cyclosporine A prodrug, both of which must be characterized as large lipophilic compounds. The reason for these two compounds being actively transported across the membrane is probably because they contain a dipeptide sequence facilitating transport.

From more qualitative point of view, it appears that most naturally occurring PTR2/POT/NPF substrates contain a formal negative charge, whether part of a zwitterion or a single charge (±); exceptions, so far, are chloramphenicol and the cyanogenic glucosides linnemarin and prunasin, which does not possess neither C nor N termini. The transportability of chloramphenicol may be explained by the presence of a peptide bond, however interestingly, for the rest of the substrates fit well with results from independent structural studies performed on the POTs, as discussed below.

3. 3D-structure and transport mechanism

Several crystal structures of the family members have been determined so far (Minhas & Newstead, 2020), by far most from bacterial species and two from *A. thaliana* (Newstead, 2011, 2014, 2017a, 2017b). They all exhibited a canonical overall fold with twelve transmembrane (TM) helices organized in to two six-helix domains interconnected by a loop, as observed in the AtNPFs, or with two additional transmembrane helices situated in all bacterial structures determined so far (Fig. 6A). This two-domain six-helix architecture is a consensus structure seen in all MFS members. Another feature of the MFS also observed among the PTR2/POT/NPFs, is that the substrate binding site is located in between the two six-helix domains (Fig. 6B).

Most of the crystallized PTR2/POT/NPFs have been shown to be monomers, however NPF 6.3 from *A. thaliana* exists as a dimer, which has functional implications as well (Parker & Newstead, 2014). Some bacterial POTs are multimeric (Fatma Guettou et al., 2013) however the functional significance of this (if any) is not clear, and rabbit PEPT1 has been proposed to be a multimer of either four or five monomers (Panitsas, Boyd, & Meredith, 2006).

The amino acid sequences of PTR2/POT/NPFs all contain three highly conserved signature motifs: Motif 1 (ExxER, lesser in NPFs), Motif 2

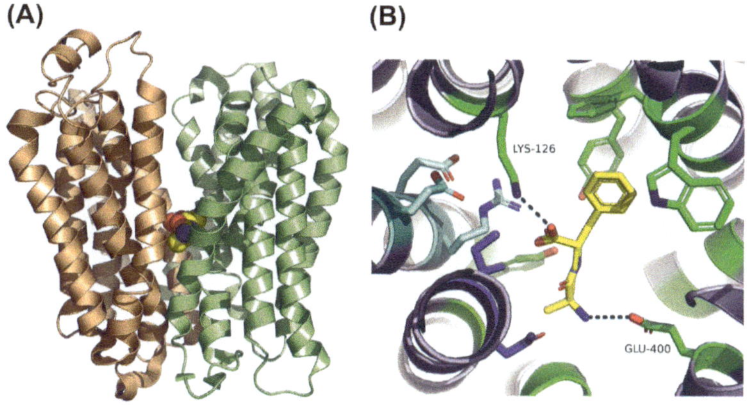

Fig. 6 General 3D structure of POTs with ligand in active site (A) and Phe-Ala at the active site of PepTst (B).

(GxxADxxxGKxxTI) and Motif 3 (FYxxINxG) (Jørgensen et al., 2015). Motifs 1 and 3 are located such that they face the interdomain space i.e. the within or in closeness, proximity of the substrate binding site. Apart from these motifs, a positively charged residue on helix 4, mostly lysine and rarely arginine, and a glutamate on helix 10, or rarely an aspartate, are located in this space, and found in most PTR2/POT/NPFs (Fig. 7).

3.1 A mechanism for peptide transport

During its transport cycle, a PTR2/POT/NPF member will go through several steps (Fig. 8): the empty transporter outward-facing conformation with both substrate and proton bound, it is not yet clear whether the proton or substrate binds first; the loaded structure will go through an occluded state to the inward-facing conformation where deprotonation and substrate

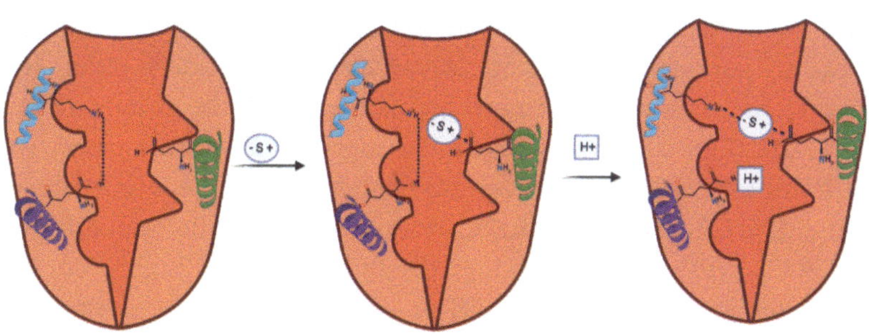

Fig. 7 Illustration of putative salt bridge interactions between substrate and conserved residues in active site (created in biorender.com).

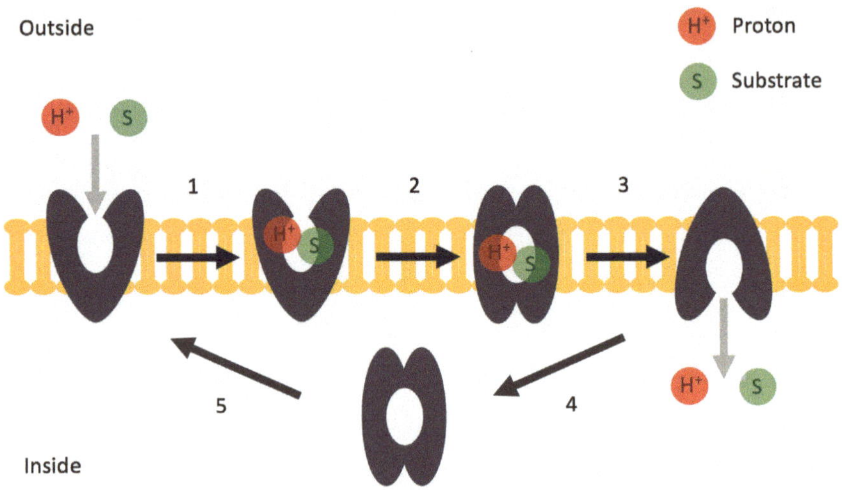

Fig. 8 Putative transport cycle.

release will occur, again it is undetermined which leaves first, finally the empty transporter will go through an occluded conformation in to the outward-facing empty conformation.

It has been shown that for lactose permease from *E. coli*, protonation of the transporter leads to the formation of a sugar binding site with higher affinity (Abramson et al., 2003); taken together with the fact that sugar alone cannot be transported and neither can protons, it can be speculated that the presence of sugar and proton both lead to formation of a high-affinity site but also the one that triggers transport. Recently it was suggested that the same may apply to POTs as well: *in vitro* and *in silico* studies of bacterial POTs from *E. coli* showed that when the EXXER motif is protonated it allows the lysine residue of helix 4 (TM4) (Aduri et al., 2015) to interact with the C-terminus of a bound dipeptide (Aduri et al., 2015). In unprotonated state the lysine remains attached to the EXXER motif, unable to interact with bound peptide. In this state, however, the peptide remains loosely attached to the transporter through interactions with the glutamate residue on helix 10 (Fig. 7). This mechanism suggests that protonation of the EXXER motif requires loosely bound peptide so that the lysine can switch from the EXXER motif to the C-terminus. It may be applicable to all PTR2/POT/NPFs, as the residues involved are very highly conserved. However, for some PTR2/POT/NPFs other residues in the substrate binding space have also been suggested to become protonated such as His57 of the mammalian PEPT1, which is not conserved among other PTR2/POT/

NPF members (Chen, Steel, & Hediger, 2000). It could be speculated that variations in protonation exist among PTR2/POT/NPFs, but that the proton always has to pass by the EXXER motif for transport to occur.

Several investigations were made to determine the substrate:proton coupling ratios and it was found that for PEPT1, the peptide:proton ratio was one for neutral substrates, two for anionic substrates and less than one for the cationic substrates. On the contrary for PEPT2, two proton for neutral substrates, three for anionic and 1.5–2.5 for cationic substrates (Chen, Zhu, Smith, & Hediger, 1999; Irie, Terada, Katsura, Matsuoka, & Inui, 2005; Parker, Mindell, & Newstead, 2014). PEPT2 is speculated to be having one substrate active site and two proton active sites which could be the reason for additional number of protons required for substrate translocation (Chen et al., 1999). While the studies in bacterial peptide transporter ($PepT_{St}$) reported five to six protons for dipeptides and three protons for tripeptides. This could also mean that different substrates translocate with a different number of protons. This could possible also be one of the reasons for the substrate promiscuity of POTs (Parker et al., 2014).

About 15%–20% of the plant NPFs do not contain an EXXER motif in any form, i.e. both glutamates are absent, and yet exhibit substrate transport. Although not entirely clear, it may be that these transporters energize their activity by some other sub-specialized mechanism.

3.2 Do the non-peptide substrates fit in the transport mechanism?

The "working" mechanism outlined in Fig. 8 contains a "capture" step where the peptide is bound to the conserved glutamate residue on helix 9; this glutamate (or a negative charge) is almost completely conserved among the POT and NPFs. Additionally, the lysine on the helix 4 are also conserved and well within the 5 Å of the active site.

How do we explain that substrates without a positive charge but with a negative charge, or without any charges are able to get transported? An example of the latter is observed for chloramphenicol uptake by YdgR (Prabhala et al., 2018). Chloramphenicol does not contain any charged groups and thus cannot form ionic interactions with the conserved glutamate or lysines at the active site. It does however contain a peptide bond and mutating residues that interacts with the peptide bond also led to lowered chloramphenicol transport (Prabhala et al., 2018). This indicates that chloramphenicol is bound in the same position as the peptide, and the binding is able to trigger conformational changes even though the glutamate and lysine are apparently not directly affected by binding.

For substrates with a negative charged group interaction with the lysine (or in some cases arginine) is possible and thereby proton coupling. As most naturally occurring POT/NPTF substrates contain a negatively charged group, interactions with the lysine or in some cases arginine) appears to be of utmost importance. For such substrates, it is observed that the binding pocket has evolved to enable capture and to bring the substrate in place, as observed for glucosinolates (Jørgensen et al., 2017) for example.

References

Abramson, J., Smirnova, I., Kasho, V., Verner, G., Kaback, H. R., & Iwata, S. (2003). Structure and mechanism of the lactose permease of *Escherichia coli*. *Science, 301*(5633), 610–615. https://doi.org/10.1126/science.1088196.

Aduri, N. G., Prabhala, B. K., Ernst, H. A., Jørgensen, F. S., Olsen, L., & Mirza, O. (2015). Salt bridge swapping in the EXXERFXYY motif of proton-coupled oligopeptide transporters. *Journal of Biological Chemistry, 290*(50), 29931–29940.

Anand, B. S., Patel, J., & Mitra, A. K. (2003). Interactions of the dipeptide ester prodrugs of acyclovir with the intestinal oligopeptide transporter: Competitive inhibition of glycylsarcosine transport in human intestinal cell line-Caco-2. *Journal of Pharmacology and Experimental Therapeutics, 304*(2), 781–791. https://doi.org/10.1124/jpet.102.044313.

Beale, J. H., Parker, J. L., Samsudin, F., Barrett, A. L., Senan, A., Bird, L. E., et al. (2015). Crystal structures of the extracellular domain from PepT1 and PepT2 provide novel insights into mammalian peptide transport. *Structure, 23*(10), 1889–1899. https://doi.org/10.1016/j.str.2015.07.016.

Berthold, M. R., Cebron, N., Dill, F., Gabriel, T. R., Kötter, T., Meinl, T., et al. (2011). Data analysis, machine learning and applications. In C. Preisach, H. Burkhardt, L. Schmidt-Thieme, & R. Decker (Eds.), *Knowledge organization*. https://doi.org/10.1007/978-3-642-76307-6.

Bhardwaj, R. K., Herrera-Ruiz, D., Eltoukhy, N., Saad, M., & Knipp, G. T. (2006). The functional evaluation of human peptide/histidine transporter 1 (hPHT1) in transiently transfected COS-7 cells. *European Journal of Pharmaceutical Sciences, 27*(5), 533–542. https://doi.org/10.1016/j.ejps.2005.09.014.

Biegel, A., Knutter, I., Hartrodt, B., Gebauer, S., Theis, S., Luckner, P., et al. (2006). The renal type H+/peptide symporter PEPT2: Structure-affinity relationships. *Amino Acids, 31*(2), 137–156. https://doi.org/10.1007/s00726-006-0331-0.

Bissa, B., Beedle, A. M., & Govindarajan, R. (2016). Lysosomal solute carrier transporters gain momentum in research. *Clinical Pharmacology & Therapeutics, 100*(5), 431–436. https://www.ncbi.nlm.nih.gov/pmc/articles/PMC5056150/pdf/CPT-100-431.pdf.

Boggavarapu, R., Jeckelmann, J. M., Harder, D., Ucurum, Z., & Fotiadis, D. (2015). Role of electrostatic interactions for ligand recognition and specificity of peptide transporters. *BMC Biology, 13*, 58. https://doi.org/10.1186/s12915-015-0167-8.

Casagrande, F., Harder, D., Schenk, A., Meury, M., Ucurum, Z., Engel, A., et al. (2009). Projection structure of DtpD (YbgH), a prokaryotic member of the peptide transporter family. *Journal of Molecular Biology, 394*(4), 708–717. https://doi.org/10.1016/j.jmb.2009.09.048.

Chen, M., Singh, A., Xiao, F., Dringenberg, U., Wang, J., Engelhardt, R., et al. (2010). Gene ablation for PEPT1 in mice abolishes the effects of dipeptides on small intestinal fluid absorption, short-circuit current, and intracellular pH. *American Journal of Physiology - Gastrointestinal and Liver Physiology, 299*(1), G265–G274. https://doi.org/10.1152/ajpgi.00055.2010.

Chen, X. Z., Steel, A., & Hediger, M. A. (2000). Functional roles of histidine and tyrosine residues in the H(+)-peptide transporter PepT1. *Biochemical and Biophysical Research Communications, 272*(3), 726–730. https://doi.org/10.1006/bbrc.2000.2851.

Chen, X. Z., Zhu, T., Smith, D. E., & Hediger, M. A. (1999). Stoichiometry and kinetics of the high-affinity H+-coupled peptide transporter PepT2. *Journal of Biological Chemistry, 274*(5), 2773–2779. https://doi.org/10.1074/jbc.274.5.2773.

Chiba, Y., Shimizu, T., Miyakawa, S., Kanno, Y., Koshiba, T., Kamiya, Y., et al. (2015). Identification of *Arabidopsis thaliana* NRT1/PTR FAMILY (NPF) proteins capable of transporting plant hormones. *Journal of Plant Research, 128*(4), 679–686. https://doi.org/10.1007/s10265-015-0710-2.

Corratgé-Faillie, C., & Lacombe, B. (2017). Substrate (un)specificity of Arabidopsis NRT1/PTR FAMILY (NPF) proteins. *Journal of Experimental Botany, 68*(12), 3107–3113. https://doi.org/10.1093/jxb/erw499.

Dalmasso, G., Nguyen, H. T. T., Charrier-Hisamuddin, L., Yan, Y., Laroui, H., Demoulin, B., et al. (2010). PepT1 mediates transport of the proinflammatory bacterial tripeptide L-Ala-{gamma}-D-Glu-meso-DAP in intestinal epithelial cells. *American Journal of Physiology - Gastrointestinal and Liver Physiology, 299*(3), G687–G696. https://doi.org/10.1152/ajpgi.00527.2009.

Dietrich, D., Hammes, U., Thor, K., Suter-Grotemeyer, M., Fluckiger, R., Slusarenko, A. J., et al. (2004). AtPTR1, a plasma membrane peptide transporter expressed during seed germination and in vascular tissue of Arabidopsis. *The Plant Journal, 40*(4), 488–499. https://doi.org/10.1111/j.1365-313X.2004.02224.x.

Doki, S., Kato, H. E., Solcan, N., Iwaki, M., Koyama, M., Hattori, M., et al. (2013). Structural basis for dynamic mechanism of proton-coupled symport by the peptide transporter POT. *Proceedings of the National Academy of Sciences of the United States of America, 110*(28), 11343–11348. https://doi.org/10.1073/pnas.1301079110.

Doring, F., Will, J., Amasheh, S., Clauss, W., Ahlbrecht, H., & Daniel, H. (1998). Minimal molecular determinants of substrates for recognition by the intestinal peptide transporter. *Journal of Biological Chemistry, 273*(36), 23211–23218. https://doi.org/10.1074/jbc.273.36.23211.

Du, Y., Tian, C., Wang, M., Huang, D., Wei, W., Liu, Y., et al. (2018). Dipeptide-modified nanoparticles to facilitate oral docetaxel delivery: New insights into PepT1-mediated targeting strategy. *Drug Delivery, 25*(1), 1403–1413. https://doi.org/10.1080/10717544.2018.1480675.

El-Gebali, S., Mistry, J., Bateman, A., Eddy, S. R., Luciani, A., Potter, S. C., et al. (2019). The Pfam protein families database in 2019. *Nucleic Acids Research, 47*(D1). https://doi.org/10.1093/nar/gky995.

Ernst, H. A., Pham, A., Hald, H., Kastrup, J. S., Rahman, M., & Mirza, O. (2009). Ligand binding analyses of the putative peptide transporter YjdL from *E. coli* display a significant selectivity towards dipeptides. *Biochemical and Biophysical Research Communications, 389*(1), 112–116. https://doi.org/10.1016/j.bbrc.2009.08.098.

Foley, D. W., Bermudez, I., Bailey, P. D., & Meredith, D. (2016). A cyclosporine derivative is a substrate of the oligopeptide transporter PepT1. *MedChemComm, 7*, 999–1002. https://doi.org/10.1039/c5md00575b.

Foley, D. W., Pathak, R. B., Phillips, T. R., Wilson, G. L., Bailey, P. D., Pieri, M., et al. (2018). Thiodipeptides targeting the intestinal oligopeptide transporter as a general approach to improving oral drug delivery. *European Journal of Medicinal Chemistry, 156*, 180–189. https://doi.org/10.1016/j.ejmech.2018.06.064.

Friedrichsen, G. M., Chen, W., Begtrup, M., Lee, C. P., Smith, P. L., & Borchardt, R. T. (2002). Synthesis of analogs of L-valacyclovir and determination of their substrate activity for the oligopeptide transporter in Caco-2 cells. *European Journal of*

Pharmaceutical Sciences, 16(1–2), 1–13. http://www.sciencedirect.com/science/article/pii/S0928098702000477.

Frommer, W. B., Hummel, S., & Rentsch, D. (1994). Cloning of an Arabidopsis histidine transporting protein related to nitrate and peptide transporters. *FEBS Letters, 347*(2–3), 185–189.

Ganapathy, M. E., Huang, W., Wang, H., Ganapathy, V., & Leibach, F. H. (1998). Valacyclovir: A substrate for the intestinal and renal peptide transporters PEPT1 and PEPT2. *Biochemical and Biophysical Research Communications, 246*(2), 470–475. https://doi.org/10.1006/bbrc.1998.8628.

Geissler, S., Zwarg, M., Knutter, I., Markwardt, F., & Brandsch, M. (2010). The bioactive dipeptide anserine is transported by human proton-coupled peptide transporters. *FEBS Journal, 277*(3), 790–795. https://doi.org/10.1111/j.1742-4658.2009.07528.x.

Gourdon, B., Chemin, C., Moreau, A., Arnauld, T., Baumy, P., Cisternino, S., et al. (2017). Functionalized PLA-PEG nanoparticles targeting intestinal transporter PepT1 for oral delivery of acyclovir. *International Journal of Pharmaceutics, 529*(1–2), 357–370. https://doi.org/10.1016/j.ijpharm.2017.07.024.

Guettou, F., Quistgaard, E. M., Raba, M., Moberg, P., Low, C., & Nordlund, P. (2014). Selectivity mechanism of a bacterial homolog of the human drug-peptide transporters PepT1 and PepT2. *Nature Structural & Molecular Biology, 21*(8), 728–731. https://doi.org/10.1038/nsmb.2860.

Guettou, F., Quistgaard, E. M., Trésaugues, L., Moberg, P., Jegerschöld, C., Zhu, L., et al. (2013). Structural insights into substrate recognition in proton-dependent oligopeptide transporters. *EMBO Reports, 14*(9), 804–810. https://doi.org/10.1038/embor.2013.107.

Han, H., de Vrueh, R. L. A., Rhie, J. K., Covitz, K.-M. Y., Smith, P. L., Lee, C.-P., et al. (1998). 5′-Amino acid esters of antiviral nucleosides, acyclovir, and AZT are absorbed by the intestinal PEPT1 peptide transporter. *Pharmaceutical Research, 15*(8), 1154–1159. https://doi.org/10.1023/A:1011919319810.

Harder, D., Stolz, J., Casagrande, F., Obrdlik, P., Weitz, D., Fotiadis, D., et al. (2008). DtpB (YhiP) and DtpA (TppB, YdgR) are prototypical proton-dependent peptide transporters of *Escherichia coli*. *FEBS Journal, 275*(13), 3290–3298. https://doi.org/10.1111/j.1742-4658.2008.06477.x.

Irie, M., Terada, T., Katsura, T., Matsuoka, S., & Inui, K. (2005). Computational modelling of H(+)-coupled peptide transport via human PEPT1. *The Journal of Physiology, 565*(Pt 2), 429–439. https://doi.org/10.1113/jphysiol.2005.084582.

Jensen, J. M., Ismat, F., Szakonyi, G., Rahman, M., & Mirza, O. (2012). Probing the putative active site of YjdL: An unusual proton-coupled oligopeptide transporter from *E. coli*. *PLoS One,* 7(10), e47780. https://doi.org/10.1371/journal.pone.0047780.

Jensen, J. M., Simonsen, F. C., Mastali, A., Hald, H., Lillebro, I., Diness, F., et al. (2012). Biophysical characterization of the proton-coupled oligopeptide transporter YjdL. *Peptides, 38*(1), 89–93. https://doi.org/10.1016/j.peptides.2012.08.012.

Jiang, Q., Zhang, J., Tong, P., Gao, Y., Lv, Y., Wang, C., et al. (2019). Bioactivatable pseudotripeptidization of cyclic dipeptides to increase the affinity toward oligopeptide transporter 1 for enhanced oral absorption: An application to Cyclo(l-hyp-l-ser) (JBP485). *Journal of Medicinal Chemistry, 62*(17), 7708–7721. https://doi.org/10.1021/acs.jmedchem.9b00358.

Jørgensen, M. E., Olsen, C. E., Geiger, D., Mirza, O., Halkier, B. A., & Nour-Eldin, H. H. (2015). A functional EXXEK motif is essential for proton coupling and active glucosinolate transport by NPF2.11. *Plant and Cell Physiology, 56*(12), 2340–2350. https://doi.org/10.1093/pcp/pcv145.

Jørgensen, M. E., Xu, D., Crocoll, C., Ernst, H. A., Ramírez, D., Motawia, M. S., et al. (2017). Origin and evolution of transporter substrate specificity within the NPF family. *Elife, 6*, e19466. https://doi.org/10.7554/eLife.19466.

Kanno, Y., Hanada, A., Chiba, Y., Ichikawa, T., Nakazawa, M., Matsui, M., et al. (2012). Identification of an abscisic acid transporter by functional screening using the receptor complex as a sensor. *Proceedings of the National Academy of Sciences, 109*(24), 9653–9658. https://doi.org/10.1073/pnas.1203567109.

Knutter, I., Wollesky, C., Kottra, G., Hahn, M. G., Fischer, W., Zebisch, K., et al. (2008). Transport of angiotensin-converting enzyme inhibitors by H+/peptide transporters revisited. *Journal of Pharmacology and Experimental Therapeutics, 327*(2), 432–441. https://doi.org/10.1124/jpet.108.143339.

Kottra, G., Spanier, B., Verri, T., & Daniel, H. (2013). Peptide transporter isoforms are discriminated by the fluorophore-conjugated dipeptides beta-Ala- and d-Ala-Lys-N-7-amino-4-methylcoumarin-3-acetic acid. *Physics Reports, 1*(7), e00165. https://doi.org/10.1002/phy2.165.

Krouk, G., Lacombe, B., Bielach, A., Perrine-Walker, F., Malinska, K., Mounier, E., et al. (2010). Nitrate-regulated auxin transport by NRT1.1 defines a mechanism for nutrient sensing in plants. *Developmental Cell, 18*(6), 927–937. https://doi.org/10.1016/j.devcel.2010.05.008.

Leran, S., Varala, K., Boyer, J. C., Chiurazzi, M., Crawford, N., Daniel-Vedele, F., et al. (2014). A unified nomenclature of NITRATE TRANSPORTER 1/PEPTIDE TRANSPORTER family members in plants. *Trends in Plant Science, 19*(1), 5–9. https://doi.org/10.1016/j.tplants.2013.08.008.

Li, B., Byrt, C., Qiu, J., Baumann, U., Hrmova, M., Evrard, A., et al. (2016). Identification of a stelar-localized transport protein that facilitates root-to-shoot transfer of chloride in Arabidopsis. *Plant Physiology, 170*(2), 1014–1029. https://doi.org/10.1104/pp.15.01163.

Li, B., Qiu, J., Jayakannan, M., Xu, B., Li, Y., Mayo, G. M., et al. (2016b). AtNPF2.5 modulates chloride (Cl(-)) efflux from roots of *Arabidopsis thaliana*. *Frontiers in Plant Science, 7*, 2013. https://doi.org/10.3389/fpls.2016.02013.

Lipinski, C. A., Lombardo, F., Dominy, B. W., & Feeney, P. J. (2001). Experimental and computational approaches to estimate solubility and permeability in drug discovery and development settings. *Advanced Drug Delivery Reviews, 46*(1–3), 3–26. https://doi.org/10.1016/s0169-409x(00)00129-0.

Liu, K. H., & Tsay, Y. F. (2003). Switching between the two action modes of the dual-affinity nitrate transporter CHL1 by phosphorylation. *The EMBO Journal, 22*(5), 1005–1013. https://doi.org/10.1093/emboj/cdg118.

Mandal, A., Pal, D., & Mitra, A. K. (2016). Circumvention of P-gp and MRP2 mediated efflux of lopinavir by a histidine based dipeptide prodrug. *International Journal of Pharmaceutics, 512*(1), 49–60. https://doi.org/10.1016/j.ijpharm.2016.08.027.

Meredith, D., Temple, C. S., Guha, N., Sword, C. J., Boyd, C. A. R., Collier, I. D., et al. (2000). Modified amino acids and peptides as substrates for the intestinal peptide transporter PepT1. *European Journal of Biochemistry, 267*(12), 3723–3728. https://doi.org/10.1046/j.1432-1327.2000.01405.x.

Merlin, D., Steel, A., Gewirtz, A. T., Si-Tahar, M., Hediger, M. A., & Madara, J. L. (1998). hPepT1-mediated epithelial transport of bacteria-derived chemotactic peptides enhances neutrophil-epithelial interactions. *Journal of Clinical Investigation, 102*(11), 2011–2018. https://doi.org/10.1172/jci4179.

Miao, Q., Liu, Q., Wang, C., Meng, Q., Guo, X., Peng, J., et al. (2011). Inhibitory effect of zinc on the absorption of JBP485 via the gastrointestinal oligopeptide transporter (PEPT1) in rats. *Drug Metabolism and Pharmacokinetics, 26*(5), 494–502. https://www.jstage.jst.go.jp/article/dmpk/26/5/26_DMPK-11-RG-014/_pdf.

Minhas, G. S., & Newstead, S. (2019). Structural basis for prodrug recognition by the SLC15 family of proton coupled peptide transporters. *Proceedings of the National Academy of Sciences of the United States of America, 116*(3), 804–809. https://doi.org/10.1101/454116.

Minhas, G. S., & Newstead, S. (2020). Recent advances in understanding prodrug transport through the SLC15 family of proton-coupled transporters. *Biochemical Society Transactions, 48*(2), 337–346. https://doi.org/10.1042/BST20180302.
Neumann, J., Bruch, M., Gebauer, S., & Brandsch, M. (2004). Transport of the phosphono-dipeptide alafosfalin by the H+/peptide cotransporters PEPT1 and PEPT2 in intestinal and renal epithelial cells. *European Journal of Biochemistry, 271*(10), 2012–2017. https://doi.org/10.1111/j.1432-1033.2004.04114.x.
Newstead, S. (2011). Towards a structural understanding of drug and peptide transport within the proton-dependent oligopeptide transporter (POT) family. *Biochemical Society Transactions, 39*(5), 1353–1358. https://doi.org/10.1042/bst0391353.
Newstead, S. (2014). Molecular insights into proton coupled peptide transport in the PTR family of oligopeptide transporters. *Biochimica et Biophysica Acta, 1850*(3), 488–499. https://doi.org/10.1016/j.bbagen.2014.05.011.
Newstead, S. (2017a). Recent advances in understanding proton coupled peptide transport via the POT family. *Current Opinion in Structural Biology, 45*, 17–24. https://doi.org/10.1016/j.sbi.2016.10.018.
Newstead, S. (2017b). Symmetry and structure in the POT family of proton coupled peptide transporters. *Symmetry, 9*(6). https://doi.org/10.3390/sym9060085.
Newstead, S., Drew, D., Cameron, A. D., Postis, V. L., Xia, X., Fowler, P. W., et al. (2011). Crystal structure of a prokaryotic homologue of the mammalian oligopeptide-proton symporters, PepT1 and PepT2. *The EMBO Journal, 30*(2), 417–426. https://doi.org/10.1038/emboj.2010.309.
Nosworthy, M. G., & Brunton, J. A. (2016). Cysteinyl-glycine reduces mucosal proinflammatory cytokine response to fMLP in a parenterally-fed piglet model. *Pediatric Research, 80*(2), 293–298. https://doi.org/10.1038/pr.2016.69.
Nour-Eldin, H. H., Andersen, T. G., Burow, M., Madsen, S. R., Jørgensen, M. E., Olsen, C. E., et al. (2012). NRT/PTR transporters are essential for translocation of glucosinolate defence compounds to seeds. *Nature, 488*(7412), 531–534. https://doi.org/10.1038/nature11285.
Panitsas, K. E., Boyd, C. A. R., & Meredith, D. (2006). Evidence that the rabbit proton-peptide co-transporter PepT1 is a multimer when expressed in *Xenopus laevis* oocytes. *Pfluegers Archiv European Journal of Physiology, 452*(1), 53–63. https://doi.org/10.1007/s00424-005-0002-0.
Parker, J. L., Mindell, J. A., & Newstead, S. (2014). Thermodynamic evidence for a dual transport mechanism in a POT peptide transporter. *Elife, 3*, e04273.
Parker, J. L., & Newstead, S. (2014). Molecular basis of nitrate uptake by the plant nitrate transporter NRT1.1. *Nature, 507*(7490), 68–72. https://doi.org/10.1038/nature13116.
Payne, R. M., Xu, D., Foureau, E., Teto Carqueijeiro, M. I., Oudin, A., Bernonville, T. D., et al. (2017). An NPF transporter exports a central monoterpene indole alkaloid intermediate from the vacuole. *Native Plants, 3*, 16208. https://doi.org/10.1038/nplants.2016.208.
Pike, S., Gao, F., Kim, M. J., Kim, S. H., Schachtman, D. P., & Gassmann, W. (2014). Members of the NPF3 transporter subfamily encode pathogen-inducible nitrate/nitrite transporters in grapevine and Arabidopsis. *Plant and Cell Physiology, 55*(1), 162–170. https://doi.org/10.1093/pcp/pct167.
Prabhala, B. K., Aduri, N. G., Jensen, J. M., Ernst, H. A., Iram, N., Rahman, M., et al. (2014). New insights into the substrate specificities of proton-coupled oligopeptide transporters from *E. coli* by a pH sensitive assay. *FEBS Letters, 588*(4), 560–565. https://doi.org/10.1016/j.febslet.2014.01.004.
Prabhala, B. K., Aduri, N., Hald, H., & Mirza, O. (2015). Investigation of the substrate specificity of the proton coupled peptide transporter PepTSo from *Shewanella oneidensis*.

International Journal of Peptide Research and Therapeutics, 1–6. https://doi.org/10.1007/s10989-014-9427-7.

Prabhala, B. K., Aduri, N. G., Iqbal, M., Rahman, M., Gajhede, M., Hansen, P. R., et al. (2017). Several hPepT1-transported drugs are substrates of the *Escherichia coli* proton-coupled oligopeptide transporter YdgR. *Research in Microbiology, 168*(5), 443–449. https://doi.org/10.1016/j.resmic.2017.01.005.

Prabhala, B. K., Aduri, N. G., Sharma, N., Shaheen, A., Sharma, A., Iqbal, M., et al. (2018). The prototypical proton-coupled oligopeptide transporter YdgR from *Escherichia coli* facilitates chloramphenicol uptake into bacterial cells. *Journal of Biological Chemistry, 293*(3), 1007–1017. https://doi.org/10.1074/jbc.M117.805960.

Saito, H., Oikawa, T., Hamamoto, S., Ishimaru, Y., Kanamori-Sato, M., Sasaki-Sekimoto, Y., et al. (2015). The jasmonate-responsive GTR1 transporter is required for gibberellin-mediated stamen development in Arabidopsis. *Nature Communications, 6*, 6095. https://doi.org/10.1038/ncomms7095.

Sala-Rabanal, M., Loo, D. D., Hirayama, B. A., Turk, E., & Wright, E. M. (2006). Molecular interactions between dipeptides, drugs and the human intestinal H+ -oligopeptide cotransporter hPEPT1. *The Journal of Physiology, 574*(Pt 1), 149–166. https://doi.org/10.1113/jphysiol.2006.107904.

Sharma, N., Aduri, N. G., Iqbal, A., Prabhala, B. K., & Mirza, O. (2016). Peptide selectivity of the proton-coupled oligopeptide transporter from *Neisseria meningitidis*. *Journal of Molecular Microbiology and Biotechnology, 26*(5), 312–319. http://www.karger.com/DOI/10.1159/000447129.

Sugawara, M., Huang, W., Fei, Y. L., Leibach, F. H., Ganapathy, V., & Ganapathy, M. E. (2000). Transport of valganciclovir, a ganciclovir prodrug, via peptide transporters PEPT1 and PEPT2. *Journal of Pharmaceutical Sciences, 89*(6), 781–789. https://doi.org/10.1002/(sici)1520-6017(200006)89:6<781::aid-jps10>3.0.co;2-7.

Sun, Y., Gan, W., Lei, M., Jiang, W., Cheng, M., He, J., et al. (2018). PEPT1-mediated prodrug strategy for oral delivery of peramivir. *Asian Journal of Pharmaceutical Sciences, 13*(6), 555–565. https://doi.org/10.1016/j.ajps.2018.05.008.

Sun, Y., Sun, J., Shi, S., Jing, Y., Yin, S., Chen, Y., et al. (2009). Synthesis, transport and pharmacokinetics of 5'-amino acid ester prodrugs of 1-beta-D-arabinofuranosylcytosine. *Molecular Pharmaceutics, 6*(1), 315–325. https://doi.org/10.1021/mp800200a.

Takeo, N., & Ikumi, T. (2013). Transporters in drug development: Discovery, optimization, clinical study and regulation. In S. Yuichi, & B. Steffansen (Eds.), *Advances in the Pharmaceutical Sciences Series* 7 (pp. 179–200). Springer New York LLC. https://doi.org/10.1007/978-1-4614-8229-1.

Tamai, I., Nakanishi, T., Nakahara, H., Sai, Y., Ganapathy, V., Leibach, F. H., et al. (1998). Improvement of L-dopa absorption by dipeptidyl derivation, utilizing peptide transporter PepT1. *Journal of Pharmaceutical Sciences, 87*(12), 1542–1546. https://doi.org/10.1021/js980186o.

Taochy, C., Gaillard, I., Ipotesi, E., Oomen, R., Leonhardt, N., Zimmermann, S., et al. (2015). The Arabidopsis root stele transporter NPF2.3 contributes to nitrate translocation to shoots under salt stress. *The Plant Journal, 83*(3), 466–479. https://doi.org/10.1111/tpj.12901.

Thomsen, A. E., Friedrichsen, G. M., Sorensen, A. H., Andersen, R., Nielsen, C. U., Brodin, B., et al. (2003). Prodrugs of purine and pyrimidine analogues for the intestinal di/tri-peptide transporter PepT1: Affinity for hPepT1 in Caco-2 cells, drug release in aqueous media and in vitro metabolism. *Journal of Controlled Release, 86*(2–3), 279–292. http://www.sciencedirect.com/science/article/pii/S0168365902004133.

Tsay, Y. F., Schroeder, J. I., Feldmann, K. A., & Crawford, N. M. (1993). The herbicide sensitivity gene CHL1 of Arabidopsis encodes a nitrate-inducible nitrate transporter. *Cell, 72*(5), 705–713. http://www.cell.com/cell/pdf/0092-8674(93)90399-B.pdf.

Tsuda, M., Terada, T., Irie, M., Katsura, T., Niida, A., Tomita, K., et al. (2006). Transport Characteristics of a novel peptide transporter 1 substrate, antihypotensive drug midodrine, and its amino acid derivatives. *Journal of Pharmacology and Experimental Therapeutics, 318*(1), 455–460. https://doi.org/10.1124/jpet.106.102830.

Tsume, Y., Hilfinger, J. M., & Amidon, G. L. (2008). Enhanced cancer cell growth inhibition by dipeptide prodrugs of floxuridine: Increased transporter affinity and metabolic stability. *Molecular Pharmaceutics, 5*(5), 717–727. https://doi.org/10.1021/mp800008c.

Ural-Blimke, Y., Flayhan, A., Strauss, J., Rantos, V., Bartels, K., Nielsen, R., et al. (2019). Structure of prototypic peptide transporter DtpA from *E. coli* in complex with valganciclovir provides insights into drug binding of human PepT1. *Journal of the American Chemical Society, 141*(6), 2404–2412. https://doi.org/10.1021/jacs.8b11343.

Verri, T., Terova, G., Dabrowski, K., & Saroglia, M. (2011). Peptide transport and animal growth: The fish paradigm. *Biology Letters,* 7(4), 597–600. https://doi.org/10.1098/rsbl.2010.1164.

Vig, B. S., Stouch, T. R., Timoszyk, J. K., Quan, Y., Wall, D. A., Smith, R. L., et al. (2006). Human PEPT1 pharmacophore distinguishes between dipeptide transport and binding. *Journal of Medicinal Chemistry, 49*(12), 3636–3644. https://doi.org/10.1021/jm0511029.

Wang, C.-L., Fan, Y.-B., Lu, H.-H., Tsai, T.-H., Tsai, M.-C., & Wang, H.-P. (2010). Evidence of D-phenylglycine as delivering tool for improving L-dopa absorption. *Journal of Biomedical Science, 17*(1), 71. https://doi.org/10.1186/1423-0127-17-71.

Wang, Y., Li, P., Song, F., Yang, X., Weng, Y., Ma, Z., et al. (2019). Substrate transport properties of the human peptide/histidine transporter PHT2 in transfected MDCK cells. *Journal of Pharmaceutical Sciences, 108*(10), 3416–3424. https://doi.org/10.1016/j.xphs.2019.06.016.

Wang, R., Liu, D., & Crawford, N. M. (1998). The Arabidopsis CHL1 protein plays a major role in high-affinity nitrate uptake. *Proceedings of the National Academy of Sciences of the United States of America, 95*(25), 15134–15139. https://www.ncbi.nlm.nih.gov/pmc/articles/PMC24588/pdf/pq015134.pdf.

Wenzel, U., Thwaites, D. T., & Daniel, H. (1995). Stereoselective uptake of beta-lactam antibiotics by the intestinal peptide transporter. *British Journal of Pharmacology, 116*(7), 3021–3027. https://www.ncbi.nlm.nih.gov/pmc/articles/PMC1909205/pdf/brjpharm00180-0242.pdf.

Wulff, N., Ernst, H. A., Jørgensen, M. E., Lambertz, S., Maierhofer, T., Belew, Z. M., et al. (2019). An optimized screen reduces the number of GA transporters and provides insights into nitrate transporter 1/peptide transporter family substrate determinants. *Frontiers in Plant Science, 10,* 1106. https://doi.org/10.3389/fpls.2019.01106.

Yamashita, T., Shimada, S., Guo, W., Sato, K., Kohmura, E., Hayakawa, T., et al. (1997). Cloning and functional expression of a brain peptide/histidine transporter. *Journal of Biological Chemistry, 272*(15), 10205–10211. http://www.jbc.org/content/272/15/10205.full.pdf.

Zhao, Y., Mao, G., Liu, M., Zhang, L., Wang, X., & Zhang, X. C. (2014). Crystal structure of the *E. coli* peptide transporter YbgH. *Structure, 22*(8), 1152–1160. https://doi.org/10.1016/j.str.2014.06.008.

Zhu, T., Chen, X. Z., Steel, A., Hediger, M. A., & Smith, D. E. (2000). Differential recognition of ACE inhibitors in *Xenopus laevis* oocytes expressing rat PEPT1 and PEPT2. *Pharmaceutical Research, 17*(5), 526–532. https://doi.org/10.1023/a:1007556630189.

CHAPTER ELEVEN

Network analysis of transcriptomics data for the prediction and prioritization of membrane-associated biomarkers for idiopathic pulmonary fibrosis (IPF) by bioinformatics approach

Smriti Mishra[a,f,#], Mohammad Imran Shah[a,f,#], S. Udhaya Kumar[b], D. Thirumal Kumar[b], Chandrasekhar Gopalakrishnan[b], Abeer Mohammed Al-Subaie[c], R. Magesh[d], C. George Priya Doss[b,*] and Balu Kamaraj[e,*]

[a]Department of Biotechnology and Bioinformatics, Jaypee University of Information Technology, Waknaghat, Solan, Himachal Pradesh, India
[b]School of Biosciences and Technology, Vellore Institute of Technology, Vellore, Tamil Nadu, India
[c]Department of Clinical Laboratory Sciences, College of Applied Medical Sciences, Imam Abdulrahman Bin Faisal University, Dammam, Saudi Arabia
[d]Faculty of Biomedical Sciences, Technology & Research, Department of Biotechnology, Sri Ramachandra University, Chennai, Tamil Nadu, India
[e]Department of Neuroscience Technology, College of Applied Medical Sciences in Jubail, Imam Abdulrahman Bin Faisal University, Jubail, Saudi Arabia
[f]Navipoint Health India Pvt Ltd, Moula-Ali, Hyderabad, Telangana, India
*Corresponding authors: E-mail: georgepriyadoss@vit.ac.in, bkranganayaki@iau.edu.sa

Contents

1. Introduction 242
2. Materials and methods 244
 2.1 Overview of the analysis 244
 2.2 RNA-seq data analyses 244
  2.2.1 edgeR 244
  2.2.2 DESeq2 246
 2.3 Enrichment analysis 246
  2.3.1 DAVID database analysis 247
  2.3.2 STRING database analysis 248

Equal first author.

Advances in Protein Chemistry and Structural Biology, Volume 123
ISSN 1876-1623
https://doi.org/10.1016/bs.apcsb.2020.10.003

3. Results and discussion 248
3.1 Differential gene expression analysis 249
3.2 Enrichment analysis 252
3.2.1 Protein-protein interaction (PPI) reticulation calibration 252
3.2.2 Pathway analysis 252
3.2.3 Gene ontology (GO) analysis 252
3.3 Network analysis for the characterization of therapeutic/biomarker 260
3.3.1 Validation through literature data 261
3.4 Comprehensive functional investigation of hub genes 264
4. Conclusion 267
Supplementary data 268
Acknowledgement 268
References 268

Abstract

Idiopathic pulmonary fibrosis (IPF) is a rare yet crucial persistent lung disorder that actuates scarring of lung tissues, which makes breathing difficult. Smoking, environmental pollution, and certain viral infections could initiate lung scarring. However, the molecular mechanism involved in IPF remains elusive. To develop an efficient therapeutic arsenal against IPF, it is vital to understand the pathology and deviations in biochemical pathways that lead to disorder. In this study, we availed network analysis and other computational pipelines to delineate the prominent membrane proteins as diagnostic biomarkers and therapeutic targets for IPF. This study yielded a significant role of glycosaminoglycan binding, endothelin, and GABA-B receptor signaling pathway in IPF pathogenesis. Furthermore, *ADCY8, CRH, FGB, GPR17, MCHR1, NMUR1,* and *SAA1* genes were found to be immensely involved with IPF, and the enrichment pathway analysis suggests that most of the pathways were corresponding to membrane transport and signal transduction functionalities. This analysis could help in better understanding the molecular mechanism behind IPF to develop an efficient therapeutic target or biomarkers for IPF.

1. Introduction

Idiopathic pulmonary fibrosis (IPF) is an acute and persistent pulmonary disorder that is peculiarized by a variable amount of progressive and irreversible inflammation and/or fibrosis (Flaherty et al., 2017). The incidence of IPF is approximately 3—20 per 1,00 ,000 people yearly throughout Europe, America, and Asia (Flaherty et al., 2017; Hutchinson, Fogarty, Hubbard, & McKeever, 2015; Richeldi, Rubin, Avdeev, Udwadia, & Xu, 2015). IPF mainly affects people with mean age 40—70 years and, is distinguished by the cumulative deterioration of dyspnea and lung efficacy (Raghu et al., 2011).

The exact phenomenon of the pathogenic mechanism and fibrosis is unknown (idiopathic-literaly means "a disease of its own kind") (Bagnato & Harari, 2015; Crystal et al., 1981; Furukawa, Oka, Shimada, Tsuchiya, & Tohma, 2015). However, occupational, environmental, and prolonged drug courses were found to be associated with IPF. Varying degrees of lung tissue inflammation and the increased amount and abnormal structure of the connective tissue are commonly seen in IPF (Camus, Kudoh, & Ebina, 2004). The origin of the fibroblast's activation is still unclear and poorly explained. One possible scenario is when a stimuli (drug, co-morbidities, environmental exposure, or genetic risk factors) is exposed to the lung alveoli; it causes epithelial damage and inflammations. These inflammations then promote the resident fibroblasts activation and differentiation that causes the interstitial thickening and fibrosis to develop mild to chronic lung disease symptoms (Bagnato & Harari, 2015; Kaur, Mathai, & Schwartz, 2017; Sheu, Chang, Tsai, Liao, Chong, & Kuo, 2019a, 2019b). Existing knowledge of these processes indicates the role of pathways, such as wound healing, signaling cascades, and pro-fibrotic protein expression play a crucial part in IPF pathology. Analysis of significant regulatory genes, including interleukin-4,7,8,10 and Chemokine Ligands, suggested the association of these genes with defense response, reaction to external stimuli, and immunological process (Mishra, Shah, Sarkar, Asati, & Rout, 2018).

Although many such investigations have advocated the role of many genetic factors and biomolecular pathways that could refurbish the existing knowledge of IPF, still the understanding of therapeutic targets for severe/end-stage IPF is insufficient. More importantly, effective treatment of advanced IPF is still limited, and reversal of fibrosis is yet to be unattainable (Justice et al., 2019). Presently, lung transplantation is the only viable option for IPF patients who are in a dire situation. Abundant studies on inflammatory and tissue remodeling pathways have not come out with positive outcomes due to insufficient understanding of the disease and its molecular markers (Caminati, Cassandro, Torre, & Harari, 2017). However, subsequent research in the area advocated the differential gene expression analysis comparing normal and pathogenic lung tissues (Sheu et al., 2019a,b). Still the mechanism of IPF pathogenesis and progression is still limited, and further investigations are essential to identify critical regulatory gene networks and their genetic association. In recent years, next-generation sequencing (NGS) technologies has gained significant advancement and improved the analysis of genome-wide mRNA expression data (Udhaya Kumar, Thirumal Kumar, Siva, George Priya Doss, & Zayed, 2019; Udhaya Kumar et al., 2020a, 2020b; Wang et al., 2019). The identification of associated molecular function, biological processes, and critical pathways of significantly expressed genes could lead us to a better understanding of IPF. This is based on the premise that proteins interact with each other as a part of a giant mesh to execute a specific metabolic and cellular process. Constructing biological networks and applying computational approaches could help identify the key genes that are pertinent for IPF.

Thus, this study is designed to identify differentially expressed genes that are encoded by either transmembrane or other protein and their functional interactions for a better comprehension of disease prognosis and elucidation of diagnostics-based biomarkers. The outcomes supported the role of few genes such as *ADCY8*, *CRH*, *FGB*, *GPR17*, *MCHR1*, *NMUR1*, *SAA1*, and associated processes as significant regulatory signatures for disease prognosis. The study could help elucidate a better discernment of the bio-molecular and biochemical logistics that could cause IPF and may assist in delineating potential prominent proteo-therapeutic targets for IPF.

2. Materials and methods

2.1 Overview of the analysis

The overview of the workflow is outlined in Fig. 1. The IPF data was procured from the Gene Expression Omnibus (GEO) database (ID: GSE92592). The cohort includes lung tissue samples from IPF patients and healthy control groups with a sample size of 20 and 19, respectively.

2.2 RNA-seq data analyses

The R Bioconductor packages (Huber et al., 2015) were used for RNAseq data analysis. The egdeR and DESeq2 packages were used for gene expression analysis.

2.2.1 edgeR

The edgeR (Robinson, McCarthy, & Smyth, 2010) identifies differentially expressed genes (DEGs) by incorporating negative binomial probabilistic models in the datasets. Further, the log expression ratios' weighted trimmed

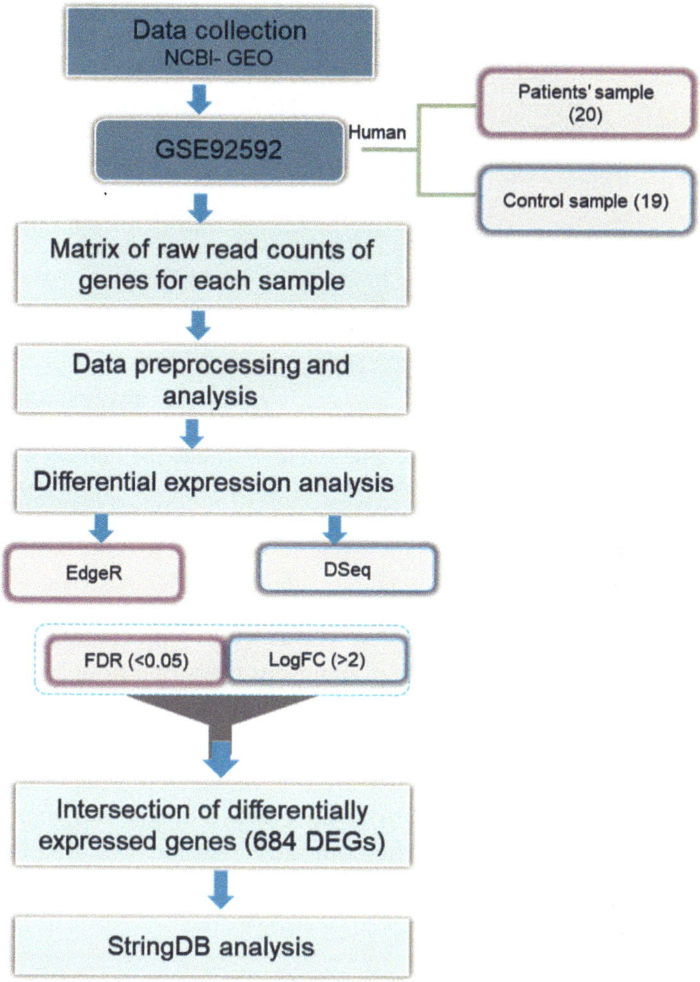

Fig. 1 A flowchart depicting the schematics of the workflow pipeline. It contains the sequential list of software and the parameters used to process the dataset.

mean was used to normalize the gene length and sequencing depth between samples. Then the model is fitted on the read count with formulation $\nu = \mu + \alpha\mu 2$, where in v is the variance, μ-arithmetic mean, and α depicts the dispersion factor. Also, the tool combines common, typical dispersion with gene-specific dispersion to compute α. Finally, Fisher's exact test with a false detection rate (FDR) correction control is implemented to find out the (DEGs).

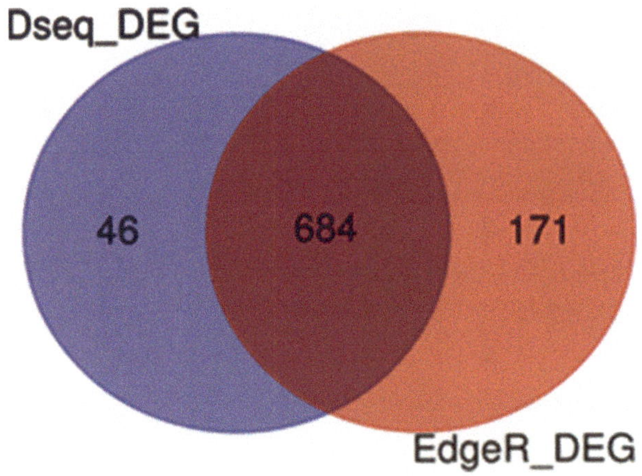

Fig. 2 A Venn diagram illustrating the intersection between the number of differentially expressed genes as per the Dseq R package(*blue circle*) and EdgeR package(*red circle*).

2.2.2 DESeq2

The DESeq2 (Love et al., 2014) is also contrived by implementing negative binomial probabilistic models. For calibrating the dispersions, it avails the gene-specific shrinkage estimation along with all genes having a similar mean expression value. The fold-change ratios were used to identify significant up or downregulation in genes. The intersection (Fig. 2) of significant differentially expressed genes from EdgeR and DESeq2 was then further filtered out for future analysis. The combined method was used to ensure the most robust gene set identification.

2.3 Enrichment analysis

The workflow of enrichment analysis is given in Fig. 3. The DAVID and StringDB were employed to integrate functional explication and protein-protein interactivity analysis for the significant DEGs. The differentially expressed genes were used as an input to DAVID and string DB tools. The detailed analysis and method are given in subsequent sections. The GOplot (Walter, Sánchez-Cabo, & Ricote, 2015) from R Studio was availed to further conceptualize the findings from DAVID tool. The GOBubble from GOplot was used to plot and annotate IPF GO terms from 210 DEGs identified from the dataset.

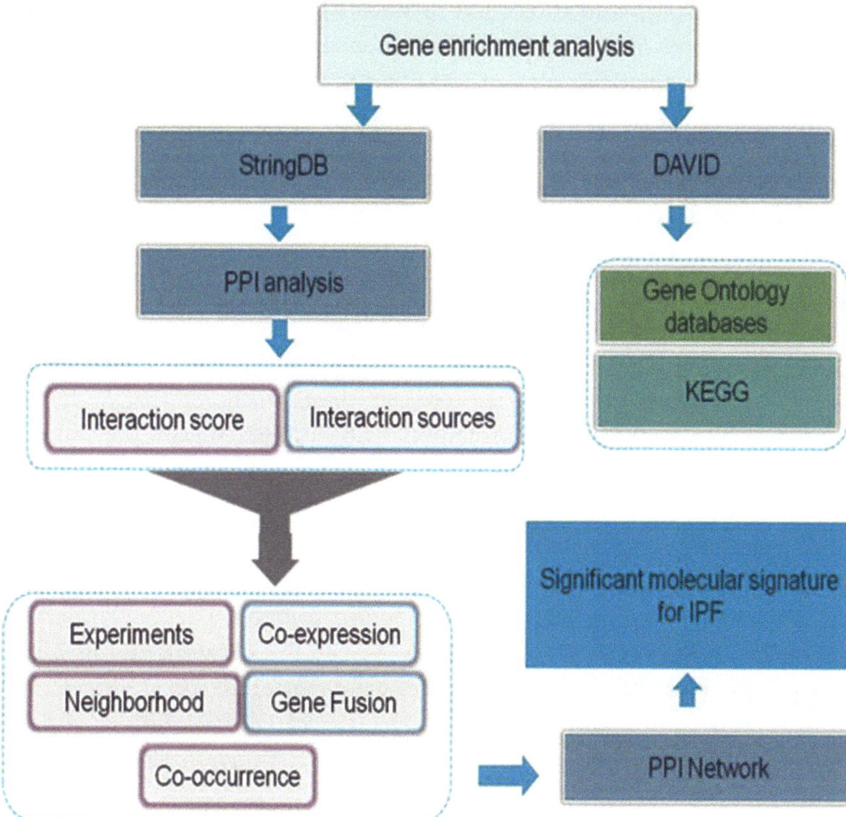

Fig. 3 A detailed illustration of the pipeline involved in gene enrichment analysis. The flowchart illustrates the following list of tools and filtering parameters used to get the final set of genes signatures that could potentially be attributed to IPF.

2.3.1 DAVID database analysis

The DEGs were uploaded to the DAVID tool for functional annotation and clustering. DAVID is an integrated analytical tool and biological knowledge base that provides a high-throughput integrated data-mining environment for gene list analysis (Sherman & Lempicki, 2009). It uses multiple functional annotation data repositories such as the KEGG and Gene Ontology (da Huang, Sherman, & Lempicki, 2009; Krämer, Green, Pollard Jr, & Tugendreich, 2014). It classifies genes into different clusters by agglomeration algorithm method, which calculates the similarity of global annotation profiles. These different clusters contain the list of genes associated with ontological terms, including "Molecular Functions (MFs), 'Cellular

Components (CCs),' and Biological Process (BP)". Further, to validate the statistical inference of the gene-category association, it computes an Expression Analysis Systematic Explorer (EASE) score. A default score = 0.1 can be extended to 1 to extend the clustering range.

2.3.2 STRING database analysis

To discern the interaction among DEGs, the STRING DB database was used. The Search Tool for the Retrieval of Interacting Genes (STRING) database (DB) is utilized to know how these proteins are functionally associated. STRING is a comprehensive repository of documented and predicted protein-protein associations for different organisms (Szklarczyk et al., 2017). These associations could be physical (direct) or functional (indirect). Associations in STRING are procured from various significant sources, including High-throughput Lab Experiment, Conserved Co-Expressions, Genomic Context Predictions, Previous Knowledge in Database, and Automated Text-mining (Szklarczyk et al., 2017).

The DEGs were uploaded and mapped with their STRING DB identifiers and analyzed using all active interaction sources. The benchmarked confidence scores of >0.4 were used to identify the network and its functional interactions. The interaction scores were exported, and an interaction plot was created using visNetwork. It is an R package for network visualization that uses a JavaScript library called "vis.js" (Almende, 2016).

3. Results and discussion

The pathology and causal mechanism behind IPF remain elusive. Previous studies (Bonanni, Frymoyer, & Jacox, 1965; Kaur, Mathai, & Schwartz, 2017) have laid the foundation for understanding the prognosis and etiology of the disease. However, an integrated and contiguous study of genes and other affiliated factors has the aptitude to comprehend the destructive nature of IPF better and delineate significant therapeutic targets. Moreover, diagnostics based on non-invasive techniques are gaining more popularity to mitigate the unpropitious consequence of radiations, biopsies, and recurring co-morbidities. Molecular or genetic biomarkers for IPF have the potential to ameliorate present diagnostic and prognostic situations (De Lauretis & Renzoni, 2014).

A significant gene expression-based biomarker discovery can be helpful for precious and viable drug discovery, development, and diagnostic research. A holistic gene expression analysis can provide significant insights into the state and nature of cells, including pharmacological response, disease progressiveness, and physiological processes: growth and development. These findings lead to the sets of genes and associated pathways which are exclusively associated with a disease condition, prognosis, and/or response to the drug. Also, in numerous cases, only a few genes in the complete genome (e.g., 20–500 genes) show a statistically significant association with the disease condition. These numbers can be further reduced by applying a more stringent threshold for statistical and biological relevance. The resulting smaller subset becomes an implicative set of targets for diagnosis of the disease state and a marker of therapeutic response. Furthermore, this smaller subset could be enriched by the different regulatory processes such as crosstalk in genes, pathways, GO terms to filter out an ultimate list of key genes that could serve as a diagnostic and therapeutic biomarker for a disease state. In this study, the implementation of an integrated analysis approach for gene expression signature discovery and enrichment analysis to identify critical genes has been done using experimental and bioinformatics expertise.

3.1 Differential gene expression analysis

The RNAseq data profile GSE92592 was used for differential gene expression calculations. To create robust and more statistically significant DEGs, a combination of DESeq2 and EdgeR packages were used. After applying the threshold of p-value and FDR, a total of 855 significant genes were extracted from EdgeR, and 730 genes were extracted from DESeq2. To understand and validate the computed normalized expression of IPF versus control group genes, an MA-plot (Fig. 4) and volcano-plot (Fig. 5) were generated. In the MA-plot, the genes with nominal differential expression values for a disease will cluster around the red lines (logFC = 0), whereas the genes with very significant differential expression are plotted toward the edges of the window (Fig. 4). The plot also shows that the expression data is normalized adequately across the samples. The volcano plot shows the distribution of the logFC and corrected p-values in study subgroups. The plot enables visual detection of most biologically significant genes (with large logFC values) that are also statistically significant (Fig. 5). Essential genes such as *MMP1, MMP11, CDH3, EMP2,* etc. were plotted with high relevance in IPF regulation. The heatmap based gene expression analysis showed clear clusters of genes showing differential expression among control versus IPF groups (Fig. 6). Gene *COL1A* (Collagen type III alpha I) is shown to have significant up-regulation and closely clustered with other significant genes

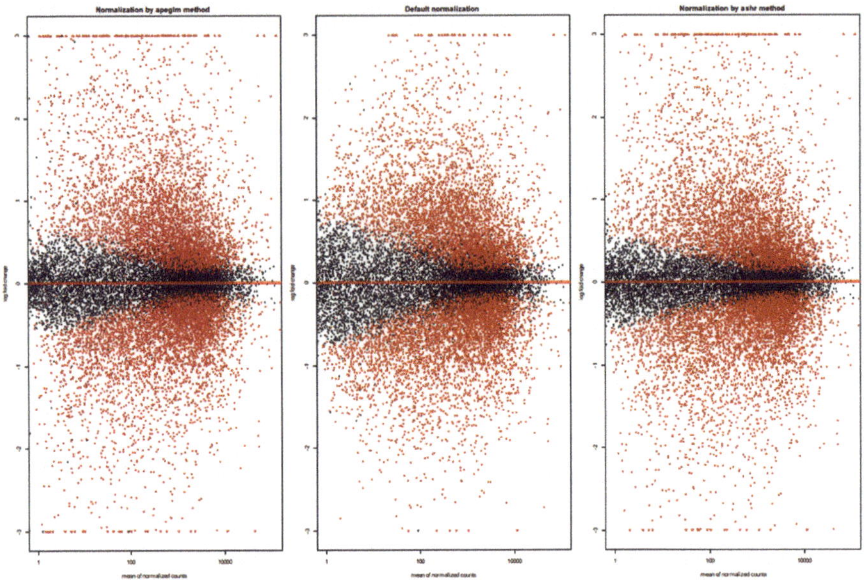

Fig. 4 An MA-plot of differentially expressed genes. The X-axis (M) and Y-axis (A) denote the log of normalized read count and the LogFC value of the RNAseq data, respectively.

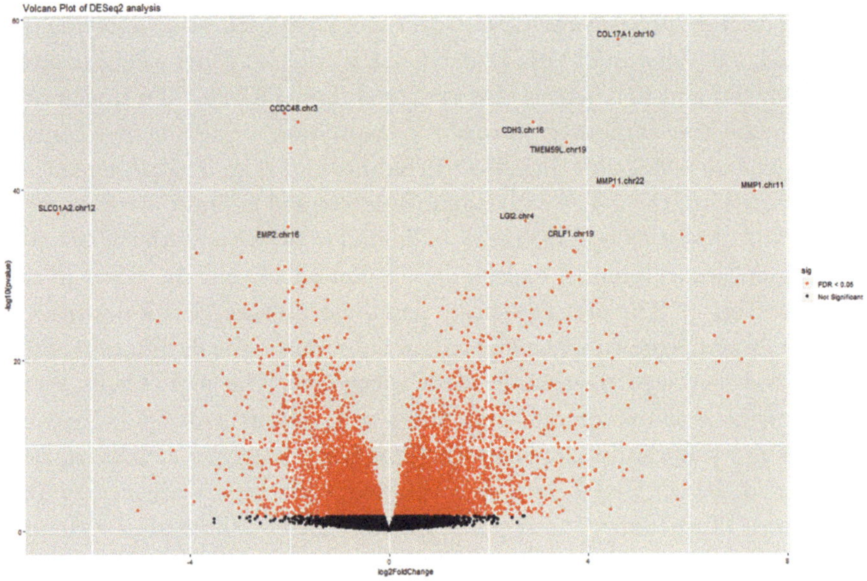

Fig. 5 A volcano plot to showcase the most statistically significant genes with biological relevance.

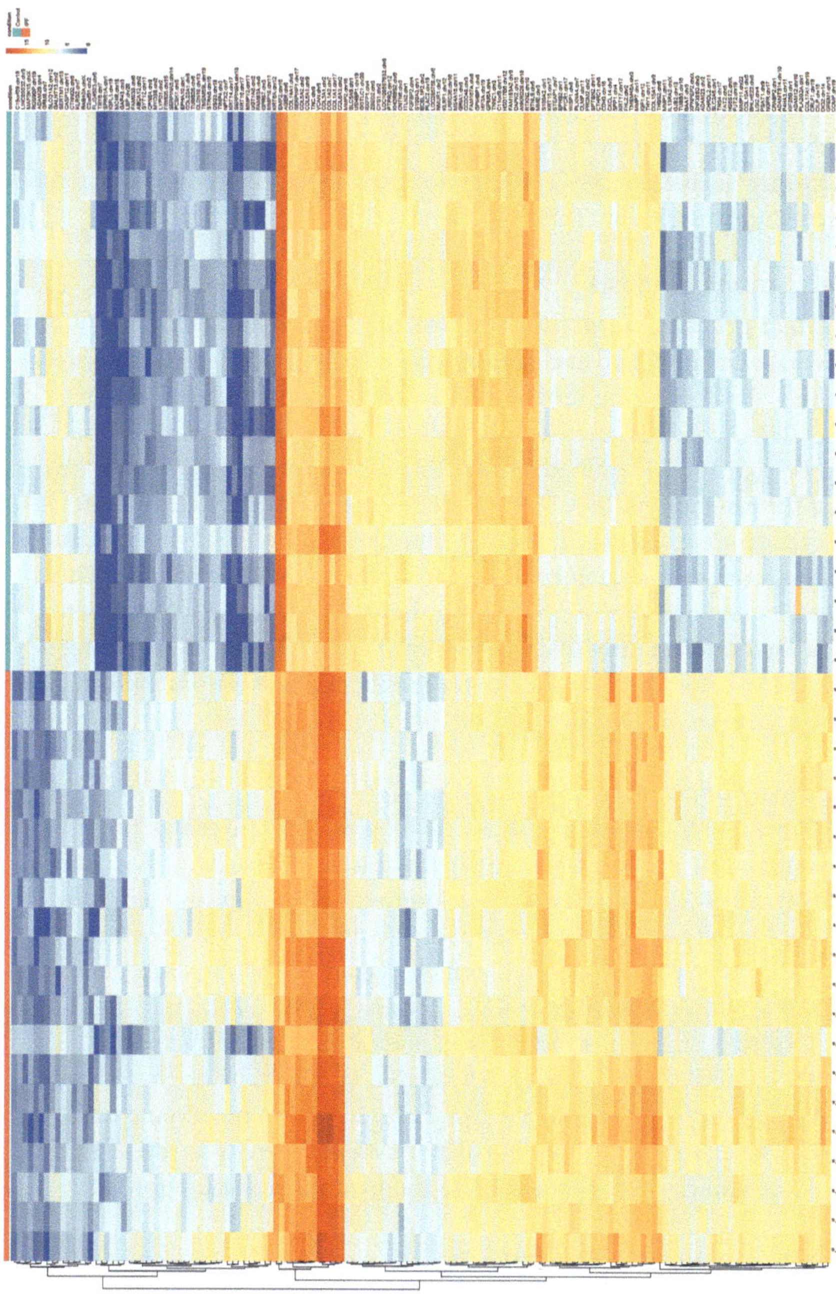

Fig. 6 Clustering heatmap of control versus IPF group gene list.

as *MMP2* etc. *COL1A* is a gastro-esophageal reflux disease susceptibility gene, which is a co-morbid disease of IPF. Similarly, gene *NMUR1* was clustered with other neighboring genes significantly shown downregulation in control versus IPF genes. *NMUR1* gene was found to be one of the salient genes in subsequent analyses. For further filtering, the intersection of results from DESeq2 and EdgeR was used, and a total of 684 genes were filtered out for further analysis (Fig. 2). The differential gene expression data for all the genes are available in Table S1.

3.2 Enrichment analysis

3.2.1 Protein-protein interaction (PPI) reticulation calibration

PPI network calibration and analysis are progressively used to decode pertinent pathways and signaling molecules, and that may propel aspects of pathology. Hence, we examined PPI to ascribe pivotal proteins transcribed by the gene signatures identified as potential DEGs with the culpability to actuate to the prognosis of IPF. The PPI network is consisting of 210 nodes. The interaction score was calculated with the threshold of 0.07 and applied all the interaction source parameters available in the StringDB database. The complete data for each pair of nodes were shown in Table S2. The visNetwork was used to create the interaction plot (Fig. 7). A total of seven hub genes were identified for their maximum number of interactions indicated by network analysis of PPI (Table 1).

3.2.2 Pathway analysis

Since IPF's pathology leads to fibrosis in lung tissues. Targeting fibrotic pathways may assist in disease therapeutics (Lear & Chen, 2016; Richeldi et al., 2014). Pathway investigation was done on the seven hub genes by availing the DAVID server and other pathway databases. This enrichment analysis has provided 43 pathways, which include 7 of the hub genes, i.e., *ADCY8, CRH, FGB, GPR17, MCHR1, NMUR1*, and *SAA1* (Fig. 2). Out of all the genes, *SAA1* was present in most of the pathways (a total of 29 pathways), suggesting its vital role in disease pathophysiology (Table 2). Most of the pathways in Table 2 correspond to membrane transport and signal transduction functionalities. Altered activity of these two classes of biochemical pathways could be the pathological cause for IPF.

3.2.3 Gene ontology (GO) analysis

To determine the relationship between disease candidate genes (DCGs) and corresponding gene product's biology, the statistically significant GO terms

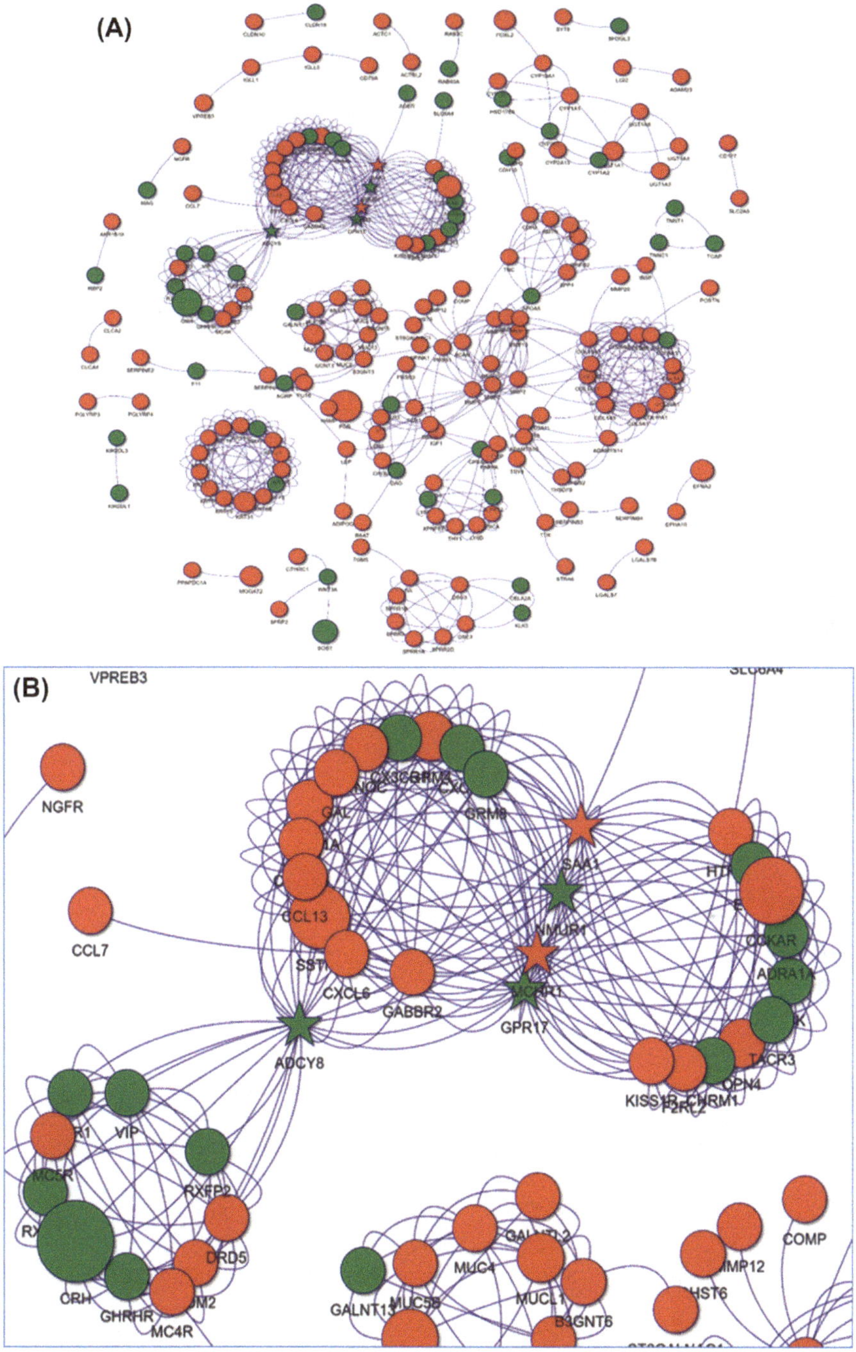

Fig. 7 (A) a comprehensive network graph constructed based on the differentially expressed genes (*red node*: down-regulated genes, *green node*: up-regulated genes) of IPF compared with the control group. The network was assembled with StringDB and visualized using the visNetwork R package. (B) Zoomed view of the network focusing on the predicted hub genes (*star-shaped nodes*) and their connected partners.

Table 1 Table containing the list of hub genes that were predicted by network analysis that could be construed as potential therapeutic targets for IFP.

Approved symbol	Approved name	Chromosome location	NCBI gene ID	Ensemble gene ID	OMIM ID	Panther family/ subfamily	PANTHER protein class
ADCY8	adenylate cyclase 8	8q24.22	114	ENSG00000155897	103070	NA	NA
CRH	corticotropin releasing hormone	8q13.1	1392	ENSG00000147571	122560	NA	NA
FGB	fibrinogen beta chain	4q31.3	2244	ENSG00000171564	134830	Fibrinogen beta chain (pthr19143: sf332)	signaling molecule (PC00207)
GPR17	G protein-coupled receptor 17	2q14.3	2840	ENSG00000144230	603071	Uracil nucleotide/ cysteinyl leukotriene receptor (pthr24232: sf44)	NA
MCHR1	melanin concentrating hormone receptor 1	22q13.2	2847	ENSG00000128285	601751	Melanin-concentrating hormone receptor 1 (pthr24229: sf91)	G-protein coupled receptor(PC00021)

NMUR1	neuromedin U receptor 1	2q37.1	10316	ENSG00000171596	604153	Neuromedin-u receptor 1 (pthr24243: sf109)	G-protein coupled receptor(PC00021)
SAA1	serum amyloid A1	11p15.1	6288	ENSG00000173432	104750	Serum amyloid a-1 protein (pthr23424: sf22)	Apolipoprotein (PC00052); defense/immunity protein (PC00090); transporter(PC00227)

Gene's NCBI, ensemble, and OMIM IDs, along with its protein classification, are given. NA- Not available.

Table 2 Table delineating the list of pathways, Gene Ontology terms, and its functional descriptions associated with the Hub genes obtained from network analysis.

S.No	Term ID	Term description	Observed gene count	Matching proteins in the network (labels)	Functional enrichments
1	GO:0007165	signal transduction	7	ADCY8, CRH, FGB, GPR17, MCHR1, NMUR1, SAA1	Biological processes
2	HSA-162582	Signal Transduction	7	ADCY8, CRH, FGB, GPR17, MCHR1, NMUR1, SAA1	Pathways
3	GO:0007186	G protein-coupled receptor signaling pathway	6	ADCY8, CRH, GPR17, MCHR1, NMUR1, SAA1	Biological processes
4	GO:0065008	regulation of biological quality	6	ADCY8, CRH, FGB, GPR17, MCHR1, SAA1	Biological processes
5	HSA-388396	GPCR downstream signaling	6	ADCY8, CRH, GPR17, MCHR1, NMUR1, SAA1	Pathways
6	GO:0005102	signaling receptor binding	5	CRH, FGB, GPR17, MCHR1, SAA1	Molecular function
7	GO:0048878	chemical homeostasis	5	ADCY8, CRH, GPR17, MCHR1, SAA1	Biological processes
8	GO:0071310	cellular response to organic substance	5	ADCY8, CRH, FGB, GPR17, SAA1	Biological processes
9	HSA-418594	G alpha (i) signaling events	5	ADCY8, GPR17, MCHR1, NMUR1, SAA1	Pathways
10	HSA-500792	GPCR ligand binding	5	CRH, GPR17, MCHR1, NMUR1, SAA1	Pathways
11	GO:0031347	regulation of defense response	4	CRH, FGB, GPR17, SAA1	Biological processes

12	GO:0050801	ion homeostasis	4	CRH, GPR17, MCHR1, SAA1	Biological processes
13	GO:0051050	positive regulation of transport	4	CRH, FGB, MCHR1, SAA1	Biological processes
14	GO:1902533	positive regulation of intracellular signal transduction	4	CRH, FGB, GPR17, SAA1	Biological processes
15	GO:0003008	system process	4	ADCY8, CRH, FGB, NMUR1	Biological processes
16	HSA-416476	G alpha (q) signaling events	4	GPR17, MCHR1, NMUR1, SAA1	Pathways
17	HSA-373076	Class A/1 (Rhodopsin-like receptors)	4	GPR17, MCHR1, NMUR1, SAA1	Pathways
18	GO:0008528	G protein-coupled peptide receptor activity	3	GPR17, MCHR1, NMUR1	Molecular function
19	GO:0007204	positive regulation of cytosolic calcium ion concentration	3	GPR17, MCHR1, SAA1	Biological processes
20	GO:0032102	negative regulation of response to external stimulus	3	FGB, GPR17, SAA1	Biological processes
21	GO:0050714	positive regulation of protein secretion	3	CRH, FGB, SAA1	Biological processes
22	GO:0050878	regulation of body fluid levels	3	ADCY8, FGB, SAA1	Biological processes
23	GO:0007610	behavior	3	ADCY8, CRH, MCHR1	Biological processes
24	GO:0032870	cellular response to hormone stimulus	3	ADCY8, CRH, FGB	Biological processes
25	GO:0001934	positive regulation of protein phosphorylation	3	CRH, FGB, SAA1	Biological processes

(Continued)

Table 2 Table delineating the list of pathways, Gene Ontology terms, and its functional descriptions associated with the Hub genes obtained from network analysis.—cont'd

S.No	Term ID	Term description	Observed gene count	Matching proteins in the network (labels)	Functional enrichments
26	GO:0071345	cellular response to cytokine stimulus	3	FGB, GPR17, SAA1	Biological processes
27	GO:0006952	defense response	3	CRH, FGB, SAA1	Biological processes
28	HSA-375276	Peptide ligand-binding receptors	3	MCHR1, NMUR1, SAA1	Pathways
29	GO:0008188	neuropeptide receptor activity	2	MCHR1, NMUR1	Molecular function
30	GO:0001664	G protein-coupled receptor binding	2	CRH, SAA1	Molecular function
31	GO:0007193	adenylate cyclase-inhibiting G protein-coupled receptor signaling pathway	2	ADCY8, MCHR1	Biological processes
32	GO:0090277	positive regulation of peptide hormone secretion	2	CRH, FGB	Biological processes
33	GO:0007200	phospholipase C-activating G protein-coupled receptor signaling pathway	2	GPR17, NMUR1	Biological processes
34	GO:0007218	neuropeptide signaling pathway	2	MCHR1, NMUR1	Biological processes
35	GO:0030168	platelet activation	2	FGB, SAA1	Biological processes
36	GO:0050728	negative regulation of inflammatory response	2	GPR17, SAA1	Biological processes

37	GO:0051928	positive regulation of calcium ion transport	2	CRH, MCHR1	Biological processes
38	GO:0007611	learning or memory	2	ADCY8, CRH	Biological processes
39	GO:0019932	second-messenger-mediated signaling	2	ADCY8, NMUR1	Biological processes
40	GO:0031349	positive regulation of defense response	2	CRH, FGB	Biological processes
41	GO:0008015	blood circulation	2	CRH, FGB	Biological processes
42	GO:0045785	positive regulation of cell adhesion	2	FGB, SAA1	Biological processes
43	HSA-168898	Toll-like Receptor Cascades	2	FGB, SAA1	Pathways

which discerns the molecular functions, cellular components, and biological processes of DCGs were ascertained. Readings of molecular function have revealed that a promiscuous number of DCGs were involved in transferring amino-acyl groups (GO:0016755), protein-glutamine gamma-glutamyl transferase activity (GO:0003810), hydrolase activity (GO:0016825), etc. The role of many hydrolases is already being studied in regulating inflammation and repair. The majority of genes were found to be mapped with extracellular structure organization (GO:0043062), extracellular matrix organization (GO:0030198), etc. Similarly, associated cellular components also indicated the importance of the collagen-containing extracellular matrix (GO:0062023) and extracellular matrix (GO:0031012). The role of the extracellular matrix and its associated physiological and developmental pathways has been established for IPF pathology in several studies (Tomos et al., 2017; Sivakumar et al., 2019). GO analysis of specific hub-genes is elucidated in the subsequent section (Table 2). Further study on these associations could unravel novel effective therapeutic targets for IPF therapeutics.

3.3 Network analysis for the characterization of therapeutic/biomarker

Biological systems' network investigation is receiving more popularity in the research community for better data integration and analysis (Eisenberg, Marcotte, Xenarios, & Yeates, 2000). Network representation of intracellular biological interaction usually considers biomolecular units present inside a cell as nodes and their congruent association (direct or indirect) between themselves as edges (Ma'ayan, 2011). Features of central nodes encompass connectivity degrees that compute the number of interactions for each unit and significant differential regulation. It also includes Gene Ontology annotations (Gene Ontology Consortium, 2010), which depicts the nodes' attributes, location, and complicity of biological functions. To perform network analysis, we have first identified the nodes which have Protein-Protein interactions with the help of STRING DB. Resulting, 210 nodes were then further analyzed with the help of visNetwork (Fig. 7A and B). Three parameters were used to provide more interactive results through nodes and edges calculations. The node color was based on differential expression, the node size was based on FDR value, and the numbers of edges were calculated using PPI interaction scores gained from STRING DB PPI analysis. A total of seven essential genes were then delineated by their differential expression values and number of interactions: *ADCY8, GPR17, MCHR1, NMUR1, SAA1, CRH,* and *FGB.*

3.3.1 Validation through literature data

Out of 7 predicted essential genes, four genes (*ADCY8*, *GPR17*, *MCHR1*, and *NMUR1*) are encoded by transport protein, and the other three genes (*SAA1*, *CRH*, and *FGB*) encoded proteins were interacting with membrane-associated protein which was involved in the biological function of IPF. We generated the bar charts (Figs. S1 and S2) using counts (CPM) to observe the regulation of predicted genes between the disease and control samples. Out of 4 transmembrane protein biomarkers, three proteins were downregulated in the disease sample relative to control. (Fig. S1A–C). Whereas, MCHR1 transmembrane protein is upregulated in disease samples (Fig. S1D). The other three predicted biomarkers, two proteins (*SAA1* and *FGB*), were downregulated in the disease sample condition, in contrast to control (Fig. S2A–B). In contrast, CRH is upregulated in the disease sample (Fig. S2C). Further, the top seven key genes have been prioritized using a two-step process, which is differential expression analysis followed by PPI network analysis. The pathways and GO terms associated with these critical genes were also identified to check their association with IPF. Furthermore, to emphasize the role of these critical genes in IPF pathogenicity, past studies that experimentally validated the role of these critical genes and/or its associated pathways, GO terms are discussed in the following section.

3.3.1.1 Adenylyl cyclase type 8 (ADCY8)

It is an enzyme that is transcribed by the *ADCY8* gene (Defer et al., 1994). ADCY8 is a membrane-fettered biocatalyst that enhances the synthesis of cAMP from ATP. Its 3D structure posses two modules with six transmembrane helices each that are required for the enzymatic activity. Further, its catalytic site is situated at the intersection of the two domains. The role of *ADY8* (Fig. 7A and B) was also studied for its role in acute lung injury and association with significant function SNPs (rs31127725, rs31127727, rs31131314, rs13482616, and rs32478823). Further, the gene was positively associated with the WNT signaling pathway, and proteins derived from *ADCY8* have associated with GTP-mediated cell signaling (Li et al., 2012). The purpose of WNT signaling pathways has been established for IPF and lung injuries in many studies (Mishra et al., 2018).

3.3.1.2 GPR17

GPR17 encodes a uracil nucleotide/cysteinyl leukotriene receptor, which is a G protein-coupled receptor whose genetic code is on chromosome 2q21 (Marucci et al., 2016). GPR17, also called a P2Y-like receptor, was first identified by Raport and co-workers (Raport et al., 1996) in an attempt to discover new chemokine receptors. It has eight amphipathic helices, seven membrane-spanning domains (TM1—TM7), and a segment of the reaction sites for N-glycosylation, which is crucial for its movement from the rough endoplasmic reticulum to the plasma membrane. G protein-coupled receptors (GPCRs) (Fig. 7A and B), are the most arduously researched therapeutic targets owing to their pharmacological docility and its significant connotation with human pathophysiology (Hauser, Attwood, Rask-Andersen, Schiöth, & Gloriam, 2017). The negative regulatory function of the *GPR17* gene has been proven to significantly increase the vascular permeability in mice tissue (Maekawa, Balestrieri, Austen, & Kanaoka, 2009). The permeability of blood vessels is a highly fluctuating and a significant indicator of inflammatory mediators. Law (1985) suggested that fibrosis, which develops after irradiation, is caused by increases in vascular permeability.

3.3.1.3 MCHR1

MCHR1 is the gene that encodes a protein called a Melanin-concentrating hormone receptor 1, found in all mammals. The biochemical role of MCH is facilitated by 2 GPCR, MCHR1 and MCHR2 (Chambers et al., 1999; Lembo et al., 1999; Philippe & Mitterhauser, 2017; Saito, Nothacker, Wang, Lin, Leslie, & Civelli, 1999). The MCHR1 protein consists of 353 amino acids in length and contains the characteristics of GPCR, including 3 possible glycosylation sites in the N-terminus, 7 transmembrane helices, and a DRY motif at the proximity of the third intracellular loop (Lakaye, Minet, Zorzi, & Grisar, 1998). MCHR1 is an integral plasma membrane protein that associates with the melanin-concentrating hormone. It inhibits cAMP accretion and initiates intracellular calcium flux and is possibly entangled in the neuronal control of food intake. (Lakaye et al., 1998). MCHR1 (Fig. 7A and B) interaction with periplakin, which is a prominent controller of lung injury and reparation in mice, is well studied, and outcomes shown the periplakin can modify the *MCHR1* to initiate signal transduction (Murdoch et al., 2005).

3.3.1.4 SAA1

Serum amyloid A1 (SAA1) is a protein transcribed by the *SAA1* gene in humans. (Glaser, Housman, Lewis, Gerhard, & Jones, 1989; Polyak et al., 1997). The *SAA1* is one of the four *SAA* genes and located on Chromosome 15. (Uhlar, Burgess, Sharp, & Whitehead, 1994). SAA1 is a main acute-phase protein primarily synthesized by liver cells in reaction to tissue damage, malignancy, and infection. The SAA1 possesses a hexameric structure with each subunit, assuming an antiparallel 4-helix bundle structure (Lu, Yu, Zhu, Cheng, & Sun, 2014). The G protein-coupled chemoattractant receptor FPR2 (formyl peptide receptor 2) (Su et al., 1999) which is thought to facilitate the chemotactic reaction of the recombinant SAA1. Experimental evidence indicates that SAA1 has a role in host defense against bacterial infection (Hari-Dass, Shah, Meyer, & Raynes, 2005). The Serum amyloid A1 (SAA1) (Fig. 7A and B) is an already well-established indicator of inflammatory reaction in patients with Systemic Sclerosis. Furthermore, *SAA* regulates the expression of the dominant influencer of fibrogenesis (Wilson et al., 2008). A recent study also indicates *SAA* as a promising identifier of IPF disease pathology (Vietri et al., 2019).

3.3.1.5 NMURI

The NMU consists of 2 receptors (NMU-R1 and NMU-R2) and belongs to the GPCR family that activates intracellular Ca^{2+} stores in reaction to NMU association. (Brighton, Szekeres, Wise, & Willars, 2004). Neuromedin-U receptor 1 (NMUR1) is coded by the NMUR1 gene (Tan et al., 1998). The NMUR1 associated pathways are signaled by GPCR and Peptide ligand-interacting receptors. Gene Ontology (GO) annotations related to this gene include neuromedin U receptor activity and GPCR activity. The NMU-R1 is sufficiently expressed in different peripheral tissues, including various immune cells (Hedrick et al., 2000). The NMU—NMUR1 (Fig. 7A and B) signaling regulates inflammatory responses, where NMUR1 is a receptor for the neuropeptide NMU. The *NMUR1* signaling elevates inflammatory responses, which imply the significance of neuro-immune crosstalk in allergic inflammation at mucosal surfaces (Moriyama et al., 2006). The more significant insights into the precision of the molecular identity and cellular components contributing to neuro-immune crosstalk have great potential for the detection of novel therapeutic targets. It also has a significant role in the effectuation of novel remedies in the framework of inflammatory, infectious, metabolic, and oncogenic conditions (Godinho-Silva, Cardoso, & Veiga-Fernandes, 2019).

3.3.1.6 Corticotropin-releasing hormone (CRH)

CRH interacts with corticotropin-releasing hormone receptor 1 (CRHR1) (Grammatopoulos et al., 1999; Gottowik et al., 1997), is a membrane-spanning protein also known as CRF1, is genetically transcribed and encoded by the *CRHR1* gene (Chen, Lewis, Perrin, Vale, & October, 1993;

Polymeropoulos, Torres, Yanovski, Chandrasekharappa, & Ledbetter, 1995). CRF1 is triggered via the binding of CRF/CRF-agonist (Grammatopoulos et al., 1999; Hauger, Grigoriadis, Dallman, Plotsky, Vale, & Dautzenberg, 2003; Hillhouse & Grammatopoulos, 2006). The ligand interacting and ensuing receptor conformational modification depends on three various sites in the second and third extracellular sites of CRF1 (Grammatopoulos et al., 1999). In the majority of tissues, CRF1 is associated to a stimulatory G-protein that triggers the adenylyl cyclase signaling pathway, and ligand-binding initiates a rise in cAMP levels (Grammatopoulos et al., 1999; Hillhouse & Grammatopoulos, 2006). However, the signal can be communicated along various signal transduction cascades, according to the confirmation of the receptor and the areas of its expression (Grammatopoulos et al., 1999). CRH (Fig. 7A and B) encodes a member from the family of the corticotropin-releasing factor. CRH primarily help in the instigation of corticosteroid-mediated pathways that usually inhibits immune function (Kim, Kayembe, Simecka, Pulse, & Jones, 2011). Significant downregulation of this gene in our network genes advocate its critical role in IPF pathogenesis. CRH receptor-mediated novel therapy could factor a substantial role in moderating immune reactions.

3.3.1.7 FGB

Fibrinogen Beta Chain is a Protein encoded by the *FGB* gene. Maladies affiliated with FGB include Congenital, Afibrinogenemia, and Dysfibrinogenemia. Among its associated pathways are RET signaling and Oncogenic MAPK signaling. Gene Ontology (GO) annotations connected to this gene include signaling receptor interaction and chaperone binding. The significant paralog of the *FGB* gene is TNR (Klose et al., 2000). FGB (fibrinogen beta chain, Fig. 7A and B) and its upregulation have been studied and verified with the iTRAQ-Based Proteomic analysis (Niu et al., 2017). The *FGB* alleles (the gene coding the β subunit) were proven to increase fibrinogen response to ambient particulates. Elevated fibrinogen levels lead to increased resistance of the clot to fibrinolysis, increased clot stiffness, and increased blood viscosity as an inflammatory response (Ariëns, 2011).

3.4 Comprehensive functional investigation of hub genes

The enrichment analysis of these central genes was performed using PANTHERs (Protein Analysis Through Evolutionary Relationships) database (Mi, Muruganujan, & Thomas, 2013) and STRING DB database. The overall classification of the genes, enrichment analysis in terms of GO terms

(CC, MF, and BP), and pathways have also been performed (Fig. 8 and Fig. S3). The molecular functions associated with these genes were binding (GO:0005488), molecular function regulator (GO:0098772), glycosaminoglycan binding (GO:0005539), catalytic activity (GO:0003824), and molecular transducer activity (GO:0060089). Suppressed transducer activity has been found to attenuate lung fibrosis significantly (Dong et al., 2015), which suggested its therapeutic competence. Furthermore, many ligand-binding facilitate transcriptional expression regulation of a few well studied IPF genes such as *MUCB5*, *FOXA2*, etc. in a variety of pathology, including cancer and IPF (Zhang, Wang, Qu, Yu, & Yang, 2019). Subsequently, direct collaboration between CXCL10 (antifibrotic Chemokine) and glycosaminoglycan binding was found to hinder fibroblast conscription and ensuing fibrosis (Jiang et al., 2010). Thus, further studies of these genes and their binding targets may characterize an innovative therapy for IPF. The pathways associated with these genes are Blood coagulation (P00011), Corticotropin-releasing factor receptor signaling pathway (P04380), GABA-B receptor II signaling (P05731), Endothelin signaling pathway (P00019), Gs alpha mediated pathway (P00026), Heterotrimeric G-protein signaling pathway-Gi alpha and Plasminogen activating cascade (P00050). The role of these pathways has been studied for the control of diverse physiological processes, with stress and immune reactions. GABA-B receptors were most often found to be associated with small-cell lung carcinoma (Zhang et al., 2013) and fetal lung development. The pathway was proven to have its role in increased cell proliferation of distal fetal lung epithelial cells (Fu, Wood, & Spindel, 2011). Similarly, the association of endolithin in the fibrosis and scarring associated with IPF have already been implicated (Abraham, 2008). Such an association can be further studied for their role in IPF and associate's phenotypes. The GPCR could upsurge cAMP and control fibroblast propagation and differentiation, advocating that they may offer novel targets to identify novel IPF treatments (Roberts, Broome, Kent, Charlton, & Rosethorne, 2018).

In addition to the comprehensive pathway analyses, we performed an analysis of the interrelation between the hub genes in order to explore the relationship between them. As seen in Fig. 9, all hub genes interacting with each other directly expect FGB protein. Nevertheless, the FGB protein is involved in the phospholipase C-activating GPCR signaling and GPCR signaling pathway. Unlike other hub genes, FGB protein might indirectly mediate the GPCR and phospholipase C-activating signaling pathway. Taken together, findings from this analysis showed that hub genes identified

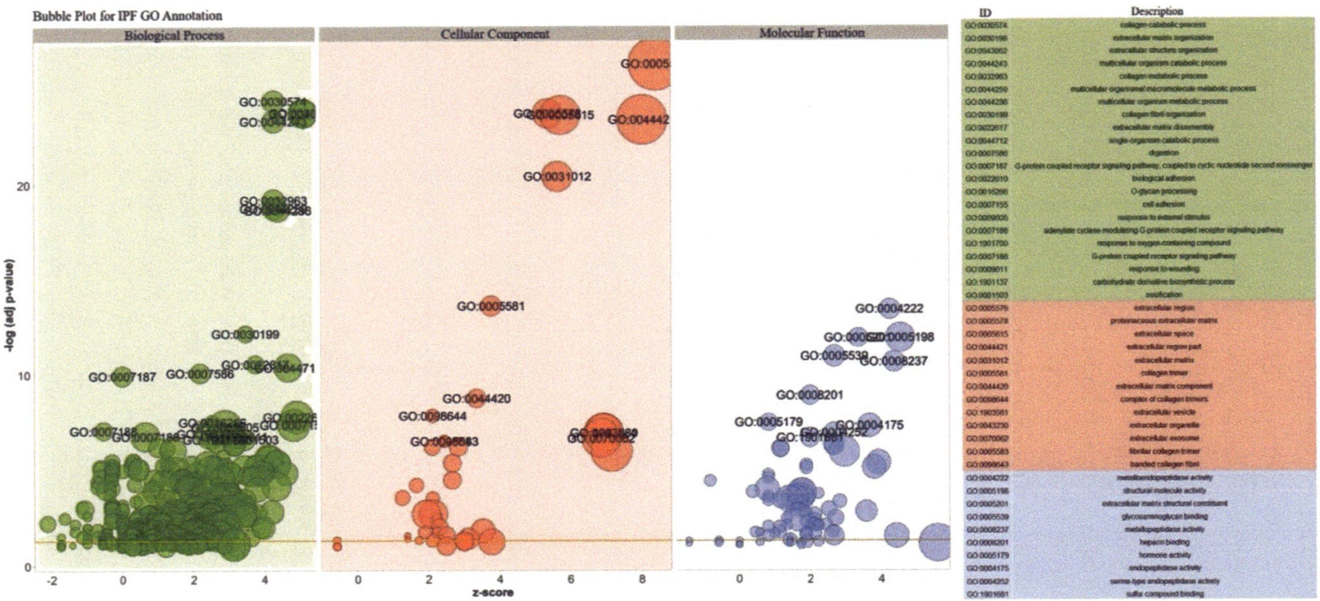

Fig. 8 GOBubble plot illustrates enrichment annotation such as biological process (*green*), cellular component (*red*), molecular function (*blue*) of the gene ontology terms from the DAVID tool. The y-axis is −log p-value, and the x-axis is Z-score. The bubbles size in each GO terms is considered proportional to the DEGs number from the dataset. The threshold (*orange line* in BP, CC, and MF) means the normalized p-value to 0.05.

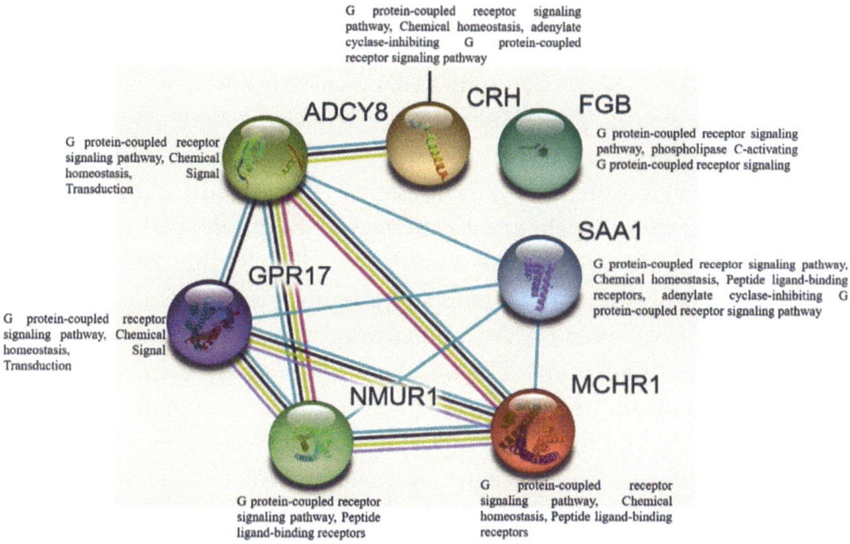

Fig. 9 Analysis of the relationship between the core genes identified from the dataset. The STRING database was utilized to establish a network and visualization. The genes such as *ADCY8, CRH, FGB, GPR17, MCHR1, NMUR1*, and *SAA1* are involved in the G protein-coupled receptor signaling pathway, whereas *FGB* and *SAA1* involved in the phospholipase C-activating and adenylate cyclase-inhibiting G protein-coupled receptor signaling pathway. The genes *ADCY8, CRH, GPR17, MCHR1*, and *SAA1* are involved in chemical homeostasis.

between IPF patients and control samples are significantly present in the critical signaling pathways and might lead to the development of IPF. Available data for these genes suggests a few of these critical genes, such as *SAA*, *FGB*, etc., are already in advance stages of experimentation to prove their role as therapeutic/biomarker targets. The results and studies validated and supported finding given in the presented research work. These findings indicate that these critical genes may further be experimentally validated to ascribe their potential role in IPF for diagnosis and prognosis. All these potential indicators warrant further endorsement to determine their diagnostic potential in clinical research.

4. Conclusion

The present investigation is a thorough analysis of differentially expressed genes and pathway systems affected in IPF and highlights salient regulatory associations. To surmise, seven prominent hub genes (clinical indicator and/or therapeutic targets) (Table 1) that are encoded by either transmembrane or other protein and ten associated pathways that correspond to membrane transport and signal transduction functionalities (Table 2) were found. These discoveries advocate the significant functionality of inflammatory reactions, cytokine-cytokine receptor interactivity in IPF disease progression, and various signal transduction pathways. The involvement of these biochemical pathways in crucial cellular mechanisms is well-known. Additionally, the outcomes ameliorate the comprehension of these markers in IPF pathogenesis and progression. The findings emphasize the vital role of seven DCGs and pathways as novel potential targets for IPF therapeutic intervention. The findings of this inquiry are anticipated to offer improved diagnostic means for IPF detection and surveilling and help in establishing a novel molecular model to discover, examine, and envision novel therapeutic outcomes. Datasets and analysis provided here will be supportive for research communities putting their efforts for the betterment of IPF.

Supplementary data

Supplementary data to this chapter can be found online at https://doi.org/10.1016/bs.apcsb.2020.10.003.

Acknowledgement

We would like to acknowledge the Bioinformatics lab at College of Applied Medical Sciences in Jubail; Imam Abdulrahman Bin Faisal University, Jubail, 35816, Saudi Arabia; Vellore Institute of Technology, Vellore India; Sri Ramachandra University, Chennai, India for their computing facility to carry out this work.

Conflict of interests

We have no conflicts of interest to disclose.

References

Abraham, D. (2008). Role of endothelin in lung fibrosis. *European Respiratory Review, 17*(109), 145–150.

Almende, B. V. (2016). *ThieurmelB.visNetwork: Network visualization using 'vis.js' Library*. R package version 1.0.3 https://CRAN.R-project.org/package=visNetwork.

Ariëns, R. A. (2011). Elevated fibrinogen causes thrombosis. *Blood, 117*(18), 4687–4688.

Bagnato, G., & Harari, S. (2015). Cellular interactions in the pathogenesis of interstitial lung diseases. *The European Respiratory Review, 24*(135), 102–114. https://doi.org/10.1183/09059180.00003214. PMID: 25726561.

Bonanni, P. P., Frymoyer, J. W., & Jacox, R. F. (1965). A family study of idiopathic pulmonary fibrosis: A possible dysproteinemic and genetically determined disease. *The American Journal of Medicine, 39*(3), 411–421.

Brighton, P. J., Szekeres, P. G., Wise, A., & Willars, G. B. (2004). Signaling and ligand binding by recombinant neuromedin U receptors: Evidence for dual coupling to galphaq/11 and galphai and an irreversible ligand-receptor interaction. *Molecular Pharmacology, 66*, 154415–154456.

Caminati, A., Cassandro, R., Torre, O., & Harari, S. (2017). Severe idiopathic pulmonary fibrosis: What can be done? *European Respiratory Review, 26*, 170047.

Camus, P., Kudoh, S., & Ebina, M. (2004). Interstitial lung disease associated with drug therapy. *British Journal of Cancer, 91*(Supply 2), S18–S23. https://doi.org/10.1038/sj.bjc.6602063. PMCID: PMC2750811.

Chambers, J., Ames, R. S., Bergsma, D., Muir, A., Fitzgerald, L. R., Hervieu, G., et al. (1999). Melanin concentrating hormone is the cognate ligand for the orphan G-protein-coupled receptor SLC-1. *Nature, 400*, 261–265.

Chen, R., Lewis, K. A., Perrin, M. H., & Vale, W. W. (1993). Expression cloning of a human corticotropin-releasing-factor receptor. *Proceedings of the National Academy of Sciences of the United States of America, 90*(19), 8967–8971.

Crystal, R. G., Gadek, J. E., Ferrans, V. J., Fulmer, J. D., Line, B. R., & Hunninghake, G. W. (1981). Interstitial lung disease: current concepts of pathogenesis, staging and therapy. *The American Journal of Medicine, 70*(3), 542–568. https://doi.org/10.1016/0002-9343(81)90577-5. PMID: 7011012.

da Huang, W., Sherman, B. T., & Lempicki, R. A. (2009). Systematic and integrative analysis of large gene lists using DAVID bioinformatics resources. *Nature Protocols, 4*, 44–57.

De Lauretis, A., & Renzoni, E. A. (2014). Molecular biomarkers in interstitial lung diseases. *Molecular Diagnosis and Therapy, 18*(5), 505–522.

Defer, N., Marinx, O., Stengel, D., Danisova, A., Iourgenko, V., Matsuoka, I., et al. (1994). Molecular cloning of the human type VIII adenylyl cyclase. *FEBS Letters, 351*(1), 109–113. https://doi.org/10.1016/0014-5793(94)00836-1

Dong, Z., Lu, X., Yang, Y., Zhang, T., Li, Y., Chai, Y., et al. (2015). IL-27 alleviates the bleomycin-induced pulmonary fibrosis by regulating the Th17 cell differentiation. *BMC Pulmonary Medicine, 15*, 13.

Eisenberg, D., Marcotte, E. M., Xenarios, I., & Yeates, T. O. (2000). Protein function in the post-genomic era. *Nature, 405*, 823–826.

Flaherty, K. R., Brown, K. K., Wells, A. U., Clerisme-Beaty, E., Collard, H. R., Cottin, V., et al. (2017). Design of the PF-ILD trial: A double-blind, randomised, placebo-controlled phase III trial of nintedanib in patients with progressive fibrosing interstitial lung disease. *BMJ Open Respiratory Research, 4*(1), e000212.

Fu, X. W., Wood, K., & Spindel, E. R. (2011). Prenatal nicotine exposure increases GABA signaling and mucin expression in airway epithelium. *American Journal of Respiratory Cell and Molecular Biology, 44*(2), 222–229. https://doi.org/10.1165/rcmb.2010-0109OC

Furukawa, H., Oka, S., Shimada, K., Tsuchiya, N., & Tohma, S. (2015). Genetics of interstitial lung disease: Vol de nuit (Night Flight). *Clinical Medicine Insights: Circulatory, Respiratory and Pulmonary Medicine, 9*(Supply 1), 1–7. https://doi.org/10.4137/CCRPM.S23283. PMCID: PMC4444491.

Gene Ontology Consortium. (2010). The gene ontology in 2010: Extensions and refinements. *Nucleic Acids Research, 38*(Database issue), D331–D335. https://doi.org/10.1093/nar/gkp1018. Epub 2009 Nov 17. PMID: 19920128; PMCID: PMC2808930.

Glaser, T., Housman, D., Lewis, W. H., Gerhard, D., & Jones, C. (1989). A fine-structure deletion map of human chromosome 11p: Analysis of J1 series hybrids. *Somatic Cell and Molecular Genetics, 15*(6), 477–501.

Godinho-Silva, C., Cardoso, F., & Veiga-Fernandes, H. (2019). Neuro—immune cell units: A new paradigm in physiology. *Annual Review of Immunology, 37*, 19—46.

Gottowik, J., Goetschy, V., Henriot, S., Kitas, E., Fluhman, B., Clerc, R. G., et al. (1997). Labelling of CRF1 and CRF2 receptors using the novel radioligand, [3H]-urocortin. *Neuropharmacology, 36*(10), 1439—1446.

Grammatopoulos, D. K., Dai, Y., Randeva, H. S., Levine, M. A., Karteris, E., Easton, A. J., et al. (1999). A novel spliced variant of the type 1 corticotropin-releasing hormone receptor with a deletion in the seventh transmembrane domain present in the human pregnant term myometrium and fetal membranes. *Molecular Endocrinology, 13*(12), 2189—2202.

Hari-Dass, R., Shah, C., Meyer, D. J., & Raynes, J. G. (2005). Serum amyloid A protein binds to outer membrane protein A of gram-negative bacteria. *Journal of Biological Chemistry, 280*(19), 18562—18567.

Hauger, R. L., Grigoriadis, D. E., Dallman, M. F., Plotsky, P. M., Vale, W. W., & Dautzenberg, F. M. (2003). International Union of Pharmacology. XXXVI. Current status of the nomenclature for receptors for corticotropin-releasing factor and their ligands. *Pharmacological Reviews, 55*(1), 21—26.

Hauser, A. S., Attwood, M. M., Rask-Andersen, M., Schiöth, H. B., & Gloriam, D. E. (2017). Trends in GPCR drug discovery: New agents, targets and indications. *Nature Reviews Drug Discovery, 16*(12), 829.

Hedrick, J. A., Morse, K., Shan, L., Qiao, X., Pang, L., Wang, S., et al. (2000). Identification of a human gastrointestinal tract and immune system receptor for the peptide neuromedin U. *Molecular Pharmacology, 58*, 870—875.

Hillhouse, E. W., & Grammatopoulos, D. K. (2006). The molecular mechanisms underlying the regulation of the biological activity of corticotropin-releasing hormone receptors: Implications for physiology and pathophysiology. *Endocrine Reviews, 27*(3), 260—286.

Huber, W., Carey, V. J., Gentleman, R., Anders, S., Carlson, M., Carvalho, B. S., et al. (2015). Orchestrating high-throughput genomic analysis with Bioconductor. *Nature Methods, 12*, 115—121.

Hutchinson, J., Fogarty, A., Hubbard, R., & McKeever, T. (2015). Global incidence and mortality of idiopathic pulmonary fibrosis: A systematic review. *European Respiratory Journal, 46*, 795—806.

Jiang, D., Liang, J., Campanella, G. S., Guo, R., Yu, S., Xie, T., et al. (2010). Inhibition of pulmonary fibrosis in mice by CXCL10 requires glycosaminoglycan binding and syndecan-4. *Journal of Clinical Investigation, 120*(6), 2049—2057. https://doi.org/10.1172/JCI38644. Epub 2010 May 17. PMID: 20484822; PMCID: PMC2877927.

Justice, J. N., Nambiar, A. M., Tchkonia, T., LeBrasseur, N. K., Pascual, R., Hashmi, S. K., et al. (2019). Senolytics in idiopathic pulmonary fibrosis: Results from a first-in-human, open-label, pilot study. *EBioMedicine, 40*, 554—563. https://doi.org/10.1016/j.ebiom.2018.12.052. Epub 2019 Jan 5. PMID: 30616998; PMCID: PMC6412088.

Kaur, A., Mathai, S. K., & Schwartz, D. A. (2017). Genetics in idiopathic pulmonary fibrosis pathogenesis, prognosis, and treatment. *Frontiers in Medicine, 4*, 154. https://doi.org/10.3389/fmed.2017.00154

Kim, B. J., Kayembe, K., Simecka, J. W., Pulse, M., & Jones, H. P. (2011). Corticotropin-releasing hormone receptor-1 and 2 activity produces divergent resistance against stress-induced pulmonary Streptococcus pneumoniae infection. *Journal of Neuroimmunology, 237*(1—2), 57—65. https://doi.org/10.1016/j.jneuroim.2011.06.016. Epub 2011 Jul 20. PMID: 21774994; PMCID: PMC5715473.

Klose, R., Fresser, F., Kochl, S., Parson, W., Kapetanopoulos, A., Fruchart-Najib, J., et al. (2000). Mapping of a minimal apolipoprotein(a) interaction motif conserved in fibrin(-ogen) beta - and gamma -chains. *Journal of Biological Chemistry, 275*(49), 38206—38212.

Krämer, A., Green, J., Pollard, J., Jr., & Tugendreich, S. (2014). Causal analysis approaches in ingenuity pathway analysis. *Bioinformatics, 30*, 523–530.
Lakaye, B., Minet, A., Zorzi, W., & Grisar, T. (1998). Cloning of the rat brain cDNA encoding for the SLC-1 G protein-coupled receptor reveals the presence of an intron in the gene. *Biochimica et Biophysica Acta, 1401*, 216–220.
Law, M. P. (1985). Vascular permeability and late radiation fibrosis in mouse lung. *Radiation Research, 103*(1), 60–76.
Lear, T., & Chen, B. B. (2016). Therapeutic targets in fibrotic pathways. *Cytokine, 88*, 193–195.
Lembo, P. M., Grazzini, E., Cao, J., Hubatsch, D. A., Pelletier, M., Hoffert, C., et al. (1999). The receptor for the orexigenic peptide melanin-concentrating hormone is a G-protein-coupled receptor. *Nature Cell Biology, 1*, 267–271.
Li, H. H., Li, Q., Liu, P., Liu, Y., Li, J., Wasserloos, K., et al. (2012). WNT1-inducible signaling pathway protein 1 contributes to ventilator-induced lung injury. *American Journal of Respiratory Cell and Molecular Biology, 47*(4), 528–535. https://doi.org/10.1165/rcmb.2012-0127OC. Epub 2012 Jun 14. PMID: 22700866; PMCID: PMC3488625.
Love, M. I., Huber, W., & Anders, S. (2014). Moderated estimation of fold change and dispersion for RNA-seq data with DESeq2. *Genome Biology, 15*(12), 550.
Lu, J., Yu, Y., Zhu, I., Cheng, Y., & Sun, P. D. (2014). Structural mechanism of serum amyloid A-mediated inflammatory amyloidosis. *Proceedings of the National Academy of Sciences of the United States of America, 111*(14), 5189–5194.
Ma'ayan, A. (2011). Introduction to network analysis in systems biology. *Science Signaling, 4*(190), tr5. https://doi.org/10.1126/scisignal.2001965
Maekawa, A., Balestrieri, B., Austen, K. F., & Kanaoka, Y. (2009). GPR17 is a negative regulator of the cysteinyl leukotriene 1 receptor response to leukotriene D4. *Proceedings of the National Academy of Sciences, 106*(28), 11685–11690.
Marucci, G., Dal Ben, D., Lambertucci, C., Santinelli, C., Spinaci, A. A., Thomas, A., et al. (2016). The G protein-coupled receptor GPR17: Overview and update. *ChemMedChem, 11*(23), 2567–2574. https://doi.org/10.1002/cmdc.201600453
Mi, H., Muruganujan, A., & Thomas, P. D. (2013). PANTHER in 2013: Modeling the evolution of gene function, and other gene attributes, in the context of phylogenetic trees. *Nucleic Acids Research, 41*(Database issue), D377–D386. https://doi.org/10.1093/nar/gks1118. Epub 2012 Nov 27. PMID: 23193289; PMCID: PMC3531194.
Mishra, S., Shah, M. I., Sarkar, M., Asati, N., & Rout, C. (2018). ILDgenDB: Integrated genetic knowledge resource for interstitial lung diseases (ILDs). *Database, 2018*, bay053. https://doi.org/10.1093/database/bay053
Moriyama, M., Fukuyama, S., Inoue, H., Matsumoto, T., Sato, T., Tanaka, K., et al. (2006). The neuropeptide neuromedin U activates eosinophils and is involved in allergen-induced eosinophilia. *American Journal of Physiology - Lung Cellular and Molecular Physiology, 290*, L971–L977.
Murdoch, H., Feng, G. J., Bächner, D., Ormiston, L., White, J. H., Richter, D., et al. (2005). Periplakin interferes with G protein activation by the melanin-concentrating hormone receptor-1 by binding to the proximal segment of the receptor C-terminal tail. *Journal of Biological Chemistry, 280*(9), 8208–8220. https://doi.org/10.1074/jbc.M405215200
Niu, R., Liu, Y., Zhang, Y., Zhang, Y., Wang, H., Wang, Y., et al. (2017). iTRAQ-based proteomics reveals novel biomarkers for idiopathic pulmonary fibrosis. *PLoS One, 12*(1), e0170741. https://doi.org/10.1371/journal.pone.0170741. PMID: 28122020; PMCID: PMC5266322.
Philippe, C., & Mitterhauser, M. (2017). The potential role of the MCHR1 in diagnostic imaging: Facts and trends. *Melanin, 27*.
Polyak, K., Xia, Y., Zweier, J. L., Kinzler, K. W., & Vogelstein, B. (1997). A model for p53-induced apoptosis. *Nature, 389*(6648), 300–305.

Polymeropoulos, M. H., Torres, R., Yanovski, J. A., Chandrasekharappa, S. C., & Ledbetter, D. H. (1995). The human corticotropin-releasing factor receptor (CRHR) gene maps to chromosome 17q12-q22. *Genomics, 28*(1), 123–124.

Raghu, G., Collard, H. R., Egan, J. J., Martinez, F. J., Behr, J., Brown, K. K., et al. (2011). An official ATS/ERS/JRS/ALAT statement: Idiopathic pulmonary fibrosis: Evidence-based guidelines for diagnosis and management. *American Journal of Respiratory and Critical Care Medicine, 183*, 788–824.

Raport, C. J., Schweickart, V. L., Chantry, D., Eddy, R. L., Jr., Shows, T. B., Godiska, R., et al. (1996). *Journal of Leukocyte Biology, 59*, 18–23.

Richeldi, L., du Bois, R. M., Raghu, G., Azuma, A., Brown, K. K., & Costabel, U. (2014). INPULSIS Trial Investigators. *Efficacy and safety of nintedanib in idiopathic pulmonary fibrosis. The New England Journal of Medicine, 370*(22), 2071–2082. https://doi.org/10.1056/NEJMoa1402584. Erratum in: The New England Journal of Medicine, 373, 2015, 782. PMID: 24836310.

Richeldi, L., Rubin, A. S., Avdeev, S., Udwadia, Z. F., & Xu, Z. J. (2015). Idiopathic pulmonary fibrosis in BRIC countries: The cases of Brazil, Russia, India, and China. *BMC Medicine, 13*, 237.

Roberts, M. J., Broome, R. E., Kent, T. C., Charlton, S. J., & Rosethorne, E. M. (2018). The magnitude of cAMP response does not predict the inhibition of human lung fibroblast proliferation and differentiation by Gs-coupled receptors. *Respiratory Research, 19*(1), 56.

Robinson, M. D., McCarthy, D. J., & Smyth, G. K. (2010). edgeR: a bioconductor package for differential expression analysis of digital gene expression data. *Bioinformatics, 26*, 139–140.

Saito, Y., Nothacker, H. P., Wang, Z., Lin, S. H., Leslie, F., & Civelli, O. (1999). Molecular characterization of the melanin-concentrating-hormone receptor. *Nature, 400*, 265–269.

Sherman, B. T., & Lempicki, R. A. (2009). Systematic and integrative analysis of large gene lists using DAVID bioinformatics resources. *Nature Protocols, 4*(1), 44.

Sheu, C. C., Chang, W. A., Tsai, M. J., Liao, S. H., Chong, I. W., & Kuo, P. L. (April 1, 2019). Bioinformatic analysis of next-generation sequencing data to identify dysregulated genes in fibroblasts of idiopathic pulmonary fibrosis. *International Journal of Molecular Medicine, 43*(4), 1643–1656.

Sheu, C. C., Chang, W. A., Tsai, M. J., Liao, S. H., Chong, I. W., & Kuo, P. L. (March 5, 2019). Gene expression changes associated with nintedanib treatment in idiopathic pulmonary fibrosis fibroblasts: A next-generation sequencing and bioinformatics study. *Journal of Clinical Medicine, 8*(3), 308. https://doi.org/10.3390/jcm8030308. PMID: 30841487; PMCID: PMC6462954.

Sivakumar, P., Thompson, J. R., Ammar, R., Porteous, M., McCoubrey, C., Cantu, E., 3rd, et al. (2019). RNA sequencing of transplant-stage idiopathic pulmonary fibrosis lung reveals unique pathway regulation. *ERJ Open Research, 5*(3), 00117–02019. https://doi.org/10.1183/23120541.00117-2019

Su, S. B., Gong, W., Gao, J. L., Shen, W., Murphy, P. M., Oppenheim, J. J., et al. (1999). A seven-transmembrane, G protein-coupled receptor, FPRL1, mediates the chemotactic activity of serum amyloid A for human phagocytic cells. *Journal of Experimental Medicine, 189*(2), 395–402.

Szklarczyk, D., Morris, J. H., Cook, H., Kuhn, M., Wyder, S., Simonovic, M., et al. (2017). The string database in 2017: Quality-controlled protein-protein association networks, made broadly accessible. *Nucleic Acids Research, 45*, D362–D368.

Tan, C. P., McKee, K. K., Liu, Q., Palyha, O. C., Feighner, S. D., Hreniuk, D. L., et al. (1998). Cloning and characterization of a human and murine T-cell orphan G-

protein-coupled receptor similar to the growth hormone secretagogue and neurotensin receptors. *Genomics, 52*(2), 223–229.
Tomos, I. P., Tzouvelekis, A., Aidinis, V., Manali, E. D., Bouros, E., Bouros, D., et al. (2017). Extracellular matrix remodeling in idiopathic pulmonary fibrosis. It is the 'bed' that counts and not 'the sleepers'. *Expert Review of Respiratory Medicine, 11*(4), 299–309.
Udhaya Kumar, S., Thirumal Kumar, D., Siva, R., George Priya Doss, C., & Zayed, H. (2019). Integrative bioinformatics approaches to map potential novel genes and pathways involved in ovarian cancer. *Frontiers in Bioengineering and Biotechnology, 7*, 391.
Udhaya Kumar, S., Thirumal Kumar, D., Siva, R., George Priya Doss, C., Younes, S., et al. (2020). Dysregulation of signaling pathways due to differentially expressed genes from the B-cell transcriptomes of systemic lupus erythematosus patients—A bioinformatics approach. *Frontiers in Bioengineering and Biotechnology, 8*, 276.
Udhaya Kumar, S., Thirumal Kumar, D., Bithia, R., Sankar, S., Magesh, R., Sidenna, M., et al. (2020). Analysis of differentially expressed genes and molecular pathways in familial hypercholesterolemia involved in atherosclerosis: A systematic and bioinformatics approach. *Frontiers in Genetics, 11*, 734.
Uhlar, C. M., Burgess, C. J., Sharp, P. M., & Whitehead, A. S. (1994). Evolution of the serum amyloid A (SAA) protein superfamily. *Genomics, 19*(2), 228–235.
Vietri, L., Bennett, D., Cameli, P., Bergantini, L., Cillis, G., Sestini, P., et al. (2019). Serum amyloid A in patients with idiopathic pulmonary fibrosis. *Respiratory Investigation, 57*(5), 430–434. https://doi.org/10.1016/j.resinv.2019.03.010
Walter, W., Sánchez-Cabo, F., & Ricote, M. (2015). GOplot: an R package for visually combining expression data with functional analysis. *Bioinformatics, 31*, 2912–2914. https://doi.org/10.1093/bioinformatics/btv300
Wang, T., Li, B., Nelson, C. E., & Nabavi, S. (2019). Comparative analysis of differential gene expression analysis tools for single-cell RNA sequencing data. *BMC Bioinformatics, 20*(1), 40.
Wilson, P. G., Thompson, J. C., Webb, N. R., de Beer, F. C., King, V. L., et al. (2008). Serum amyloid A, but not C-reactive protein, stimulates vascular proteoglycan synthesis in a pro-atherogenic manner. *American Journal Of Pathology, 173*, 1902–1910. https://doi.org/10.2353/ajpath.2008.080201
Zhang, X., Zhang, R., Zheng, Y., Shen, J., Xiao, D., Li, J., et al. (2013). Expression of gamma-aminobutyric acid receptors on neoplastic growth and prediction of prognosis in non-small cell lung cancer. *Journal of Translational Medicine, 11*, 102. https://doi.org/10.1186/1479-5876-11-102
Zhang, Q., Wang, Y., Qu, D., Yu, J., & Yang, J. (2019). The possible pathogenesis of idiopathic pulmonary fibrosis considering *MUC5B*. *BioMed Research International, 2019*, 9712464. https://doi.org/10.1155/2019/9712464. PMID: 31309122; PMCID: PMC6594326.

FSC
www.fsc.org
MIX
Papier | Fördert gute Waldnutzung
FSC® C083411

Kolibri 360 GmbH
Ferdinand-Jühlke-Straße 7
99095 Erfurt, Deutschland
produktsicherheit@kolibri360.de